T0299728

Formal Methods for Control of Nonlinear Systems

Formal methods is a field of computer science that emphasizes the use of rigorous mathematical techniques for verification and design of hardware and software systems. Analysis and design of nonlinear control systems plays an important role across many disciplines of engineering and applied sciences, ranging from the control of an aircraft engine to the design of genetic circuits in synthetic biology.

While linear control is a well-established subject, analysis and design of nonlinear control systems remains a challenging topic due to some of the fundamental difficulties caused by nonlinearity. ***Formal Methods for Control of Nonlinear Systems*** provides a unified computational approach to analysis and design of nonlinear systems.

Features

- Constructive approach to nonlinear control.
- Rigorous specifications and validated computation.
- Suitable for graduate students and researchers who are interested in learning how formal methods and validated computation can be combined together to tackle nonlinear control problems with complex specifications from an algorithmic perspective.
- Combines mathematical rigor with practical applications.

Formal Methods for Control of Nonlinear Systems

Yinan Li
OTTO Motors, Clearpath Robotics, Canada

Jun Liu
University of Waterloo

CRC Press is an imprint of the
Taylor & Francis Group, an **informa** business
A CHAPMAN & HALL BOOK

First edition published 2023

by CRC Press
6000 Broken Sound Parkway NW, Suite 300, Boca Raton, FL 33487-2742

and by CRC Press
4 Park Square, Milton Park, Abingdon, Oxon, OX14 4RN

CRC Press is an imprint of Taylor & Francis Group, LLC

ISBN: 978-0-367-21999-4 (hbk)
ISBN: 978-1-032-42816-1 (pbk)
ISBN: 978-0-4292-7025-3 (ebk)

DOI: 10.1201/9780429270253

Typeset in MLM Roman
by KnowledgeWorks Global Ltd.

Publisher's note: This book has been prepared from camera-ready copy provided by the authors.

To our families

Contents

Preface

Formal methods is a field of computer science that emphasizes the use of rigorous mathematical techniques for verification and design of hardware and software systems. Among formal methods techniques, *formal verification* and *model checking* of finite transition systems have achieved considerable success. A landmark result was due to Pnueli [185], who introduced temporal logic to computer science for formal program verification in the late 1970s. This also laid the foundation for the later development of model checking [48], an automated formal verification technique now widely regarded as a mature technology with many tools available [47, 49, 105].

While formal verification seeks to verify a system (or its component) that has already been designed, *formal synthesis* targets the ambitious goal of formally designing systems (or system components such as a controller) that are provably correct by design. Seminal work was done in the late 1980s by Pnueli and Rosner [186, 207] in the context of *reactive synthesis* and, separately in the control community, by Ramadge and Wonham [200, 243] for control of *discrete-event systems*.

For its ability to handle complex specifications, formal methods captured the attention of researchers in the robotics and control communities over the past two decades. The readers are referred to recent surveys and monographs [25, 126, 150, 160, 184, 187, 222] and references therein for a more complete overview of this research trend.

The main aim of this book is to present formal methods for control of *nonlinear* dynamical systems with theoretical guarantees. We consider both continuous-time and discrete-time dynamical systems whose states evolve in continuous state spaces. Such models are paramount to control and robotics applications. The specifications we consider are general ω-regular properties that concern the infinite time-horizon behaviors of dynamical systems, which include relatively simple but useful properties of safety, invariance, reachability, reach-avoid (reachability with safety), as well as more complex specifications described by linear temporal logic (LTL). We embrace the fact that, in general, it is impossible to exactly determine if an ω-regular property is satisfiable by a nonlinear control system. Indeed, even reachability is undecidable for low-dimensional continuous or simple hybrid system models (see, e.g., [17, 104, 119, 172, 215]). Instead, we seek to answer the question whether a robust controller can be found to enforce the property in the presence of small perturbations to the dynamical system under control, provided that certain robust controllers do exist. We call such algorithms *robustly complete.* This can be seen as a *delta-decision procedure* [80] for nonlinear controller synthesis, which extends previous work on bounded-horizon reachability analysis [122] and observations that "*robustness implies decidability*" in the context of reachability analysis [14, 72, 73] to the more challenging situation of controller synthesis for nonlinear systems with respect to ω-regular specifications.

In the presented framework, there are two key enablers for automated formal synthesis of controllers for nonlinear systems with theoretical guarantees. One is validated computation for dynamical systems, notably *interval analysis* [173]. The other is the notion of discrete *control abstractions* [154, 203, 247] for transforming a continuous control problem into a discrete synthesis problem. The latter is amenable to automated formal synthesis technique developed for finite transition systems. Interval analysis was introduced by Moore in the 1960s [173] to provide rigorous enclosures of solutions to numerical problems. When applied to validated computation for dynamical systems, it enables computation of arbitrarily precise over-approximations of one-step forward reachable sets under Lipschitz continuity assumptions on system dynamics. Such over-approximations in turn can be used to construct arbitrarily precise control abstractions [152] or compute arbitrarily precise under-approximations of controlled predecessors [139] for dynamical systems, both of which lead to robust completeness of controller synthesis.

Scope and Outline of the Book

It is intended that the book be more than just a research monograph. The targeted readers are graduate students and researchers who are interested in learning how formal methods and validated computation can be combined together to tackle nonlinear control problems with complex specifications from an algorithmic perspective.

Chapter 1 provides an introduction to the theory of continuous-time dynamical systems and discusses how Lyapunov functions can be used as a tool for certifying relatively simple properties such as stability, safety, and reachability. Chapter 2 covers similar topics for discrete-time dynamical systems.

Chapter 3 introduces formal specifications and discrete synthesis for readers without any prior knowledge of automata theory and formal methods. Notably, we present the discrete synthesis problems and their solutions in the context of nondeterministic transition systems, in contrast with their usual presentation in the language of two-player infinite games. We believe this exposition is closer to the formulation of control problems usually seen in control and robotics applications.

Chapter 4 presents the basic machinery of interval computation and uses it to compute one-step forward and backward reachable sets for both discrete-time and continuous-time dynamical systems. Such computation is central for controller synthesis in later chapters.

Chapter 5 discusses abstraction-based controller synthesis. The focus is on constructing finite abstractions with theoretical guarantees of soundness and approximate completeness via a robustness argument. In other words, such abstractions can be used for algorithmic synthesis of provably correct controllers with completeness guarantees.

Chapter 6 considers specification-guided controller synthesis without having to computing discrete abstractions first. The approach is motivated by needs to overcome the limitation of uniform discretization usually used by discrete abstractions.

Chapter 7 is devoted to a set of case studies, including control of a simple DC-DC boost converter, Moore-Greitzer jet engine model, mobile robots, robot manipulator, and bipedal locomotion. Examples in this chapter are solved using the the robust controller synthesis tool ROCS[1] [141, 146] for nonlinear systems.

The appendices include several chapters of background and supplementary materials that accompany the main text.

Acknowledgment

The origin of this book can be traced back to the collaborative work the second author had with Necmiye Ozay, Ufuk Topcu, Pavithra Prabhakar, and Richard Murray about a decade ago. The second author is indebted to their collaboration and influence in this line of research. The authors are also grateful to Yiming Meng, Zhibing Sun, Max Fitzsimmons, Ye Zhao, Luis Sentis, and Ebrahim Shahrivar for joint research related to the contents of this book.

A good part of the book (Chapters 6 and 7) is based on the first author's PhD thesis [135]. The first author would like to express her heartfelt thanks to those who encouraged and supported her during her research. The second author is grateful to his mentors, students, colleagues, and collaborators who have supported him throughout his career. Both authors thank colleagues at the Department of Applied Mathematics, University of Waterloo for creating a pleasant working environment. Funding support from the Natural Sciences and Engineering Research Council of Canada, the Canada Research Chairs Program, and the Ontario Early Researcher Award Program is gratefully acknowledged. Comments and corrections are welcome (`j.liu@uwaterloo.ca`).

Cambridge, Ontario, Canada — Yinan Li
Waterloo, Ontario, Canada — Jun Liu
June, 2022

[1]`https://git.uwaterloo.ca/hybrid-systems-lab/rocs`

About the Authors

Yinan Li is a senior autonomy developer at OTTO Motors—the industrial division of Clearpath Robotics—in Kitchener, Ontario, Canada. She works on navigation algorithms for industrial mobile robots, multi-robot traffic control and coordination. Before joining OTTO Motors, she was a postdoctoral fellow at Hybrid Systems Laboratory, shortly after she received her Ph.D. degree in applied mathematics from University of Waterloo in 2019. Her Ph.D. and postdoctoral research focused on formal control methods for nonlinear systems, and the joint research with Georgia Institute of Technology and Clearpath Robotics was about applications of formal control methods in humanoid and mobile robots. She was awarded the 2020 Huawei prize for best research paper for her research in this area.

Jun Liu received the B.S. degree in applied mathematics from Shanghai Jiao-Tong University in 2002, the M.S. degree in mathematics from Peking University in 2005, and the Ph.D. degree in applied mathematics from the University of Waterloo in 2011. Following an NSERC Postdoctoral Fellowship in Control and Dynamical Systems at Caltech, he became a Lecturer in Control and Systems Engineering at the University of Sheffield in 2012. He joined the Faculty of Mathematics of the University of Waterloo in 2015, where he currently is an Associate Professor of Applied Mathematics and directs the Hybrid Systems Laboratory. Dr. Liu's main research interests are in the theory and applications of hybrid systems and control, including rigorous computational methods for control design with applications in cyber-physical systems and robotics. He was awarded a Marie-Curie Career Integration Grant from the European Commission in 2013, a Canada Research Chair from the Government of Canada in 2017 and 2022, an Early Researcher Award from the Ontario Ministry of Research, Innovation and Science in 2018, and an Early Career Award from the Canadian Applied and Industrial Mathematics Society and Pacific Institute for the Mathematical Sciences in 2020. His best paper awards include the Zhang Si-Ying Outstanding Youth Paper Award and the Nonlinear Analysis: Hybrid Systems Paper Prize. Dr. Liu is a senior member of IEEE, a member of SIAM, and a lifetime member of CAIMS. He has served on the editorial boards and program committees of several journals and conferences, including Automatica, Nonlinear Analysis: Hybrid Systems, Systems & Control Letters, the ACM International Conference on Hybrid Systems: Computation and Control (HSCC), the IEEE Conference on Decision and Control (CDC), and the American Control Conference (ACC).

List of Symbols

$\mathbb{R}$	the set of all real numbers
$\mathbb{R}^n$	the n-dimensional Euclidean space
$\mathbb{R}_{\geq 0}$	the set of all nonnegative real numbers
$\mathbb{R}_{>0}$	the set of all positive real numbers
$[x]$	a compact interval (box, hyperrectangle) in $\mathbb{R}$ or $\mathbb{R}^n$
$\mathbb{IR}$	the set of all compact intervals in $\mathbb{R}$
$\mathbb{IR}^n$	the set of all compact intervals (boxes, hyperrectangles) in $\mathbb{R}^n$
$\mathbf{0}^n$	the n-dimensional zero vector, i.e., all components are zero
$\mathbf{1}^n$	the n-dimensional one vector, i.e., all components are one
$\mathbb{Z}$	the set of all integers
$\mathbb{Z}_{\geq 0}$, $\mathbb{N}$	the set of all nonnegative integers (natural numbers with zero)
$\mathbb{Z}_{>0}$	the set of all positive integers
$\mathbb{Z}^k$	the k-dimensional integer lattice, i.e., the set of all k-tuples of integers
$\lvert x\rvert$	the (component-wise) absolute value of a vector $x \in \mathbb{R}^n$
$\lVert x\rVert_2$, $\lVert x\rVert$	the 2-norm (Euclidean norm) of a vector $x \in \mathbb{R}^n$
$\lVert x\rVert_\infty$	the infinity (maximum) norm of a vector $x \in \mathbb{R}^n$
$x \leq y$	the inequality holds component-wise for vectors $x, y \in \mathbb{R}^n$ $x < y$, $x \geq y$, and $x > y$ are similarly defined
$\lvert A\rvert$	the cardinality of a finite set A
$A \setminus B$	the set difference defined by $A \setminus B := \{x \mid x \in A,\, x \notin B\}$
$A + B$	the Minkowski sum defined by $A + B := \{x + y \mid x \in A,\, y \in B\}$
$A - B$	the Minkowski/Pontryagin difference defined by $A - B := \{x \mid x + y \in A,\, \forall y \in B\}$

$A \times B$	the Cartesian product defined by $A \times B := \{(x, y) \mid x \in A,\ y \in B\}$
2^A	the power set, i.e., the set of all subsets, of A
$\overline{A}$	the closure of the set A
A^c	the complement of the set A
$\mathrm{int}(A)$	the interior of the set A
$\mathbb{B}_r(x)$, $\mathbb{B}_r$, $\mathbb{B}$	$\mathbb{B}_r(x)$ is the closed ball with radius r with respect to the infinity norm centered at x; we also write $\mathbb{B}_r(0)$ as $\mathbb{B}_r$ and $\mathbb{B}_1$ as $\mathbb{B}$ for simplicity
$d(x, A)$, $\|x\|_A$	the distance from a point x to a set A
$\mathbb{B}_r(A)$	the r-neighborhood of the set A defined by $\mathbb{B}_r(A) := \{x \mid \|x\|_A \leq r\}$

Chapter 1

Continuous-Time Dynamical Systems

An important class of dynamical systems is modeled by differential equations. The purpose of this chapter is to introduce differential equations with inputs for modeling continuous-time control systems. We also discuss how classical tools from control theory such as Lyapunov functions can be used to establish relatively simple solution properties such as stability, boundedness, invariance, safety, and reachability.

1.1 Continuous-Time Control System

A continuous-time control system can be written in the form

$$x'(t) = f(t, x(t), u(t), w(t)), \tag{1.1}$$

where $x(t) \in D \subseteq \mathbb{R}^n$ is the system state and $u(t) \in U \subseteq \mathbb{R}^m$ and $w(t) \in W \subseteq \mathbb{R}^p$ are input signals. In this book, we regard $u(t)$ as a *control input* that can be designed to drive the system to have desired behaviors. In contrast, $w(t)$ is treated as a *disturbance input* that cannot be controlled. We also adopt a worst-case scenario approach to control design; i.e., the control signal is responsible for controlling the system to have desired behaviors for all possible choices of disturbance inputs.

In the absence of disturbance, the system reduces to

$$x'(t) = f(t, x(t), u(t)), \tag{1.2}$$

which we sometimes call a *nominal control system*. Furthermore, in the absence of control and without explicit dependence on t by f, the system reduces to

$$x'(t) = f(x(t)), \tag{1.3}$$

which is called an *autonomous system* in the dynamical systems literature.

The primary goal of **control**[1] **synthesis** is to design appropriate controls u such that the behaviors (formally defined as *solutions* in Section 1.2) of (1.1)

[1]We shall use the words "control" and "controller" interchangeably, both meaning a mechanism to regulate the operation of a system.

DOI: 10.1201/9780429270253-1

or (1.2) have desired properties (formally defined as *specifications* in Chapter 3). By an *open-loop* control, we mean a time-dependent function $u(t)$. By a *feedback* control, we mean a choice of u that may depend on the history of the states. A common form of feedback control is called (memoryless) *state feedback*, where $u = \kappa(x)$ is a function of the state x.

System **verification** is concerned with analyzing whether the behaviors of a system without control, or for which a control is already given, such as the autonomous system (1.3) or a closed-loop system of the form

$$x'(t) = f(x(t), \kappa(x(t))), \tag{1.4}$$

meet desired requirements. Formal definitions of verification and control synthesis problems will be defined in Chapter 3.

1.2 Existence of Local and Global Solutions

By a *solution* to (1.1), we mean a function $x(t)$ such that (1.1) is satisfied for given functions $u(t)$ and $w(t)$. To make this precise, consider the initial value problem (IVP) for (1.1) with the initial condition

$$x(t_0) = x_0 \in D \subseteq \mathbb{R}^n, \tag{1.5}$$

where $t_0 \geq 0$ and D is an open set.

Due to the existence of both control and disturbance inputs, we may not want to restrict too much on the regularity of these inputs. We shall assume that both $u(t)$ and $w(t)$ are locally essentially bounded measurable[2] functions, which are defined as follows.

Let $J \subseteq \mathbb{R}$ be an interval. A (Lebesgue) measurable function $v : J \to \mathbb{R}^n$ is said to be *essentially bounded* if there exists a constant M such that $\|v(t)\| \leq M$ for almost all $t \in J$. In other words, the function is bounded except possibly on a measure zero set. We say a function is *locally essentially bounded* on a certain domain if it is essentially bounded on every compact interval in the domain. It is clear that every piecewise continuous function is locally essentially bounded.

Because of the change in regularity requirements for the inputs, we also need to relax the requirement for the regularity of solutions. The requirement of differentiability or piecewise differentiability is replaced with absolute continuity, which entails almost everywhere differentiability.

Let $J \subseteq \mathbb{R}$ be a compact interval. A function $v : J \to \mathbb{R}^n$ is said to be *absolutely continuous* on J, if for every $\varepsilon > 0$, there exists a $\delta > 0$ such that

[2]Readers who are not familiar with measure theory can assume that inputs are piecewise continuous functions when reading this section.

for every finite collection of disjoint open subintervals $\{(a_i, b_i)\}_{i=1}^k$ of J, we have

$$\sum_{i=1}^{k}(b_i - a_i) < \delta \quad \text{implies} \quad \sum_{i=1}^{k}\|f(b_k) - f(a_k)\| < \varepsilon.$$

It is clear that absolute continuity implies uniform continuity because the definition of uniform continuity is captured by the above definition with $k = 1$. We say a function is *locally absolutely continuous* on a certain domain, if it is absolutely continuous on every compact interval in the domain.

Definition 1.1 (Local solution) *Let $J \subseteq \mathbb{R}$ be an interval containing t_0. Suppose that $u : J \to \mathbb{R}^m$ and $w : J \to \mathbb{R}^m$ are locally essentially bounded functions. A* ***solution*** *to the IVP (1.1) and (1.5) on J is a locally absolutely continuous function x such that*

$$x(t) = x_0 + \int_{t_0}^{t} f(s, x(s), u(s), w(s))ds \tag{1.6}$$

for all $t \in J$. The solution is sometimes denoted by $x(t; t_0, x_0, u, w)$.

For the integral (1.6) to make sense, we need to show that $f(t, x(t), u(t), w(t))$ is measurable in t. This is guaranteed by the following assumptions on f.

Assumption 1.1 (Carathéodory function) *Let $J \subseteq \mathbb{R}$ be an interval and $D \subseteq \mathbb{R}^n$ be an open set. The function $f : J \times D \times U \times W \to \mathbb{R}^n$ satisfies the following properties:*

- *$f(\cdot, x, u, w) : J \to \mathbb{R}^n$ is measurable for each $(x, u, w) \in D \times U \times W$; and*
- *$f(t, \cdot, \cdot, \cdot) : D \times U \times W \to \mathbb{R}^n$ is continuous for each $t \in J$.*

Under this assumption, we can show that if $x(t)$, $u(t)$, and $w(t)$ are measurable on J, then $t \mapsto f(t, x(t), u(t), w(t))$ is also measurable on J (see Appendix A). If this condition is met, the integral equation (1.6) is equivalent to that the differential equation (1.1) holds for almost every $t \in J$. Since all solutions to differential equations will be understood in this sense, we shall drop the phrase "almost every" for brevity if no confusion arises.

We now state an existence and uniqueness theorem for the IVP (1.1) and (1.5).

Proposition 1.1 (Local existence and uniqueness) *Let $J \subseteq \mathbb{R}$ be an interval and $D \subseteq \mathbb{R}^n$ be an open set. Suppose that $f : J \times D \times U \times W \to \mathbb{R}^n$ satisfies Assumption 1.1 and the following conditions:*

1. *(local integrable) For every compact set $K \subseteq J \times D \times U \times W$, there exists a Lebesgue integrable function $m_K(t)$ such that $\|f(t, x, u, w)\| \le m_K(t)$ for all $(t, x, u, w) \in K$.*

2. *(local Lipschitz) For every compact set $K \subseteq J \times D \times U \times W$, there exists a Lebesgue integrable function $l_K(t)$ such that*

$$\|f(t,x,u,w) - f(t,y,u,w)\| \leq l_K(t)\,\|x - y\|,$$

for all $(t,x,u,w) \in K$.

Then, for every $(t_0, x_0) \in J \times D$ and every pair of locally essentially bounded inputs $u(t)$ and $w(t)$ on J, there exists a unique solution $x(t; t_0, x_0, u, w)$ to the IVP (1.1) and (1.5), defined on a maximal interval of existence $J^* \subseteq J$ that is open relative to J.

Proof: See Appendix A.2. ■

Remark 1.1 *A standing assumption of this book is that the right-hand side of a differential equation of the form (1.1), or more special forms, satisfies the assumptions of Proposition 1.1.*

We now discuss conditions under which there exists a global solution to the IVP (1.1) and (1.5). Throughout this section, suppose that the assumptions in Proposition 1.1 hold with $J = \mathbb{R}$ and $D = \mathbb{R}^n$. By Proposition 1.1, for every pair of (t_0, x_0) and every pair of locally essentially bounded inputs $u(t)$ and $w(t)$ defined on $\mathbb{R}$, there exists a unique solution $x(t; t_0, x_0, u, w)$ to the IVP (1.1) and (1.5), defined on the maximal interval of existence $J^* = (\alpha, \beta)$. Let $x(t) = x(t; t_0, x_0, u, w)$.

Proposition 1.2 (Boundedness of solutions implies global existence) *We have the following:*

1. *If, for every $c > t_0$, $x(t)$ is bounded on $J^* \cap [t_0, c)$, then $\beta = \infty$.*

2. *If, for every $c < t_0$, $x(t)$ is bounded on $J^* \cap (c, t_0]$, then $\alpha = -\infty$.*

3. *If, for every $c > 0$, $x(t)$ is bounded on $J^* \cap (-c, c)$, then $J^* = (-\infty, \infty)$.*

Proof: See Appendix A.3. ■

If $\beta = \infty$, the solution is said to be *forward complete.* We say that the system (1.1) is forward complete if every solution to the IVP (1.1) and (1.5) is forward complete. Similarly, $\alpha = -\infty$ defines *backward completeness.* For control systems, forward completeness is conceptually more important, because we are interested in system behaviors in forward time.

To apply Proposition 1.2, we need to establish boundedness of solutions. We will introduce such results in Section 1.3. Nonetheless, based on Proposition 1.2, we can also formulate global existence results by imposing conditions on the vector field f. For example, the following result states that if the vector field satisfies a (sub)linear growth condition, then solutions are defined globally.

Corollary 1.1 (Sublinearity implies global existence) *If there exist locally integral functions $a(t)$ and $b(t)$ defined on $\mathbb{R}$ such that*

$$\|f(t,x,u,w)\| \leq a(t)\,\|x\| + b(t)$$

for all $(t,x,u,w) \in \mathbb{R} \times \mathbb{R}^n \times U \times W$, then $J^* = \mathbb{R}$.

Proof: See Appendix A.3. ■

1.3 Stability and Boundedness

An important class of techniques to analyze nonlinear systems such as (1.1) is Lyapunov analysis. The intuitive idea is to analyze the system via an auxiliary function called a *Lyapunov function.* In this section, we review two types of results, namely results on boundedness and stability of solutions. Since the focus of this section is on analysis, we consider a system without control input of the following form

$$x'(t) = f(t, x(t), w(t)), \tag{1.7}$$

which can be obtained from (1.1), e.g., when the control is already given by some state feedback $u = \kappa(x)$. We assume that the disturbance signal $w(t)$ is an arbitrary locally essentially bounded function, taking value in W, and denote by $\mathcal{W}$ the set of all such disturbance signals.

Stability is a central notion in systems and control. We define stability for solutions of (1.7) with respect to a compact invariant set as follows.

Definition 1.2 *A set $A \subseteq \mathbb{R}^n$ is said to be a* ***positively invariant set*** *for system (1.7), if it is nonempty and, for any $x_0 \in A$, all solutions of (1.7) starting from $x(t_0) = x_0$ stay in A for all $t \geq t_0$.*

When there is no ambiguity, we shall simply call a positively invariant set an *invariant set.* Given $x \in \mathbb{R}^n$ and a set $A \subseteq \mathbb{R}^n$, let $\|x\|_A$[3] denote the distance from x to A defined by

$$\|x\|_A := \inf_{y \in A} \|x - y\|.$$

[3]Another commonly used notation for this point-to-set distance $\|x\|_A$ is $d(x, A)$. The former makes more sense in a normed vector space, while the latter is for a general metric space. Since we work mostly in $\mathbb{R}^n$, we use both notations interchangeably.

Definition 1.3 *Let $A \subseteq \mathbb{R}^n$ be a compact invariant set for system (1.7). We say that A is* **uniformly asymptotically stable** *(UAS) for system (1.7), if the following two conditions hold:*

1. *(Uniform stability) For every $\varepsilon > 0$, there exists $\delta = \delta(\varepsilon) > 0$ such that, if $\|x_0\|_A \le \delta$, then $\|x(t)\|_A \le \varepsilon$ for all $t \ge t_0$, where $x(t)$ is any solution of (1.7) starting from $x(t_0) = x_0$.*

2. *(Uniform attraction) There exists some $\rho > 0$ such that, for any $\eta > 0$, there exists some $T = T(\rho, \eta) \ge 0$ such that $\|x(t)\|_A \le \eta$, whenever $\|x_0\|_A \le \rho$ and $t \ge t_0 + T$, where $x(t)$ is any solution of (1.7) starting from $x(t_0) = x_0$.*

We say that A is **globally uniformly asymptotically stable** *(GUAS) for system (1.7) if the above conditions hold for δ chosen such that $\lim_{\varepsilon\to\infty} \delta(\varepsilon) = \infty$ and any $\rho > 0$.*

We present a Lyapunov theorem on the stability analysis of system (1.7) with respect to a compact invariant set. The proof is also fairly standard (see, e.g., texts [98, 114]). Here we preset the proof for stability analysis with respect to a compact invariant set A.

Definition 1.4 *Let $D \subseteq \mathbb{R}^n$ be an open set and A be a compact set for the system (1.7) contained in D. We say that a function $V : D \to \mathbb{R}$ is* **positive definite with respect to** *A if $V(x) = 0$ for all $x \in A$ and $V(x) > 0$ for all $x \in D \setminus A$.*

Definition 1.5 *We say that a function $\alpha : \mathbb{R}^n \to \mathbb{R}$ is* **radially unbounded** *if $\alpha(x) \to \infty$ as $\|x\| \to \infty$.*

Theorem 1.1 (Lyapunov theorem for stability) *Let A be a compact invariant set of system (1.7). Let $D \subseteq \mathbb{R}^n$ be an open set containing A and $V : [0, \infty) \times D \to \mathbb{R}$ be a continuously differentiable function. Suppose that there exist continuous functions α_i ($i = 1, 2, 3$) that are defined on D and positive definite with respect to A such that*

$$\alpha_1(x) \le V(t, x) \le \alpha_2(x), \quad \forall x \in D, \quad \forall t \ge 0, \tag{1.8}$$

and

$$\frac{dV}{dt} + \frac{dV}{dx} f(t, x, w) \le -\alpha_3(x), \quad \forall x \in D \setminus A, \quad \forall t \ge 0, \quad \forall w \in W. \tag{1.9}$$

Then A is uniformly asymptotically stable for the system (1.7). If the above conditions hold for $D = \mathbb{R}^n$ and α_1 is radially unbounded, then A is globally uniformly asymptotically stable for system (1.7).

Proof: (**Uniform stability**) Fix an arbitrary $\varepsilon > 0$. Without loss of generality, assume that ε is sufficiently small such that $\mathbb{B}_\varepsilon(A) \subseteq D$. Choose c such that

$$0 < c < \min_{\|x\|_A = \varepsilon} \alpha_1(x).$$

Then the set

$$\Omega_1^c := \{x \in \mathbb{B}_\varepsilon(A) : \alpha_1(x) \le c\}$$

is contained in the interior of $\mathbb{B}_\varepsilon(A)$. Define

$$\Omega_2^c := \{x \in \mathbb{B}_\varepsilon(A) : \alpha_2(x) \le c\}.$$

Then $\Omega_2^c \subseteq \Omega_1^c$. Pick $\delta \in (0, \varepsilon)$ sufficiently small such that $\mathbb{B}_\delta(A) \subseteq \Omega_2^c$. This is always possible, because $\alpha_2(x)$ is continuous on D and $\alpha_2(x) = 0$ for all $x \in A$. Hence, for sufficiently small $\delta > 0$, $\alpha_2(x) \le c$ for all $x \in \mathbb{B}_\delta(A)$, which implies $\mathbb{B}_\delta(A) \subseteq \Omega_2^c$. We claim that solutions of (1.7) starting from any initial state $x_0 \in \mathbb{B}_\delta(A)$ and any initial time $t_0 \ge 0$ will remain in $\Omega_1^c \subseteq \mathbb{B}_\varepsilon(A)$ for all $t \ge t_0$. This would imply uniform stability.

Pick any $x_0 \in \mathbb{B}_\delta(A) \subseteq \Omega_2^c$. Let $x(t)$ be any solution of (1.7) satisfying $x(t_0) = x_0$. Then

$$V(t_0, x(t_0)) = V(t_0, x_0) \le \alpha_2(x_0) \le c.$$

We have

$$\frac{dV(t, x(t))}{dt} = \frac{\partial V}{\partial t} + \frac{dV}{dx} f(t, x(t), w(t)) \le -\alpha_3(x(t)) \le 0 \tag{1.10}$$

for almost all[4] $t \ge t_0$, provided that $x(t)$ remains in D. To escape $\mathbb{B}_\varepsilon(A)$, the solution needs to cross the boundary of $\mathbb{B}_\varepsilon(A)$, on which $\alpha_1(x) > c$. This implies that $V(t, x(t)) \ge \alpha_1(x(t)) > c$ for some t. This is impossible since (1.10) implies that $V(t, x(t))$ is non-increasing for $x(t) \in D$.

(**Uniform attraction**) From the proof of uniform stability, we have shown that, for every $\epsilon > 0$ such that $\mathbb{B}_\varepsilon(A) \subseteq D$, there exists some $\delta > 0$ such that solutions of (1.7) starting from $\mathbb{B}_\delta(A)$ will stay in $\mathbb{B}_\varepsilon(A)$ for all $t \ge t_0$. We fix a choice of such ϵ and δ for the following argument. Let $\rho = \delta$. Fix any $\eta \in (0, \varepsilon)$ (without loss of generality). We show that there exists $T = T(\eta)$ such that any solution $x(t)$ of (1.7) starting in $\mathbb{B}_\delta(A)$ will reach and stay in $\mathbb{B}_\eta(A)$ for all $t \ge t_0 + T$. By the argument of uniform stability again, there exists some $\delta' = \delta'(\eta) > 0$ such that solutions of (1.7) starting in $\mathbb{B}_{\delta'}(A)$ will stay in $\mathbb{B}_\eta(A)$ for all future time. Hence we only need to show that solutions starting in $\mathbb{B}_\delta(A)$ will reach $\mathbb{B}_{\delta'}(A)$ within some finite time T.

Let

$$\lambda = \min_{\delta' \le \|x\|_A \le \varepsilon} \alpha_3(x) > 0,$$

[4]Note that $x(t)$ is absolutely continuous and $x'(t)$ exists almost everywhere.

and

$$c = \max_{x \in \mathbb{B}_\varepsilon(A)} \alpha_2(x).$$

Choose $T > \frac{c}{\lambda}$. Let $x(t)$ be any solution of (1.7) starting from $x(t_0) = x_0 \in \mathbb{B}_\rho(A)$. Without loss of generality, assume $x_0 \notin \mathbb{B}_{\delta'}(A)$. Otherwise, $x(t) \in \mathbb{B}_\eta(A)$ for all $t \geq t_0$.

Suppose that $x(t)$ never reaches $\mathbb{B}_{\delta'}(A)$ on $[t_0, t_0 + T]$. That is, $\delta' < \|x(t)\| \leq \varepsilon$ for all $t \in [t_0, t_0 + T]$. Then we have

$$\begin{aligned}
&\alpha_1(x(t_0 + T)) \\
&\quad \leq V(t_0 + T, x(t_0 + T)) \\
&\quad = V(t_0, x_0) + \int_{t_0}^{t_0+T} \frac{dV(s, x(s))}{ds} ds \\
&\quad = V(t_0, x_0) + \int_{t_0}^{t_0+T} \left[\frac{\partial V}{\partial t}(s, x(s)) + \frac{dV}{dx}(s, x(s)) f(s, x(s), w(s)) \right] ds \\
&\quad \leq V(t_0, x_0) - \int_{t_0}^{t_0+T} \alpha_3(x(s)) ds \\
&\quad \leq \alpha_2(x_0) - \lambda T \leq c - \lambda T < 0,
\end{aligned} \tag{1.11}$$

which is a contradiction, because $\alpha_1(x)$ cannot be negative. Hence, $x(t)$ must have reached $\mathbb{B}_{\delta'}(A)$ for some $t \in [t_0, t_0 + T]$. It follows that $x(t) \in \mathbb{B}_\eta(A)$ for all $t \geq t_0 + T$.

(**Global stability**) When $D = \mathbb{R}^n$ and $\alpha_1(x)$ is radially unbounded, we can pick $\delta = \delta(\varepsilon)$ for uniform stability such that $\lim_{\varepsilon \to \infty} \delta(\varepsilon) = \infty$. This is because $c \to \infty$, as $\varepsilon \to \infty$, and Ω_2^c can contain any given $\mathbb{B}_\delta(A)$ for c sufficiently large. Furthermore, $\rho = \delta$ in the proof of uniform attraction can be made arbitrarily large. ■

Another important property for solutions of (1.7) is boundedness. We introduce two closely related notions of boundedness as follows.

Definition 1.6 *We say that solutions of (1.7) are* ***uniformly bounded*** *(UB) if there exists a constant $c > 0$ such that, for all $t_0 \geq 0$ and $\rho \in (0, c)$, $\|x(t_0)\| \leq \rho$ implies $\|x(t)\| \leq b$, where $b = b(\rho)$ only depends on ρ. We say that solutions of (1.7) are* ***globally uniformly bounded*** *(GUB), if the above holds for all $\rho > 0$.*

Definition 1.7 *We say that solutions of (1.7) are* ***uniformly ultimately bounded*** *(UUB) if there exist constants $c > 0$ and $b > 0$ such that, for every $\rho \in (0, c)$, there exists $T = T(\rho, b) \geq 0$ such that $\|x(t_0)\| \leq \rho$ implies $\|x(t)\| \leq b$ for all $t \geq t_0 + T$. We say that solutions of (1.7) are* ***globally uniformly ultimately bounded*** *(GUUB), if the above holds for all $\rho > 0$. The constant b is called the ultimate bound of solutions.*

We present a Lyapunov theorem for certifying the boundedness of the solutions of (1.7).

Theorem 1.2 (Lyapunov theorem for boundedness) *Let $D \subseteq \mathbb{R}^n$ be an open set and $V : [0, \infty) \times D \to \mathbb{R}$ be a continuously differentiable function. Suppose that there exist continuous functions α_i $(i = 1, 2, 3)$ defined on D and positive on $D \setminus \mathbb{B}_\mu$, for some $\mu > 0$, such that*

$$\alpha_1(x) \leq V(t, x) \leq \alpha_2(x), \quad \forall x \in D \setminus \mathbb{B}_\mu, \quad \forall t \geq 0, \tag{1.12}$$

and

$$\frac{dV}{dt} + \frac{dV}{dx} f(t, x, w) \leq -\alpha_3(x), \quad \forall x \in D \setminus \mathbb{B}_\mu, \quad \forall t \geq 0, \quad \forall w \in W. \tag{1.13}$$

Suppose there exists some $\rho > 0$ such that $\mathbb{B}_\rho \subseteq D$ and

$$\max_{x \in \mathbb{B}_\mu} \alpha_2(x) < \min_{\|x\|_\infty = \rho} \alpha_1(x). \tag{1.14}$$

Then solutions of (1.7) are both uniformly bounded and uniformly ultimately bounded. If the conditions (1.12) and (1.13) hold for $D = \mathbb{R}^n$ and $\alpha_1(x)$ is radially unbounded, then (1.14) is satisfied and solutions of (1.7) are both globally uniformly bounded and globally uniformly ultimately bounded.

The proof of this result is similar to that for Theorem 1.1 on set stability. The only difference is that the argument mainly takes place outside the ball $\mathbb{B}_\mu$.

Proof: Let

$$c_2 = \max_{x \in \mathbb{B}_\mu} \alpha_2(x), \quad c_1 = \min_{\|x\|_\infty = \rho} \alpha_1(x).$$

By the theorem condition, we have $c_2 < c_1$. For any $c \in [c_2, c_1]$, define

$$\Omega_1^c := \{x \in \mathbb{B}_\rho : \alpha_1(x) \leq c\},$$

and

$$\Omega_2^c := \{x \in \mathbb{B}_\rho : \alpha_2(x) \leq c\}.$$

Then $\Omega_1^c \subseteq \mathbb{B}_\rho$ and $\mathbb{B}_\mu \subseteq \Omega_2^c$. By the same argument as in the proof of Theorem 1.1, we can show that solutions of (1.7) starting from Ω_2^c will not leave Ω_1^c. Using the same argument as in (1.11), we can then show that solutions of (1.7) starting from $\Omega_1^{c_1}$ will reach $\Omega_2^{c_2}$ in finite time. Combing these arguments, we conclude that all solutions of (1.7) starting from $\Omega_2^{c_1}$ will remain in $\Omega_1^{c_1}$ for all future time (hence bounded), reach $\Omega_2^{c_2}$ in finite time, and from there stay in $\Omega_1^{c_2}$ for all time after (hence ultimately bounded). We can take $b = \max_{x \in \Omega_1^{c_2}} \|x\|$ to be the ultimate bound. When $D = \mathbb{R}^n$ and $\alpha_1(x)$ is radially unbounded, we can take $\rho > 0$ sufficiently large to contain any initial condition with $\Omega_2^{c_1}$. ■

Remark 1.2 *Clearly, the choice of norm $\|\cdot\|_\infty$ is not essential for Lyapunov analysis since all norms in $\mathbb{R}^n$ are equivalent.*

We illustrate the Lyapunov analysis using an example.

Example 1.1 *Consider the three-dimensional Lorenz system given by*

$$\begin{cases} x_1' = \sigma(x_2 - x_1), \\ x_2' = rx_1 - x_2 - x_1x_3, \\ x_3' = x_1x_2 - bx_3, \end{cases} \tag{1.15}$$

where the parameters σ, r, and b are all positive. For $r > 1$, this system has three equilibrium points, including the origin. For $r \leq 1$, the system has a unique equilibrium point at the origin.

(Stability) *We first analyze stability of the origin using the following Lyapunov function*

$$V(x_1, x_2, x_3) = \frac{x_1^2}{\sigma} + x_2^2 + x_3^2.$$

Let $f : \mathbb{R}^3 \to \mathbb{R}^3$ denote the right-hand side of (1.15). Then

$$\begin{aligned} \dot{V}(x) &= \frac{\partial V}{\partial x} f(x) \\ &= \frac{2x_1}{\sigma}\sigma(x_2 - x_1) + 2x_2(rx_1 - x_2 - x_1x_3) + 2x_3(x_1x_2 - bx_3) \\ &= -2x_1^2 + 2(1+r)x_1x_2 - 2x_2^2 - 2bx_3^2. \end{aligned} \tag{1.16}$$

Clearly, conditions (1.8) and (1.9) of Theorem 1.1 are satisfied when $r < 1$. Indeed, we can take $D = \mathbb{R}^n$ and $\alpha_1(x) = \alpha_2(x) = V(x)$ and $\alpha_3(x)$ to the right-hand side of (1.16). Then we can conclude that the origin is globally uniformly asymptotically stable for (1.15) when $r < 1$.

(Boundedness) *We further analyze boundedness of solutions for (1.15). Consider the Lyapunov function*

$$V(x_1, x_2, x_3) = \frac{x_1^2}{\sigma} + x_2^2 + x_3^2 - 2(1+r)x_3.$$

Then, following (1.16), we obtain

$$\begin{aligned} \dot{V}(x) &= \frac{\partial V}{\partial x} f(x) \\ &= -2x_1^2 + 2(1+r)x_1x_2 - 2x_2^2 - 2bx_3^3 - 2(1+r)x_1x_2 + 2b(1+r)x_3 \\ &= -2x_1^2 - 2x_2^2 - 2bx_3^2 + 2b(1+r)x_3. \end{aligned}$$

To obtain an ultimate bound, note that

$$\begin{aligned} V(x) &= \frac{x_1^2}{\sigma} + x_2^2 + x_3^2 - 2(1+r)x_3 \\ &\geq \frac{x_1^2}{\sigma} + x_2^2 + x_3^2 - \varepsilon_1 x_3^2 - \frac{1}{\varepsilon_1}(1+r)^2 \\ &\geq \min(\frac{1}{\sigma}, 1 - \varepsilon_1) \left\|x\right\|_\infty^2 - \frac{1}{\varepsilon_1}(1+r)^2, \end{aligned}$$

where $\varepsilon_1 \in (0,1)$, *and*

$$\begin{aligned}\dot{V}(x) &= -2x_1^2 - 2x_2^2 - 2bx_3^2 + 2b(1+r)x_3 \\ &\leq -2x_1^2 - 2x_2^2 - 2bx_3^2 + b\varepsilon_2 x_3^2 + \frac{1}{\varepsilon_2}(1+r)^2 \\ &\leq -\min(2, 2b-\varepsilon_2)\left\|x\right\|_\infty^2 + \frac{1}{\varepsilon_2}(1+r)^2,\end{aligned}$$

where $\varepsilon_2 \in (0, 2b)$. *Pick*

$$\mu > \max\left(\sqrt{\frac{(1+r)^2}{\min(\frac{1}{\sigma}, 1-\varepsilon_1)\varepsilon_1}}, \sqrt{\frac{(1+r)^2}{\min(2, 2b-\varepsilon_2)\varepsilon_2}}\right).$$

Then conditions (1.12) and (1.13) are satisfied for all $\left\|x\right\|_\infty \geq \mu$ *with* $\alpha_1(x) = \alpha_2(x) = V(x)$ *and* $\alpha_3(x) = \dot{V}(x)$. *Clearly,* $\alpha_1(x) = V(x)$ *is radially unbounded. Hence, by Theorem 1.2, solutions of (1.15) are globally uniformly bounded and globally uniformly ultimately bounded.*

(Boundedness under bounded disturbances) *Finally, we consider the case that the system is perturbed by a bounded disturbance of the form*

$$x' = f(x) + w, \quad w \in \mathbb{B}_\delta, \tag{1.17}$$

where $\delta > 0$. *Using the same Lyapunov function, we obtain*

$$\begin{aligned}\dot{V}(x) = \frac{\partial V}{\partial x}[f(x)+w] &= -2x_1^2 - 2x_2^2 - 2bx_3^2 + 2b(1+r)x_3 \\ &\quad + \frac{2x_1}{\sigma}w_1 + 2x_2w_2 + 2x_3w_3 - 2(1+r)w_3 \\ &\leq -\min(2, 2b-\varepsilon_2)\left\|x\right\|_\infty^2 + \frac{1}{\varepsilon_2}(1+r)^2 \\ &\quad + \max(1, \frac{1}{\sigma^2})\varepsilon_3\left\|x\right\|_\infty^2 + \frac{\delta^2}{\varepsilon_3} + 2(1+r)\delta.\end{aligned}$$

We can pick

$$\mu > \max\left(\sqrt{\frac{(1+r)^2}{\min(\frac{1}{\sigma}, 1-\varepsilon_1)\varepsilon_1}}, \sqrt{\frac{\varepsilon_3(1+r)^2 + \varepsilon_2\delta^2 + 2\varepsilon_2\varepsilon_3(1+r)\delta}{(\min(2, 2b-\varepsilon_2) - \max(1, \frac{1}{\sigma^2})\varepsilon_3)\varepsilon_2\varepsilon_3}}\right)$$

such that conditions (1.12) and (1.13) are satisfied for all $\left\|x\right\|_\infty \geq \mu$ *with* $\alpha_1(x) = \alpha_2(x) = V(x)$ *and* $\alpha_3(x) = \dot{V}(x)$. *Hence, by Theorem 1.2, solutions of (1.15), subject to bounded disturbances described by (1.17), are globally uniformly bounded and globally uniformly ultimately bounded.*

As illustrated by the above example, despite the seemingly complicated analytical bounds, Lyapunov analysis proves to be one of the most powerful tools for analyzing nonlinear systems. There are computational techniques, such as linear matrix inequalities [36], that can be used to replace the tedious (and error-prone) manual process for deriving the Lyapunov conditions. There also exist techniques that can be used to search for Lyapunov functions using certain templates, such as sums of squares programming [195].

1.4 Safety and Reachability

In this section, we introduce Lyapunov theorems for certifying safety and reachability. We establish safety using set invariance and reachability using asymptotic stability, which are unified by Lyapunov analysis.

Definition 1.8 *Given an unsafe set $\Lambda \subseteq \mathbb{R}^n$ and an initial set $\Upsilon \subseteq \mathbb{R}^n$, we say that the system (1.7) satisfies a* ***safety*** *specification (Υ, Λ), if every solution $x(t)$ of (1.7) starting from $x(t_0) = x_0 \in \Upsilon$ will not enter Λ; that is, $x(t) \notin \Lambda$ for all $t \geq t_0$.*

We can also specify safety in conjunction with stability and reachability.

Definition 1.9 (Domain of attraction) *Suppose that a compact set $A \subseteq \mathbb{R}^n$ is UAS for system (1.7). The domain of attraction $\mathcal{D}(A)$ is defined as*

$$\mathcal{D}(A) = \left\{ x_0 \in \mathbb{R}^n : \lim_{t \to \infty} \|x(t)\|_A = 0, \forall x(\cdot) \in \Phi(x_0) \right\},$$

where $\Phi(x_0)$ denotes the set of all solutions for system (1.7) starting from x_0.

Definition 1.10 *Given an unsafe set $\Lambda \subseteq \mathbb{R}^n$ and a compact set $A \subseteq \mathbb{R}^n$, we say that system (1.7) satisfies a* ***stability with safety guarantee*** *specification (Υ, Λ, A), if the set A is UAS for system (1.7), $\Upsilon \subseteq \mathcal{D}(A)$, and system (1.7) satisfies a safety specification (Υ, Λ).*

Definition 1.11 *Given an unsafe set $\Lambda \subseteq \mathbb{R}^n$ and a target set $\Omega \subseteq \mathbb{R}^n$, we say that system (1.7) satisfies a* ***reach-avoid-stay*** *specification $(\Upsilon, \Lambda, \Omega)$, if system (1.7) satisfies a safety specification (Υ, Λ) and, for every solution $x(t)$ of (1.7) starting from $x(t_0) = x_0 \in \Upsilon$, there exists some $T \geq 0$ such that $x(t) \in \Omega$ for all $t \geq t_0 + T$.*

It is clear from Definitions 1.10 and 1.11 that, if system (1.7) satisfies a stability with safety guarantee specification (Υ, Λ, A), then for every $\eta > 0$, system (1.7) satisfies a reach-avoid-stay specification $(\Upsilon, \Lambda, \Omega)$ for $\Omega = \mathbb{B}_\eta(A)$. For this reason, we shall focus on discussing stability with safety guarantee. It is shown in [164] in the context of converse Lyapunov theorems that, as far as *robust* satisfaction of reach-avoid-stay specifications is concerned, this is without loss of generality.

We next present two sets of Lyapunov sufficient conditions, one for safety and one for stability with safety guarantee.

Theorem 1.3 (Lyapunov theorem for safety) *Given an unsafe set $\Lambda \subseteq \mathbb{R}^n$ and an initial set $\Upsilon \subseteq \mathbb{R}^n$, if there exists a continuously differentiable function $B : \mathbb{R}^n \to \mathbb{R}$ such that the following conditions are satisfied:*

1. *$B(x) \leq 0$ for all $x \in \Upsilon$;*

2. *$B(x) > 0$ for all $x \in \Lambda$; and*

3. *$\nabla B(x) \cdot f(t, x, w) < 0$ for all $t \geq 0$, all $w \in W$, and all x such that $B(x) = 0$,*

then system (1.7) satisfies the safety specification (Υ, Λ).

Proof: If we can show that the set

$$C = \{x \in \mathbb{R}^n : B(x) \leq 0\}, \tag{1.18}$$

is an invariant set for system (1.7), then solutions of (1.7) starting from Υ remain in C. Clearly, $C \cap \Lambda = \emptyset$. Hence, system (1.7) satisfies the safety specification (Υ, Λ). Forward invariance of C follows as a special case of Proposition 1.3(2). ∎

The function B in the above theorem is often called a **barrier function.** The following theorem combines a barrier function with a Lyapunov function for certifying stability properties with safety guarantees.

Theorem 1.4 (Lyapunov theorem for stability with safety) *Suppose that A is a compact invariant set for system (1.7) and $A \cap \Lambda = \emptyset$. If there exists an open set D such that $(A \cup \Upsilon) \subseteq D$ and continuously differentiable functions $V : D \to \mathbb{R}_{\geq 0}$ and $B : \mathbb{R}^n \to \mathbb{R}$ such that*

1. *V is positive definite on D with respect to A;*

2. *$\nabla V(x) \cdot f(t, x, w) \leq -\alpha_3(x)$ for all $t \geq 0$, $x \in D \setminus A$, and $w \in W$, where α_3 is a continuous function defined on D that is positive definite with respect to A;*

3. *$\Upsilon \subseteq C = \{x \in \mathbb{R}^n : B(x) \leq 0\} \subseteq D$ and $B(x) > 0$ for all $x \in \Lambda$;*

4. *$\nabla B(x) \cdot f(t, x, w) \leq 0$ for all $t \geq 0$, $w \in W$, and $x \in D$,*

then system (1.7) satisfies the stability with safety guarantee specification (Υ, Λ, A).

Proof: We can easily show that the set $C = \{x \in \mathbb{R}^n : B(x) \geq 0\}$ is forward invariant. Indeed, if C is not forward invariant, then there exists some $x_0 \in C$, a solution x starting from x_0, and some $\tau > 0$ such that $B(x(\tau)) < 0$. Define

$$\bar{t} = \sup\{t \geq 0 : \varphi(t) \in C\}.$$

Then $\bar{t}$ is well defined and finite. By continuity of $B(x(t))$, we have $B(x(\bar{t})) = 0$. Hence $x(\bar{t}) \in D$. Since D is open, for $\varepsilon > 0$ sufficiently small, we have $x(t) \in D$ for all $t \in [\bar{t}, \bar{t} + \varepsilon]$. This implies that, for almost all $t \in [\bar{t}, \bar{t} + \varepsilon]$,

$$\frac{dB(x(t))}{dt} = \nabla B(x(t)) \cdot f(t, x(t), w(t)) \leq 0.$$

Hence we have $B(x(t)) \leq B(x(\bar{t})) = 0$ for all $t \in [\bar{t}, \bar{t} + \varepsilon]$. This contradicts the definition of $\bar{t}$. Hence C must be forward invariant. Since $\Upsilon \subseteq C$ and $C \cap U = \emptyset$, the system satisfies the safety specification (Υ, Λ).

It remains to show that $\Upsilon \subseteq \mathcal{D}(A)$. For any $x_0 \in W$ and any solution x starting from x_0, we have $x(t) \in C \subseteq D$ for all $t \geq 0$. Hence

$$\frac{dV(x(t))}{dt} = \nabla V(x(t)) \cdot f(t, x(t), w(t)) \leq -\alpha_3(x)$$

as long as $x(t) \notin A$. A standard Lyapunov argument (similar to that in the proof of Theorem 1.1) shows that $\|x(t)\|_A \to 0$ as $t \to \infty$. ■

Remark 1.3 *The idea behind Theorem 1.4 is that the barrier conditions (3)–(4) and Lyapunov conditions (1)–(2) can be unified. In fact, it was shown in [164] in the context of converse Lyapunpov-barrier theorems that if a robust stability with safety requirement is satisfied, then a single Lyapunov function can be used to certify both stability and safety. In other words, the functions B and V in Theorem 1.4 can be theoretically chosen to be the same function. Nonetheless, Theorem 1.4 as a set of sufficient conditions may seem less conservative (at least from a practical point of view) with separately chosen functions B and V.*

Remark 1.4 *Condition (3) in Theorem 1.3 and condition (4) in Theorem 1.4 provide two slightly different technical conditions to ensure forward invariance of the set $C = \{x \in \mathbb{R}^n : B(x) \leq 0\}$ with respect to solutions of (1.7). Note that a barrier condition like (3) in Theorem 1.3 is only stipulated at the boundary of the set C, whereas (4) in Theorem 1.4 is seemingly more restrictive as it requires a non-increasing condition on B (along solutions of (1.7)) on an open set D containing C. The latter condition may offer some computational advantages, especially when one is seeking to design appropriate control inputs to enforce the barrier condition, because detecting the boundary set $\{x \in \mathbb{R}^n : B(x) = 0\}$ may not be numerically robust. To relax (4) in Theorem 1.4, while still formulating the condition on a set containing C, the following condition (cf. [9]) can be used:*

$$\nabla B(x) \cdot f(t, x, w) \leq \alpha(-B(x)), \quad \forall x \in D, \forall t \geq 0, \forall w \in W, \tag{1.19}$$

where α is a strictly increasing function defined on some interval $(-b, a)$, with $a, b > 0$ and $\alpha(0) = 0$. Condition (1.19) as stated, however, does not ensure forward invariance of the set C as shown in the following example.

Example 1.2 *Consider a special case of a one-dimensional system where $f(x) = 1$. Let $B(x) = x^3$. Then the set $C = \{x \in \mathbb{R} : B(x) \leq 0\} = \{x \in \mathbb{R} : x \leq 0\}$ in clearly not invariant with respect to the flow defined by f. Nonetheless, we have $\nabla B(x) \cdot f(x) = 3x^2 = 3(-B(x))^{2/3} = \alpha(-B(x))$ for all $x \in C$, where $\alpha(s) = 3s^{2/3}$ for $s \geq 0$. We can easily extend the definition of α to $(-\infty, \infty)$ to make it strictly increasing on $\mathbb{R}$, e.g., with $\alpha(s) = -3s^{2/3}$ for $s \leq 0$. Condition (1.19) is satisfied on $D = C$, but C is not forward invariant.*

There are several ways to rectify the situation. We could assume that (1.19) is satisfied on an open set D containing C (note that in the above example $D = C$). We could also put an additional assumption on B, which is $\nabla B(x) \neq 0$ whenever $B(x) = 0$.

We summarize this result in a slightly more general setting as follows. The following proposition provides criteria for the forward invariance of the closed set C defined in (1.18) with respect to solutions of system (1.7).

Proposition 1.3 *If one of the following conditions is satisfied, then C defined in (1.18) is forward invariant for system (1.7).*

1. *$\nabla B(x) \cdot f(t, x, w) \leq 0$ for all $t \geq 0$, all $w \in W$, and all $x \in D \setminus C$, where D is some open set containing $\mathcal{C}$.*

2. *$\nabla B(x) \cdot f(t, x, w) \leq 0$ and $\nabla B(x) \neq 0$ for all x such that $B(x) = 0$.*

Proof: 1. If $\mathcal{C}$ is not forward invariant, then there exists a solution $x(t)$ of (1.7) starting from $x(t_0) = x_0 \in C$ such that $x(t) \notin C$ for some $t > t_0$. By continuity of $x(t)$ and $B(x)$ and openness of D, there exists an interval $[t_1, t_2]$, where $t_2 > t_1 \geq t_0$, such that $B(x(t_1)) = 0$, $x(t) \in D \setminus C$ and $B(x(t)) > 0$ for all $t \in (t_1, t_2]$. It follows that

$$\frac{dB(x(t))}{dt} = \nabla B(x(t)) \cdot f(t, x(t), w(t)) \leq 0$$

for almost all $t \in (t_1, t_2]$. This contradicts $B(x(t_1)) = 0$ and $B(x(t)) > 0$ for all $t \in (t_1, t_2]$. Hence C must be forward invariant.

2. Suppose that $\mathcal{C}$ is not forward invariant, we arrive at the same conclusion as in item 1 above; that is, there exists an interval $[t_1, t_2]$, where $t_2 > t_1 \geq t_0$, such that $B(x(t_1)) = 0$ and $B(x(t)) > 0$ for all $t \in (t_1, t_2]$. By the assumption, we have $\nabla B(x(t_1)) \neq 0$. Define $\delta(t) = d(x(t), \mathcal{C})$. Then $\delta(t_1) = 0$ and $\delta(t) > 0$ for all $t \in (t_1, t_2]$. For $t \in (t_1, t_2)$, let $y(t) \in \mathcal{C}$ be such that $\delta(t) = \|x(t) - y(t)\|$. Clearly, $y(t)$ lies on the boundary $\partial C = \{x \in \mathbb{R}^n : B(x) = 0\}$, i.e., $B(y(t)) = 0$.

Note that $\delta(t) = d(x(t), C)$ is absolutely continuous[5] It follows that

$$\delta(t + \Delta) \leq \|x(t + \Delta) - y(t)\|$$

and

$$\begin{aligned}
\delta(t + \Delta) - \delta(t) &\leq \|x(t + \Delta) - y(t)\| - \|x(t) - y(t)\| \\
&\leq \frac{\|x(t + \Delta) - y(t)\|^2 - \|x(t) - y(t)\|^2}{\|x(t + \Delta) - y(t)\| + \|x(t) - y(t)\|} \\
&\leq \frac{\|x(t + \Delta) - x(t) + x(t) - y(t)\|^2 - \|x(t) - y(t)\|^2}{\|x(t) - y(t)\|} \\
&= \frac{\|x(t + \Delta) - x(t)\|^2 + 2(x(t) - y(t)) \cdot (x(t + \Delta) - x(t))}{\|x(t) - y(t)\|},
\end{aligned}$$

[5]This is because the distance function can be shown to be Lipschitz continuous (by definition and the triangle inequality), $x(t)$ is absolutely continuous, and the composition of a Lipschitz function and an absolutely continuous function can be shown to be absolutely continuous (by definition).

which implies

$$\delta'(t) = \limsup_{\Delta \to 0} \frac{\delta(t+\Delta) - \delta(t)}{\Delta}$$
$$\leq \frac{\lim_{\Delta \to 0} \Delta \left\| \frac{x(t+\Delta)-x(t)}{\Delta} \right\|^2 + 2(x(t) - y(t)) \cdot \lim_{\Delta \to 0} \frac{x(t+\Delta)-x(t)}{\Delta}}{\|x(t) - y(t)\|}$$
$$= \frac{2(x(t) - y(t)) \cdot f(t, x(t), w(t))}{\|x(t) - y(t)\|},$$

for almost all $t \in (t_1, t_2)$. From this, we obtain

$$\delta'(t) \leq \frac{2(x(t) - y(t)) \cdot (f(t, x(t), w(t)) - f(t, y(t), w(t)))}{\|x(t) - y(t)\|} \tag{1.20}$$
$$+ \frac{2(x(t) - y(t)) \cdot f(t, y(t), w(t))}{\|x(t) - y(t)\|}$$
$$\leq 2l(t)\delta(t) + \frac{2(x(t) - y(t)) \cdot f(t, y(t), w(t))}{\|x(t) - y(t)\|}, \tag{1.21}$$

where $l(t)$ is integrable and we used the Lipschitz continuity (cf. Proposition 1.1 and Remark 1.1) of f. Since $y(t) = \arg\min_{z \in C} \|z - x(t)\|$, $x(t) - y(t)$ is a normal vector to ∂C at $y(t)$. Since $B(x) \leq 0$ for $x \in C$, we have $x(t) - y(t) = c\nabla B(y(t))$ for some $c > 0$. We can do this because $\nabla B(y(t)) \neq 0$ for $B(y(t)) = 0$. Hence,

$$(x(t) - y(t)) \cdot f(t, y(t), w(t)) = c\nabla B(y(t)) \cdot f(t, y(t), w(t)) \leq 0. \tag{1.22}$$

Equations (1.21) and (1.22) lead to

$$\delta'(t) \leq 2l(t)\delta(t),$$

for almost all $t \in (t_1, t_2)$.

This differential inequality implies (e.g., as a special case of Grönwall's inequality (Lemma A.2 in differential form))

$$\delta(t) \leq e^{2\int_{t_1}^{t} l(s)ds}\delta(t_1), \quad \forall t \in (t_1, t_2).$$

Since $\delta(t_1) = 0$, we have $\delta(t) = 0$ for all $t \in (t_1, t_2)$. This contradicts that $\delta(t) > 0$ for all $t \in (t_1, t_2]$. ■

Condition (1) in Proposition 1.3 is nothing but a Lyapunov condition for stability of the set C (cf. condition (1.9) with $\alpha_3 \equiv 0$). This condition is weaker than (1.19) being satisfied on an open set D containing C, because for $x \in D \setminus C$, $B(x) > 0$ and $\alpha(-B(x)) < 0$ for $\alpha(s)$ strictly increasing on $(-b, a)$ with $a, b > 0$ and $\alpha(0) = 0$. Note that condition (4) in Theorem 1.4 is stronger than condition (1) in Proposition 1.3, while condition (3) in Theorem 1.3 is stronger than both conditions (1) and (2) in Proposition 1.3. Hence Proposition 1.3 is indeed a generalization of the barrier conditions discussed earlier. Theorems 1.3 and 1.4 remain true with those barrier conditions replaced by either of the conditions in Proposition 1.3.

1.5 Control Lyapunov Functions

In the previous sections, we discussed how to use Lyapunov functions for verifying stability, safety, and reachability properties for a system of the form (1.7) without a control input. In this section, we discuss how the ideas of Lyapunov functions can be extended to deal with control systems of the form (1.1). Under a state feedback control of the form $u = \kappa(x)$, where $\kappa : D \to U$ is assumed to be locally Lipschitz, system (1.1) becomes

$$x'(t) = f(t, x(t), \kappa(x(t)), w(t)). \tag{1.23}$$

Define $F : J \times D \times W \to \mathbb{R}^n$ by $F(t, x, w) = f(t, x, \kappa(x), w)$. Clearly, if f satisfies Assumption 1.1, then so does F (in the absence of u). Furthermore, if f satisfies the assumptions of Proposition 1.1 and additionally that f is locally Lipschitz in u (a condition similar to item 2 of Proposition 1.1 but for the argument u), then F also satisfies the assumptions of Proposition 1.1. The basic existence and uniqueness result (Proposition 1.1) applies to (1.23). The results of Section 1.4 can be readily applied to analyze the system

$$x'(t) = F(t, x(t), w(t)). \tag{1.24}$$

Take stability analysis as an example. If a Lyapunov function exists such that (1.9) holds for F; that is

$$\frac{dV}{dt} + \frac{dV}{dx}F(t, x, w) = \frac{dV}{dt} + \frac{dV}{dx}f(t, x, \kappa(x), w) \leq -\alpha_3(x),$$

for all $x \in D \setminus A$, $t \geq 0$, and $w \in W$. This clearly implies that

$$\inf_{u \in U} \sup_{t \geq 0, w \in W} \left\{ \frac{dV}{dt} + \frac{dV}{dx} f(t, x, u, w) \right\} \leq -\alpha_3(x), \quad \forall x \in D \setminus A. \tag{1.25}$$

The above condition defines a key property for what is referred to as a *control Lyapunov function*, which is the existence of a control input for every state such that the Lyapunov stability condition is satisfied. Control Lyapunov functions play a central role in designing feedback control solutions for stabilization problems (see, e.g., [75]).

Similarly, a barrier condition for set invariance, such as the ones given in Proposition 1.3, can be modified to define a *control barrier function*, which satisfies

$$\inf_{u \in U} \sup_{t \geq 0, w \in W} \{\nabla B(x) \cdot f(t, x, u, w)\} \leq 0, \tag{1.26}$$

for all x such that $B(x) = 0$. In view of condition (2) in Proposition 1.3, we assume that $\nabla B(x) \neq 0$ for all x such that $B(x) = 0$. Condition (1.26) is equivalent to the non-emptiness of the following set

$$U(x) = \{u \in U : B(x) \cdot f(t, x, u, w) \leq 0, \forall t \geq 0, \forall w \in W\}, \tag{1.27}$$

for each x such that $B(x) = 0$. If this set $U(x)$ is non-empty for any such x, one can, in principle, choose the an appropriate control u for each x by letting

$$\kappa(x) := \arg\min_{u \in U} \sup_{t \geq 0, w \in W} \nabla B(x) \cdot f(t, x, u, w). \tag{1.28}$$

There are several technical difficulties associated with this approach. Theoretically, it often requires extra technical assumptions to ensure that $\kappa(x)$ chosen this way is locally Lipschitz so that solutions for the closed-loop system are well defined in the sense of Proposition 1.1. On a more practical note, the nature of this min-max optimization may not be amenable to computationally efficient solutions and its feasibility, especially when the control inputs have to satisfy certain constraints, is not always guaranteed. Further assumptions on the right-hand side f are required to make the problem more tractable. For example, one special case is when f is affine in the control input, a quadratic programming approach can be taken [9]. We shall not discuss this in detail as the focus of the book is not on optimization-based control design.

1.6 Summary

In this chapter, we have seen a general continuous-time dynamical system model and discussed a range of properties related to its solutions, including existence, uniqueness, stability, boundedness, invariance, safety, and reachability. Such properties are analyzed indirectly using an auxiliary function known as the Lyapunov function, which has been an indispensable tool for the analysis and control of dynamical systems. Most materials are standard. The main references used in writing this chapter are [114] for stability analysis and [218] for the basic theory of ordinary differential equations.

The notion of barrier functions (or certificates) was proposed in [193, 194] as a computationally convenient tool to verify safety properties of nonlinear and hybrid systems. The main idea behind barrier functions is set invariance. Necessary and sufficient conditions for a closed set to be forward invariant with respect to the flow of a continuous-time dynamical system, however, date back to 1940s, when a now classical result known as the Nagumo theorem [176] was developed. Similar results were later re-discovered around 1970s by Bony [34] and Brezis [38] (see [201] for an exposition and [32] for a more recent survey). Proposition 1.3 can be seen as a slight generalization of these results to systems with a disturbance input. The sufficient Lyapunov-barrier conditions for stability with a safety guarantee are taken from [164], where both sufficient and necessary conditions were discussed.

The notions of control Lyapunov functions [13, 216, 217] and barrier certificates [193, 194] motivated the development of control barrier functions [9, 238, 245], which have been applied to a range of application domains [8]. The integration of control barrier functions with high-level specifications is a topic of ongoing investigation.

Chapter 2

Discrete-Time Dynamical Systems

In this chapter, we introduce a class of models for discrete-time dynamical systems. Arguably one prominent reason to consider discrete-time control systems is to cope with digital and sampled-data control applications. From a theoretical point of view, discrete-time control systems can be modeled by difference equations with inputs. We briefly discuss such models in this chapter.

2.1 Discrete-Time Control Systems

A discrete-time control system can be written as

$$x(t+1) = f(t, x(t), u(t), w(t)), \tag{2.1}$$

where $x(t) \in \mathbb{R}^n$ is the system state, $u(t) \in U \subseteq \mathbb{R}^m$ is the control input, and $w(t) \in W \subseteq \mathbb{R}^p$ is the disturbance input. The main difference between the discrete-time system (2.1) and the continuous-time system (1.1) is that in (2.1) we assume the time variable $t \in \mathbb{N}$, the set of natural numbers including zero, and we have a difference equation instead of a differential equation.

Definition 2.1 *Given $t_0 \in \mathbb{N}$, a sequence of control input $\{u(t)\}_{t=t_0}^{\infty}$, and a sequence of disturbance input $\{w(t)\}_{t=t_0}^{\infty}$, a* ***solution*** *to (2.1) is a sequence $\{x(t)\}_{t=t_0}^{\infty}$ such that (2.1) is satisfied for all $t \geq t_0$.*

2.2 Stability and Boundedness

In the absence of a control input, (2.1) becomes

$$x(t+1) = f(t, x(t), w(t)). \tag{2.2}$$

We present similar results on Lyapunov analysis of discrete-time dynamical systems of the form (2.2). The definitions are almost identical with that for continuous-time systems.

DOI: 10.1201/9780429270253-2

Definition 2.2 *A set $A \subseteq \mathbb{R}^n$ is said to be* **invariant set** *for system (2.2), if it is nonempty and, for any $x_0 \in A$, all solutions of (2.2) starting from $x(t_0) = x_0$ stay in A for all $t \geq t_0$.*

Definition 2.3 *Let $A \subseteq \mathbb{R}^n$ be a compact invariant set for system (2.2). We say that A is* **uniformly asymptotically stable** *(UAS) for system (2.2), if the following two conditions hold:*

1. *(Uniform stability) For every $\varepsilon > 0$, there exist $\delta = \delta(\varepsilon) > 0$ such that, if $\|x_0\|_A \leq \delta$, then $\|x(t)\|_A \leq \varepsilon$ for all $t \geq t_0$, where $x(t)$ is any solution of (2.2) starting from $x(t_0) = x_0$.*
2. *(Uniform attraction) There exists some $\rho > 0$ such that, for any $\eta > 0$, there exists some $T = T(\rho, \eta) \geq 0$ such that $\|x(t)\|_A \leq \eta$, whenever $\|x_0\|_A \leq \rho$ and $t \geq t_0 + T$, where $x(t)$ is any solution of (2.2) starting from $x(t_0) = x_0$.*

We say that A is **globally uniformly asymptotically stable** *(GUAS) for system (2.2), if the above conditions hold for δ chosen such that $\lim_{\varepsilon \to \infty} \delta(\varepsilon) = \infty$ and any $\rho > 0$.*

Lyapunov functions can also be used to analyze stability and boundedness properties of discrete-time systems.

Theorem 2.1 (Lyapunov theorem for stability) *Let A be a compact invariant set of system (2.2). Let $D \subseteq \mathbb{R}^n$ be an open set containing A and $V : \mathbb{N} \times D \to \mathbb{R}$ be a continuous function. Suppose that there exist continuous functions α_i $(i = 1, 2, 3)$ that are defined on D and positive definite with respect to A such that*

$$\alpha_1(x) \leq V(t, x) \leq \alpha_2(x), \quad \forall x \in D, \quad \forall t \geq 0, \tag{2.3}$$

and

$$V(t+1, f(t, x, w)) - V(t, x) \leq -\alpha_3(x), \quad \forall x \in D, \quad \forall t \geq 0, \quad \forall w \in W. \tag{2.4}$$

Then A is uniformly asymptotically stable for system (2.2). If the above conditions hold for $D = \mathbb{R}^n$ and α_1 is radially unbounded, then A is globally uniformly asymptotically stable for system (2.2).

Proof: The proof follows almost verbatim the proof of Theorem 1.1, with (1.10) replaced by

$$V(t + 1, x(t + 1)) - V(t, x(t)) \leq -\alpha_3(x(t)) \leq 0, \tag{2.5}$$

and (1.11) replaced by

$$\begin{aligned}
\alpha_1(x(t_0+T)) &\leq V(t_0+T, x(t_0+T)) \\
&= V(t_0, x_0) + \sum_{s=t_0}^{t_0+T-1} [V(s+1, x(s+1)) - V(s, x(s))] \\
&\leq V(t_0, x_0) - \sum_{s=t_0}^{t_0+T-1} \alpha_3(x(s))ds \\
&\leq \alpha_2(x_0) - \lambda T \leq c - \lambda T < 0.
\end{aligned} \tag{2.6}$$

■

Similarly, we can state and prove boundedness results for system (2.2).

Definition 2.4 *We say that solutions of (1.7) are* ***uniformly bounded*** *(UB), if there exists a constant $c > 0$ such that, for all $t_0 \geq 0$ and $\rho \in (0, c)$, $\|x(t_0)\| \leq \rho$ implies $\|x(t)\| \leq b$, where $b = b(\rho)$ only depends on ρ. We say that solutions of (2.2) are* ***globally uniformly bounded*** *(GUB), if the above holds for all $\rho > 0$.*

Definition 2.5 *We say that solutions of (2.2) are* ***uniformly ultimately bounded*** *(UUB), if there exist constants $c > 0$ and $b > 0$ such that, for every $\rho \in (0, c)$, there exists $T = T(\rho, b) \geq 0$ such that $\|x(t_0)\| \leq \rho$ implies $\|x(t)\| \leq b$ for all $t \geq t_0 + T$. We say that solutions of (2.2) are* ***globally uniformly ultimately bounded*** *(GUUB), if the above holds for all $\rho > 0$. The constant b is called the ultimate bound of solutions.*

Theorem 2.2 (Lyapunov theorem for boundedness) *Let $D \subseteq \mathbb{R}^n$ be an open set and $V : \mathbb{N} \times D \to \mathbb{R}$ be a continuous function. Suppose that there exist continuous functions α_i $(i = 1, 2, 3)$ defined on D and positive on $D \setminus B_\mu$, for some $\mu > 0$, such that*

$$\alpha_1(x) \leq V(t, x) \leq \alpha_2(x), \quad \forall x \in D \setminus \mathbb{B}_\mu, \quad \forall t \geq 0, \tag{2.7}$$

and

$$V(t+1, f(t, x, w)) - V(t, x) \leq -\alpha_3(x), \quad \forall x \in D \setminus \mathbb{B}_\mu, \quad \forall t \geq 0, \quad \forall w \in W. \tag{2.8}$$

Suppose there exists some $\rho > 0$ such that $B_\rho \subseteq D$ and

$$\max_{x \in \mathbb{B}_\mu} \alpha_2(x) < \min_{\|x\|_\infty = \rho} \alpha_1(x). \tag{2.9}$$

Then solutions of (2.2) are both uniformly bounded and uniformly ultimately bounded. If the conditions (2.7) and (2.8) hold for $D = \mathbb{R}^n$ and $\alpha_1(x)$ is radially unbounded, then (2.9) is satisfied and solutions of (2.2) are both globally uniformly bounded and globally uniformly ultimately bounded.

The proof is similar to the proof for Theorem 1.2.

2.3 Safety and Reachability

Safety and reachability of discrete-time systems can also be analyzed using a Lyapunov-like function. In this section, we present discrete-time versions of the results in Section 1.4.

Definition 2.6 *Given an unsafe set $\Lambda \subseteq \mathbb{R}^n$ and an initial set $\Upsilon \subseteq \mathbb{R}^n$, we say that system (2.2) satisfies a **safety** specification (Υ, Λ), if every solution $x(t)$ of (2.2) starting from $x(t_0) = x_0 \in \Upsilon$ will not enter Λ; that is, $x(t) \notin \Lambda$ for all $t \geq t_0$.*

Definition 2.7 (Domain of attraction) *Suppose that a compact set $A \subseteq \mathbb{R}^n$ is UAS for system (2.2). The domain of attraction $\mathcal{D}(A)$ is defined as*

$$\mathcal{D}(A) = \left\{ x_0 \in \mathbb{R}^n : \lim_{t\to\infty} \|x(t)\|_A = 0, \forall x(\cdot) \in \Phi(x_0) \right\},$$

where $\Phi(x_0)$ denotes the set of solutions for system (2.2) starting from x_0.

Definition 2.8 *Given an unsafe set $\Lambda \subseteq \mathbb{R}^n$ and a compact set $A \subseteq \mathbb{R}^n$, we say that system (2.2) satisfies a **stability with safety guarantee** specification (Υ, Λ, A), if the set A is UAS for system (2.2), $\Upsilon \subseteq \mathcal{D}(A)$, and system (2.2) satisfies a safety specification (Υ, Λ).*

Theorem 2.3 (Lyapunov theorem for safety) *Given an unsafe set $\Lambda \subseteq \mathbb{R}^n$ and an initial set $\Upsilon \subseteq \mathbb{R}^n$, if there exists a continuously differentiable function $B : \mathbb{R}^n \to \mathbb{R}$ such that the following conditions are satisfied:*

1. *$B(x) \leq 0$ for all $x \in \Upsilon$;*
2. *$B(x) > 0$ for all $x \in \Lambda$; and*
3. *$B(f(t, x, w)) \leq 0$ for all $t \in \mathbb{N}$, all $w \in W$, and all x such that $B(x) \leq 0$,*

then system (2.2) satisfies a safety specification (Υ, Λ).

Note that the derivative condition in Theorem 1.3 is replaced with a direct condition on $f(t, x, w)$ for discrete-time systems. This condition has to be enforced for all x such that $B(x) \leq 0$, instead of only when $B(x) = 0$, because a solution sequence of (2.2) can leave the set $\{x \in \mathbb{R}^n : B(x) \leq 0\}$ without hitting the boundary. This change makes the proof almost trivial.

Proof: The conclusion follows from the fact that the set $C = \{x \in \mathbb{R}^n : B(x) \leq 0\}$ is forward invariant for system (2.2). ■

Theorem 2.4 (Lyapunov theorem for stability with safety) *Suppose that A is compact and $A \cap \Lambda = \emptyset$. If there exists an open set D such that $(A \cup \Upsilon) \subseteq D$ and continuously differentiable functions $V : D \to \mathbb{R}_{\geq 0}$ and $B : \mathbb{R}^n \to \mathbb{R}$ such that*

1. *V is positive definite on D with respect to A;*

2. *$V(f(t,x,w)) - V(x) \le -\alpha_3(x)$ for all $t \ge 0$, $x \in D \setminus A$, and $w \in W$, where α_3 is a continuous function defined on D that is positive definite with respect to A;*

3. *$\Upsilon \subseteq C = \{x \in \mathbb{R}^n : B(x) \le 0\} \subseteq D$ and $B(x) > 0$ for all $x \in \Lambda$;*

4. *$B(f(t,x,w)) \le 0$ for all $t \ge 0$, $w \in W$, and $x \in D$,*

then system (1.7) satisfies the stability with safety guarantee specification (Υ, Λ, A).

Proof: The conclusion follows from the fact that the set $C = \{x \in \mathbb{R}^n : B(x) \le 0\}$ is forward invariant for system (2.2) and that V is a Lyapunov function as used in Theorem 2.1. ■

2.4 Summary

This chapter provides a brief discussion on the analysis of discrete-time dynamical systems using Lyapunov functions. Some dynamical systems are inherently discrete in time. For example, population dynamics where the population has discrete generations. Discrete-time dynamical systems can also be used to analyze or approximate continuous-time systems. For example, the Poincaré map of a continuous-time dynamical system can be interpreted as a discrete-time dynamical system and used to analyze the original continuous-time system in a simpler way. More importantly, in the context of this book, every continuous-time control system can be controlled via a discrete-time control strategy; that is, to sample the state of the continuous-time system at discrete times, known as the sampling times, and apply a constant control accordingly and hold it for the duration of the inter-sample interval, known as *sampled-data control*. This leads to a natural time-discretization of the continuous-time control system by only considering the evolution of the states at discrete sampling times. This has become the paradigm of controlling dynamical systems using digital controllers. Control systems using such a paradigm are known as sampled-data systems, or more recently, under the names of hybrid systems or cyber-physical systems. We shall mainly take this approach of controlling continuous-time systems in this book.

Chapter 3

Formal Specifications and Discrete Synthesis

In this chapter, we introduce discrete models given by transition systems and formal specifications given by linear temporal logic (LTL) and ω-regular properties. We then discuss how to solve discrete synthesis problems for nondeterministic transition systems with reachability, safety, and general ω-regular specifications.

3.1 Transition Systems

Definition 3.1 *A* ***transition system*** *is a tuple*

$$\mathcal{T} = (S, A, R, \Pi, L),$$

where

- *S is a set of states;*
- *A is a set of actions;*
- *$R \subseteq S \times A \times S$ is a transition relation;*
- *Π is a set of atomic propositions;*
- *$L : S \to 2^{\Pi}$ is a labeling function.*

We say that $\mathcal{T}$ is ***finite****, if S, A, and Π are all finite sets.*

Definition 3.2 *Given a transition system $\mathcal{T} = (S, A, R, \Pi, L)$, the set of* ***successors*** *of a state $s \in S$ under an action $a \in A$ is defined by*

$$Post(s, a) = \{s' \in S \mid (s, a, s') \in R\}.$$

We also write

$$Post(s) = \bigcup_{a \in A} Post(s, a),$$

DOI: 10.1201/9780429270253-3

which are the set of successors for the state s. If Post(s) $= \emptyset$, *we call s a* ***terminal state.***

Similarly, the set of ***predecessors*** *of a state* $s \in S$ *under an action* $a \in A$ *is defined by*

$$Pre(s,a) = \{s' \in S \mid (s',a,s) \in R\},$$

and the set of predecessors for the state s is

$$Pre(s) = \bigcup_{a \in A} Pre(s,a).$$

It is sometimes more intuitive to write a transition $(s,a,s') \in R$ as $s \xrightarrow{a} s'$. A transition system is essentially a discrete-time dynamical system defined by (S,A,R), where for a given state $s \in S$ and an action $a \in A$, the next states are taken from the set of successors $\text{Post}(s,a)$. The additional elements Π and L are introduced for specifying and reasoning about properties of transition systems. Such transition systems are also called *labeled* transition systems.

Definition 3.3 *A transition system* $\mathcal{T} = (S,A,R,\Pi,L)$ *is called* ***deterministic,*** *if for every* $s \in S$ *and* $a \in A$, $|Post(s,a))| \leq 1$. *It is called* ***nondeterministic,*** *if it is not deterministic.*

Similar to solutions for control systems, we can define paths for transition systems.

Definition 3.4 *Let* $\mathcal{T} = (S,A,R,\Pi,L)$ *be a transition system. Given a sequence of actions* $\{a_i\}_{i=0}^{\infty}$ *in A, a* ***path*** *of* $\mathcal{T}$ *is a sequence of states* $\{s_i\}_{i=0}^{\infty}$ *in S such that* $(s_i,a_i,s_{i+1}) \in R$ *for all* $i \geq 0$.

Remark 3.1 *We are interested in long-term behaviors of transition systems, which are captured by infinite paths defined above. Infinite paths may not exist if there are terminal states in a transition system. In this book, we make the standing assumption that transition systems under consideration do not have terminal states.*

Furthermore, if $Post(s,a) = \emptyset$ *for some pair* (s,a), *we may want to remove a from the set of available actions for this state. We introduce* $A_{\mathcal{T}}(s)$ *to denote the set of available actions at state s for* $\mathcal{T}$, *i.e.,*

$$A_{\mathcal{T}}(s) = \{a \in A \mid Post(s,a) \neq \emptyset\}.$$

The assumption that a transition system $\mathcal{T} = (S,A,R,\Pi,L)$ *does not contain terminal states is equivalent to that* $A_{\mathcal{T}}(s) \neq \emptyset$ *for every* $s \in A$.

Specifications on transition systems will be interpreted over "observations" made on paths, which are called *traces.* This is similar to the notion of outputs or observations in control systems.

Definition 3.5 *Given a path* $\{s_i\}_{i=0}^{\infty}$ *of a transition system* $\mathcal{T} = (S,A,R,\Pi,L)$, *its* ***trace*** *is the sequence* $\{L(s_i)\}_{i=0}^{\infty}$ *in* 2^{Π}.

3.2 Linear-Time Properties

Linear-time properties are essentially sets of traces, which are also infinite sequences in 2^{Π}. A formal definition is given below.

Definition 3.6 *A* ***linear-time (LT) property*** *over a set of atomic propositions* Π *is a subset of*

$$(2^{\Pi})^{\omega} := \left\{ \{\Pi_i\}_{i=0}^{\infty} \mid \Pi_i \in 2^{\Pi}, \forall i \geq 0 \right\}.$$

In computer science terminology, a sequence in Σ is called a *word* (with Σ being the *alphabet*). If the sequence is finite, it is called a finite word. If the sequence is infinite, it is called an infinite word. A set of words is called a *language*. More specifically, a set of infinite words is called an ω-*language*[1]. As such, an LT property is simply a language of infinite words over the alphabet 2^{Π}. We can interpret satisfaction of an LT property over traces of a transition system by set membership.

Definition 3.7 (Satisfaction of LT Properties by Paths and Traces) *Let* $\rho = \{s_i\}_{i=0}^{\infty}$ *be a path of a transition system* $\mathcal{T} = (S, A, R, \Pi, L)$ *and* $L(\rho) := \{L(s_i)\}_{i=0}^{\infty}$ *be its trace. Let* φ *be a linear-time property over* Π*. We say that* ρ *satisfies* φ *and write* $\rho \models \varphi$*, if* $L(\rho) \in \varphi$*. If this holds, we may also say that the trace* $L(\rho)$ *satisfies* φ *and write* $L(\rho) \models \varphi$*.*

3.3 Linear Temporal Logic

Linear temporal logic (LTL) was introduced by Amir Pnueli to computer science in the late 1970s [185] for specifying and verifying properties for complex computer systems. LTL extends the usual propositional logic with additional temporal operators that are intuitive to understand and use. The properties specified by LTL constitute a subset of linear-time properties.

Definition 3.8 *LTL formulas over a set of atomic propositions* Π *are formed using the following grammar (called Backus–Naur form):*

$$\varphi ::= \mathbf{true} \mid \pi \mid \varphi_1 \wedge \varphi_2 \mid \neg\varphi \mid \bigcirc\varphi \mid \varphi_1 \mathbf{U} \varphi_2,$$

where $\pi \in \Pi$.

[1]The symbol ω can be seen as the set of natural numbers $\mathbb{N} = \{0, 1, 2, 3, \cdots\}$. In this sense, the notation Σ^{ω} can be understood as the set of all functions from $\mathbb{N}$ to Σ, i.e., all infinite sequences in Σ.

The above syntax (grammar) is understood as follows:

- **true** is a formula;
- any atomic proposition $\pi \in \Pi$ is a formula;
- given formulas φ_1 and φ_2, both $\varphi_1 \wedge \varphi_2$ and $\varphi_1 \mathbf{U} \varphi_2$ are formulas;
- given a formula φ, both $\neg\varphi$ and $\bigcirc\varphi$ are both formulas.

Additional logical and temporal operators can be defined as follows:

- **false** $= \neg$**true**;
- $\varphi_1 \vee \varphi_2 := \neg(\neg\varphi_1 \wedge \neg\varphi_2)$;
- $\Diamond\varphi :=$ **true**$\mathbf{U}\varphi$;
- $\Box\varphi := \neg\Diamond\neg\varphi$;
- $\varphi_1 \mathbf{R} \varphi_2 := \neg(\neg\varphi_1 \mathbf{U} \neg\varphi_2)$.

The operators **U** and **R** are called *until* and *release*, respectively. The operators $\Box$ and $\Diamond$ are read *always* and *eventually*, respectively. The semantics (meaning) of LTL formulas are interpreted over infinite sequences in 2^Π.

Definition 3.9 *Let φ be an LTL formula over Π and let $\sigma = \{\sigma_i\}_{i=0}^\infty \in (2^\Pi)^\omega$, i.e., an infinite sequence in 2^Π, the satisfaction of φ by σ, written as $\sigma \vDash \varphi$, is defined as follows:*

- *$\sigma \vDash$ **true**;*
- *$\sigma \vDash \pi$, where $\pi \in \Pi$, if and only if $\pi \in \sigma_0$;*
- *$\sigma \vDash \varphi_1 \wedge \varphi_2$ if and only if $\sigma \vDash \varphi_1$ and $\sigma \vDash \varphi_2$;*
- *$\sigma \vDash \neg\varphi$ if and only if $\sigma \nvDash \varphi$;*
- *$\sigma \vDash \bigcirc\varphi$ if and only if $\{\sigma_i\}_{i=1}^\infty \vDash \varphi$;*
- *$\sigma \vDash \varphi_1 \mathbf{U} \varphi_2$ if and only if there exists $j \geq 0$ such that $\{\sigma_i\}_{i=j}^\infty \vDash \varphi_2$ and $\{\sigma_i\}_{i=k}^\infty \vDash \varphi_1$ for all $0 \leq k < j$.*

With this definition, we can further obtain the semantics of the derived operators as follows:

- $\sigma \nvDash$ **false**;
- $\sigma \vDash \varphi_1 \vee \varphi_2$ if and only if $\sigma \vDash \varphi_1$ or $\sigma \vDash \varphi_2$;
- $\sigma \vDash \Diamond\varphi$ if and only if there exists $j \geq 0$ such that $\{\sigma_i\}_{i=j}^\infty \vDash \varphi$;

- $\sigma \models \Box\varphi$ if and only if $\{\sigma_i\}_{i=j}^{\infty} \models \varphi$ for all $j \geq 0$;
- $\sigma \models \varphi_1 \mathbf{R} \varphi_2$ if and only if either $\{\sigma_i\}_{i=j}^{\infty} \models \varphi_2$ for all $j \geq 0$ or there exists $j \geq 0$ such that $\{\sigma_i\}_{i=j}^{\infty} \models \varphi_1$ and $\{\sigma_i\}_{i=k}^{\infty} \models \varphi_2$ for all $0 \leq k \leq j$.

Clearly, an LTL formula defines an LT property. Formally, we can let

$$\mathrm{LT}(\varphi) := \{\sigma \in (2^{\Pi})^{\omega} \mid \sigma \models \varphi\}.$$

Then $\mathrm{LT}(\varphi)$ is an LT property. We can define satisfaction of an LTL formula by paths of a transition system in the same way as for an LT property.

Definition 3.10 (Satisfaction of LTL Formulas by Paths and Traces) *Let $\rho = \{s_i\}_{i=0}^{\infty}$ be a path of a transition system $\mathcal{T} = (S, A, R, \Pi, L)$ and $L(\rho) := \{L(s_i)\}_{i=0}^{\infty}$ be its trace. Let φ be an LTL formula over Π. We say that ρ satisfies φ and write $\rho \models \varphi$, if $L(\rho) \in \varphi$. If this holds, we may also say that the trace $L(\rho)$ satisfies φ and write $L(\rho) \models \varphi$.*

An important fact about LTL formulas is that any formula can be rewritten in the so-called *positive normal form* (PNF), in which negations are pushed inside and only appear in front of atomic propositions.

Definition 3.11 *LTL formulas in release* ***positive normal form*** *(PNF) are given by*

$$\varphi ::= \mathbf{true} \mid \mathbf{false} \mid \pi \mid \neg\pi \mid \varphi_1 \wedge \varphi_2 \mid \varphi_1 \vee \varphi_2 \mid \bigcirc\varphi \mid \varphi_1 \mathbf{U} \varphi_2 \mid \varphi_1 \mathbf{R} \varphi_2.$$

Proposition 3.1 *Any LTL formula φ can be equivalently written as another LTL formula ψ in release positive normal form.*

The transformation can be done easily using the following rules:

- $\neg\mathbf{true} \equiv \mathbf{false}$;
- $\neg(\varphi_1 \wedge \varphi_2) \equiv \neg\varphi_1 \vee \neg\varphi_2$;
- $\neg(\neg\varphi) \equiv \varphi$;
- $\neg(\bigcirc\varphi) \equiv \bigcirc\neg\varphi$;
- $\neg(\varphi_1 \mathbf{U} \varphi_2) \equiv \neg\varphi_1 \mathbf{R} \neg\varphi_2$.

From these rules, we can also see that there is no significant increase in the size of the equivalently transformed formula, compared with the original formula, apart from an added constant. In fact, we can show that $|\psi| = \mathcal{O}(|\varphi|)$, where $|\varphi|$ denotes the size of the formula φ in terms of the number of operators in φ.

3.4 ω-Regular Properties

In this section, we introduce a broad class of linear-time properties that are described as the language (set) of infinite words (sequences) accepted by so-called *ω-automata.* The notion of ω-automata generalizes that of finite automata (also known as finite-state machines), which only characterize languages of finite words. Let Σ be a finite set of symbols called an *alphabet.* The set Σ^* denotes the set of all finite words (sequences) over Σ and the set Σ^ω denotes the set of all infinite words over Σ.

Definition 3.12 *An **ω-automaton** is a tuple $\mathcal{A} = (Q, \Sigma, r, q_0, Acc)$, where*

- *Q is a finite set of states;*
- *Σ is a finite alphabet;*
- *$r : Q \times \Sigma \to 2^Q$ is the transition function;*
- *$q_0 \in Q$ is the initial state;*
- *$Acc \subseteq \Sigma^\omega$ stands for the acceptance condition (to be specified by a given type of ω-automata) that determines the infinite words over Σ accepted by $\mathcal{A}$.*

*An **ω-automaton** $\mathcal{A} = (Q, \Sigma, r, q_0, Acc)$ is said to be **deterministic** if the transition function is given by $r : Q \times \Sigma \to Q$.*

Definition 3.13 *Let $\mathcal{A} = (Q, \Sigma, r, q_0, Acc)$ be an ω-automaton. Let $\sigma = \{\sigma_i\}_{i=0}^\infty \in \Sigma^\omega$ be an infinite word over Σ. A **run** of $\mathcal{A}$ for σ is an infinite sequence $\rho = \{q_i\}_{i=0}^\infty$ such that q_0 is the initial state and*

$$q_{i+1} \in \delta(q_i, \sigma_i), \quad \forall i \geq 0.$$

The acceptance condition is often not given explicitly as a subset of Σ^ω, but rather by specifying the accepted runs of $\mathcal{A}$. We introduce several different types of acceptance conditions used in this monograph.

Definition 3.14 *A **Büchi automaton** (BA) is an ω-automaton of the form $\mathcal{A} = (Q, \Sigma, r, q_0, F)$, where $F \subseteq Q$ is the set of acceptance states (called the Büchi acceptance condition). An infinite word $\sigma \in \Sigma^\omega$ is said to be accepted by $\mathcal{A}$ if and only if there exists a run $\rho = \{q_i\}_{i=0}^\infty$ of $\mathcal{A}$ for σ such that $q_i \in F$ for infinitely many indices $i \geq 0$.*

Definition 3.15 *A **Rabin automaton** (RA) is an ω-automaton of the form $\mathcal{A} = (Q, \Sigma, r, q_0, \Omega)$, where $\Omega = \{(E_i, F_i)\}_{i=1}^k$, and $E_i, F_i \subseteq Q$ for $1 \leq i \leq k$. The Rabin acceptance condition is defined as follows: an infinite word $\sigma \in \Sigma^\omega$ is said to be accepted by $\mathcal{A}$ if and only if there exists a run $\rho = \{q_i\}_{i=0}^\infty$ of $\mathcal{A}$ for*

σ such that the following condition is satisfied: there exists some $(E, F) \in \Omega$ such that $q_i \in F$ for infinitely many indices $i \geq 0$ and $q_i \in E$ for only finitely many indices $i \geq 0$.

If a language $\mathcal{L}$ coincides with the language accepted by an ω-automaton $\mathcal{A}$, we say that $\mathcal{L}$ is *recognized* by $\mathcal{A}$. We shall use the letter D and N to indicate whether an ω-automaton is deterministic or nondeterministic, respectively. For example, DBA refers to deterministic Büchi automaton/automata and DRA refers to deterministic Rabin automaton/automata.

With $\Sigma = 2^{\Pi}$, where Π is a set atomic propositions, the language (set of infinite words) accepted by an ω-automaton is clearly a linear-time property over Π. It is known that the set of languages accepted by nondeterministic Büchi automata is exactly equal to the set of languages accepted by deterministic (and nondeterministic) Rabin automata [228]. Such linear-time properties are called *ω-regular properties*[2]. More precisely, an LT property over Π is called ω-regular, if it is an ω-regular language over 2^{Π}. The following relation shows the expressive power of different specification formalisms we have introduced so far:

$$\text{DBA} \subsetneq \text{NBA} = \text{DRA} = \text{NRA},$$

where the fact that NBA can be determinized to equivalent DRA is due to Safra [212] (see also [92, Chapter 3]). See [92, Chapter 1] for the fact that NRA does not add further expressive power to DRA.

The following example shows that NBA is more expressive than DBA.

Example 3.1 *Consider the ω-language $\mathcal{L}$ over $\Sigma = \{a, b\}$ defined by*

$$\mathcal{L} = \{w_1 w_2 \mid w_1 \in \Sigma^*, w_2 = b^\omega\},$$

where w_1 is any finite word over Σ and $w_2 = b^\omega = bbb\cdots$. The language $\mathcal{L}$ can also be defined by the ω-regular expression $(a+b)^* b^\omega$, where $a + b$ indicates a or b, $(a+b)^*$ is repeating a or b zero or finitely many times, b^ω means repeating b infinitely many times. It can be shown that no DBA can recognize the language $\mathcal{L}$. However, the NBA shown in Figure 3.1 recognizes $\mathcal{L}$. If we let $\Pi = \{p\}$, $a = \{\}$, $b = \{p\}$, then $\mathcal{L}$ coincides with the LT property defined by the LTL formula $\Diamond\Box p$.

3.4.1 Translating LTL to Büchi Automata

A powerful technique for handling linear temporal logic is to translate an LTL formula to an ω-automaton which accepts exactly the set of infinite words satisfying the formula. There are many algorithms and tools available

[2] See, e.g., [23, Section 4.3.1] for a formal definition of ω-regular property using ω-regular expressions. Since LT properties recognizable by NBA (or DRA) agree with ω-regular properties, here we can simply treat this as a definition.

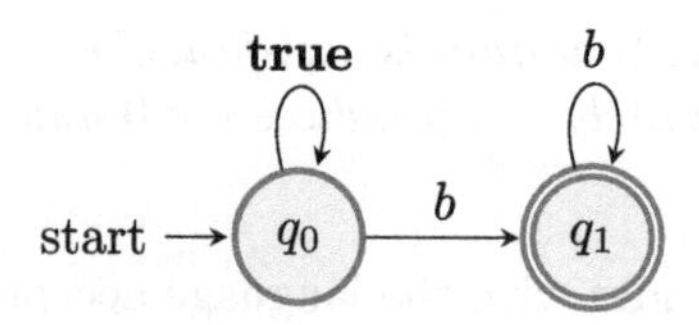

FIGURE 3.1: A nondeterministic Büchi automaton (NBA) that recognizes the ω-regular language $\mathcal{L} = (a+b)^*b^\omega$ over $\Sigma = \{a,b\}$. It can be shown that no deterministic Büchi automaton (DBA) can recognizes $\mathcal{L}$.

for translating LTL formulas to ω-automata for infinite words, e.g., LTL2BA [81], LTL3BA [21], SPOT [61], Rabinizer [79, 121, 128, 129].

The following results provide the theoretical basis for the translation of LTL formulas to ω-automata.

Theorem 3.1 *For any LTL formula over Π, there exists an NBA $\mathcal{A}_\varphi$ with alphabet $\Sigma = 2^\Pi$ such that the infinite words accepted by $\mathcal{A}_\varphi$ are exactly those satisfying φ. The resulting NBA has at most $2^{\mathcal{O}(|\varphi|)}$ states and $|\varphi|$ accepting states, where $|\varphi|$ denotes the size of φ.*

For a proof of the above result, see [23, Theorem 5.41]. While optimizations are possible for improving the size of the resulting NBA, the exponential blowup in general cannot be avoided: there exist a family of LTL formulas φ_n of size $|\varphi_n| = \mathcal{O}(\text{poly}(n))$ such that every NBA for φ_n has at least 2^n states [23, Theorem 5.42].

Example 3.2 *Consider the LTL formula $\varphi = \Diamond a \wedge \Diamond\Box b$ over $\Pi = \{a,b\}$. The formula can be translated to an NBA with four states as shown in Figure 3.2. Note that in Theorem 3.1 the alphabet is $\Sigma = 2^\Pi$, whereas in Figure 3.2, e.g., $q_0 \xrightarrow{a} q_2$ actually represents two transitions: $q_0 \xrightarrow{\{a\}} q_2$ and $q_0 \xrightarrow{\{a,b\}} q_2$.*

NBA is strictly more expressive than DBA. Example 3.1 shows that an equivalent NBA can be found for the LTL formula $\varphi = \Diamond\Box p$, but it can be shown that no DBA can express this property. Hence NBA is strictly more expressive than DBA. The expressive powers of LTL and DBA, however, are incomparable as shown in the following example.

Example 3.3 *For example, a DBA (see Figure 3.3) can be used to define the set of infinite words over $\Sigma = \{a,b\}$, where $a = \{\}, b = \{p\}$, given by*

$$\varphi = \{\sigma = \{\sigma\}_{i=0}^\infty \in \Sigma^\omega \mid \sigma_{2k} = b, \forall k \geq 0\}.$$

This linear-time property enforces every even-indexed letter of the infinite word should be b. It can be shown that no LTL formula over $\Pi = \{p\}$ can capture this property.

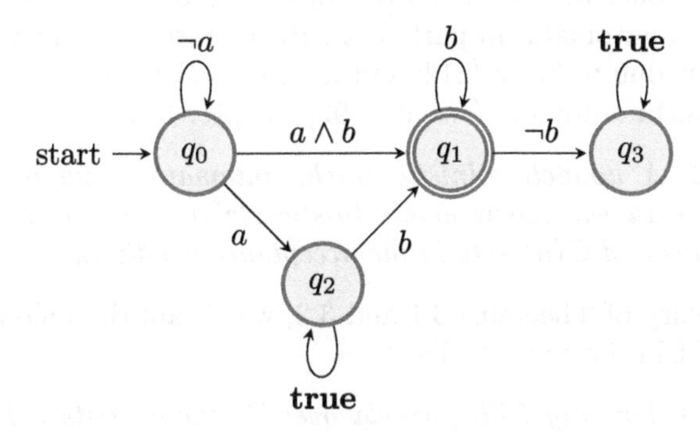

FIGURE 3.2: An equivalent NBA for the LTL formula $\varphi = \Diamond a \wedge \Diamond\Box b$. The double circled states are accepting states. The nondeterminism is caused by the subformula $\Diamond\Box b$ and shown in the automaton as the transition from q_2 to q_1, where a transition to q_1 does not need to happen as soon as the letter b is seen.

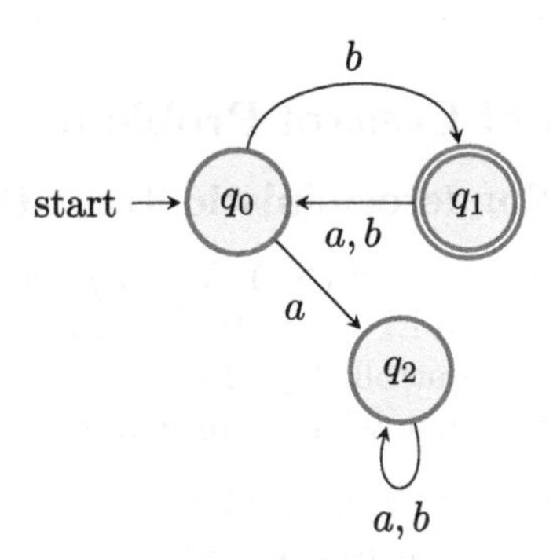

FIGURE 3.3: A deterministic Büchi automaton (DBA) for the linear-time property over $\Pi = \{p\}$ that specifies p should "happen" at every even-indexed position (starting with index 0). The DBA has the alphabet $\Sigma = 2^{\Pi} = \{a, b\}$ with $a = \{\}$ and $b = \{p\}$. It can be shown that no LTL formula over Π can specify this property.

As we will discuss in more detail in Section 3.6, for the purpose of controller synthesis, we often need to convert a nondeterministic automaton to a deterministic one. Indeed, a central topic in automata theory is on determinization of ω-automata, in particular, Büchi automata. The seminal work on this topic is due to Safra [212] and known as *Safra's construction* for determinizing Büchi automata (see also [92, Chapter 3]).

Theorem 3.2 *A nondeterministic Büchi automaton with n states can be translated into an equivalent deterministic Rabin automaton with at most $2^{\mathcal{O}(n \log n)}$ states and $\mathcal{O}(n)$ sets in the acceptance condition.*

As a corollary of Theorems 3.1 and 3.2, we obtain the following result on translation of LTL formulas to DRA.

Corollary 3.1 *For any LTL formula over Π, there exists a DRA $\mathcal{A}_\varphi$ with alphabet $\Sigma = 2^\Pi$ such that the infinite words accepted by $\mathcal{A}_\varphi$ are exactly those satisfying φ. The resulting DRA has at most $2^{2^{\mathcal{O}(|\varphi|)}}$ states and $2^{\mathcal{O}(|\varphi|)}$ pairs of sets in the acceptance condition.*

There also exist direct translations of LTL to DRA [65], which sometimes provide the benefits of preserving the semantic structures of LTL formulas. However, the worse-case complexity bounds cannot be improved in general, because there exist LTL formulas φ_n of size $|\varphi_n| = \mathcal{O}(\text{poly}(n))$ such that every DRA for φ_n has at least 2^{2^n} states [23, p. 805].

3.5 Formulation of Control Problems

3.5.1 Control of Nondeterministic Transition Systems

Given a transition system $\mathcal{T} = (S, A, R, \Pi, L)$ and a linear temporal logic formula (or more generally an LT property) φ, we are interested in controlling the system such that all controlled paths satisfy φ. To formally state the problem, we first need to define what a *controller* means.

Definition 3.16 *Let $\mathcal{T} = (S, A, R, \Pi, L)$ be a transition system. A* ***control strategy*** *for $\mathcal{T}$ is a partial function $\kappa : S^* \to A$, where S^* is the set of all finite sequences in S.*

Definition 3.17 *Let $\mathcal{T} = (S, A, R, \Pi, L)$ be a transition system and κ be a control strategy for $\mathcal{T}$. A* ***κ-controlled path*** *of $\mathcal{T}$ is a sequence $\{s_i\}_{i=0}^\infty$ such that*

$$(s_i, a_i, s_{i+1}) \in R, \quad \forall i \geq 0,$$

where $\{a_i\}_{i=0}^\infty$ is chosen according to

$$a_i = \kappa(s_0, s_1, \ldots, s_i), \quad \forall i \geq 0.$$

A control strategy of the form $\kappa : S^* \to A$ would require infinite memory, since S^* is an infinite set even if S is finite. We define memoryless (or positional) and finite-memory strategies as follows.

Definition 3.18 *Let $\mathcal{T} = (S, A, R, \Pi, L)$ be a transition system. A* ***memoryless control strategy*** *κ for $\mathcal{T}$ is a partial function $\kappa : S \to A$. For a memoryless control strategy κ, a* ***κ-controlled path*** *is a sequence $\{s_i\}_{i=0}^{\infty}$ such that*

$$(s_i, a_i, s_{i+1}) \in R, \quad \forall i \geq 0,$$

where $\{a_i\}_{i=0}^{\infty}$ is chosen according to

$$a_i = \kappa(s_i), \quad \forall i \geq 0.$$

Definition 3.19 *Let $\mathcal{T} = (S, A, R, \Pi, L)$ be a transition system. A* ***finite-memory control strategy*** *κ for $\mathcal{T}$ consists of a finite set of memory states M, a memory update function $m : S \times M \to M$, and a control function $\kappa : S \times M \to A$. For a finite-memory control strategy κ, a* ***κ-controlled path*** *is a sequence $\{s_i\}_{i=0}^{\infty}$ such that*

$$(s_i, a_i, s_{i+1}) \in R, \quad \forall i \geq 0,$$

where $\{a_i\}_{i=0}^{\infty}$ is chosen according to

$$a_i = \kappa(s_i, m_i), \quad \forall i \geq 0,$$

and $\{m_i\}_{i=0}^{\infty}$ is updated using

$$m_{i+1} = m(s_i, m_i), \quad \forall i \geq 0,$$

with m_0 being the initial state of the memory.

Definition 3.20 *Let $\mathcal{T} = (S, A, R, \Pi, L)$ be a transition system and φ be an LT property. A state $s_0 \in S$ is said to belong to the* ***winning set*** *of $\mathcal{T}$ for φ, if there exists a control strategy κ such that the set of κ-controlled paths from s_0 is nonempty and all κ-controlled paths of $\mathcal{T}$ from s_0 satisfy φ. The control strategy κ is called a* ***winning strategy*** *for s_0 with φ. The winning set of $\mathcal{T}$ for φ is denoted by $\mathbf{Win}_{\mathcal{T}}(\varphi)$.*

The following proposition shows that there exists a single winning strategy that works for every state in the winning set.

Proposition 3.2 *Let $\mathcal{T} = (S, A, R, \Pi, L)$ be a transition system, φ be an LT property, and $\mathbf{Win}_{\mathcal{T}}(\varphi)$ be the winning set. If $\mathbf{Win}_{\mathcal{T}}(\varphi) \neq \emptyset$, then there exists a control strategy κ such that all κ-controlled paths of $\mathcal{T}$ starting from $\mathbf{Win}_{\mathcal{T}}(\varphi)$ satisfy φ. Furthermore, if for every state $s_0 \in \mathbf{Win}_{\mathcal{T}}(\varphi)$, there exists a finite-memory (or memoryless) winning strategy for s_0 with φ, then there exists a single finite-memory (or memoryless) winning strategy κ that works for every $s_0 \in \mathbf{Win}_{\mathcal{T}}(\varphi)$.*

The proof of the above proposition is trivial. We just need to construct a single control strategy κ that works for all initial states in $\mathbf{Win}_{\mathcal{T}}(\varphi)$, by recording and applying the controller that corresponds to every initial state. By the same argument, it can be shown that, if from each state in $\mathbf{Win}_{\mathcal{T}}(\varphi)$, there exists a finite-memory (or memoryless) winning strategy, then we can construct an overall finite-memory (or memoryless) winning strategy.

We can now state the control problem for (possibly) nondeterministic transition systems.

Problem 3.1 (Discrete Synthesis for NTS) *Let $\mathcal{T} = (S, A, R, \Pi, L)$ be a transition system and φ be an LT property. The objectives of discrete controller synthesis of $\mathcal{T}$ with respect to φ are to:*

- *determine $\mathbf{Win}_{\mathcal{T}}(\varphi)$; and*
- *if $\mathbf{Win}_{\mathcal{T}}(\varphi)$ is nonempty, obtain a control strategy κ such that all κ-controlled paths of $\mathcal{T}$ starting from $\mathbf{Win}_{\mathcal{T}}(\varphi)$ satisfy φ.*

3.5.2 Control of Discrete-Time Dynamical Systems

Consider the discrete-time control system

$$x(t+1) = f(x(t), u(t), w(t)), \tag{3.1}$$

where $x(t) \in X \subseteq \mathbb{R}^n$, $u(t) \in U \subseteq \mathbb{R}^m$, and $w(t) \in W \subseteq \mathbb{R}^p$. Assume $f : X \times U \times W \to X$. Then we can define a transition system that captures the behaviors of the solutions of (3.1).

Definition 3.21 *The discrete-time control system (3.1) can be written as a transition system*

$$\mathcal{T} = (S, A, R, \Pi, L),$$

where

- *$S = X$ is the set of states;*
- *$A = U$ is the set of actions;*
- *$R \subseteq S \times A \times S$ is defined by $(s, a, s') \in R$ if and only if there exists $w \in W$ such that $f(s, a, w) = s'$;*
- *Π is a set of atomic propositions;*
- *$L : X \to 2^{\Pi}$ is a labeling function.*

With this definition, we can easily formulate control problems for discrete-time dynamical systems with linear-time properties in the same way as we did for nondeterministic transition systems.

Problem 3.2 (Controller Synthesis for Discrete-Time Systems) *Let $\mathcal{T} = (S, A, R, \Pi, L)$ be the transition system for (3.1) and φ be an LT property over Π. The objectives of controller synthesis of $\mathcal{T}$ with respect to φ are to:*

- *determine $\mathbf{Win}_{\mathcal{T}}(\varphi)$; and*
- *if $\mathbf{Win}_{\mathcal{T}}(\varphi)$ is nonempty, obtain a control strategy κ such that all κ-controlled paths of $\mathcal{T}$ starting from $\mathbf{Win}_{\mathcal{T}}(\varphi)$ satisfy φ.*

3.5.3 Control of Continuous-Time Dynamical Systems

Consider the continuous-time control system

$$x'(t) = f(x(t), u(t), w(t)), \tag{3.2}$$

where $x(t) \in X \subseteq \mathbb{R}^n$, $u(t) \in U \subseteq \mathbb{R}^m$, and $w(t) \in W \subseteq \mathbb{R}^p$. Assume that $f : X \times U \times W \to \mathbb{R}^n$ satisfies the usual regularity conditions such that solutions to (3.2) exist (see Proposition 1.1) for input and disturbance signals that are locally essentially bounded.

We shall mostly consider *sampled-data control* strategies where the control signal is given as a piecewise constant function. In this case, we can also easily interpret the time-discretization of (3.2) as a transition system.

Definition 3.22 *The time-discretization of control system (3.2) under sampled-data control of a sampling period $\tau > 0$ can be written as a transition system*

$$\mathcal{T} = (S, A, R, \Pi, L),$$

where

- *$S = X$ is the set of states;*
- *$A = U$ is the set of actions;*
- *$R \subseteq S \times A \times S$ is defined by $(s, a, s') \in R$ if and only if there exists a locally essentially bounded $w : [0, \tau] \to W$ such that*

  $$x'(t) = f(x(t), a, w(t))$$

  for almost all $t \in [0, \tau]$, $x(0) = s$, and $x(\tau) = s'$;
- *Π is a set of atomic propositions;*
- *$L : X \to 2^{\Pi}$ is a labeling function.*

Problem 3.3 (Sample-and-Hold Controller Synthesis) *Let $\mathcal{T} = (S, A, R, \Pi, L)$ be the transition system for (3.2) under sampled-data control of a sampling period $\tau > 0$ and φ be an LT property over Π. The objectives of controller synthesis of $\mathcal{T}$ with respect to φ are to:*

- *determine* $\mathbf{Win}_{\mathcal{T}}(\varphi)$*; and*
- *if* $\mathbf{Win}_{\mathcal{T}}(\varphi)$ *is nonempty, obtain a control strategy* κ *such that all* κ*-controlled paths of* $\mathcal{T}$ *starting from* $\mathbf{Win}_{\mathcal{T}}(\varphi)$ *satisfy* φ.

Clearly, the control strategy κ obtained for $\mathcal{T}$ can be readily implemented as a sampled-data control strategy for system (3.2).

3.6 Discrete Synthesis

In this section, we discuss control of **finite** nondeterministic transition systems (NTS). Classical algorithms for NTS controller synthesis with reachability, safety, and, more generally, ω-regular properties will be presented.

3.6.1 Safety

Consider a transition system $\mathcal{T} = (S, A, R, \Pi, L)$. A safety specification for $\mathcal{T}$ can be easily defined as $\Box p$, where p is a propositional formula. Since p is propositional formula, there exists a set $\Omega \subseteq S$ such that p is evaluated to be true at s if and only if $s \in \Omega$. This is equivalent to defining the set $U = S \setminus \Omega$ to the unsafe set. The control objective is to avoid U or remain in Ω for controlled paths of $\mathcal{T}$. With a slight abuse of notation, we can also write the specification as $\Box\Omega$.

Given $\mathcal{T} = (S, A, R, \Pi, L)$ and $\varphi = \Box\Omega$, Algorithm 1 aims to compute $Y = \mathbf{Win}_{\mathcal{T}}(\varphi)$ and output a positional control strategy $\kappa : S \to A$ such that all κ-controlled paths of $\mathcal{T}$ starting from Y satisfy φ.

Given $\mathcal{T} = (S, A, R, \Pi, L)$, the *controlled predecessors* of a set $X \subseteq S$ is defined as

$$\mathrm{CPre}(X) = \{s \in S \mid \exists a \in A \text{ s.t. } \mathrm{Post}(s, a) \subseteq X\}. \tag{3.3}$$

The following proposition establishes the soundness and completeness of Algorithm 1.

Proposition 3.3 *Given* $\mathcal{T} = (S, A, R, \Pi, L)$ *and* $\varphi = \Box\Omega$*, let* Y *and* κ *be returned by Algorithm 1. Then* $Y = \mathbf{Win}_{\mathcal{T}}(\varphi)$ *and all* κ*-controlled paths of* $\mathcal{T}$ *starting from* Y *satisfy* φ.

Proof: Algorithm 1 terminates when a fixed point is reached for the following operator $I : 2^S \to 2^S$:

$$I(X) = \mathrm{CPre}(X) \cap X.$$

This is a monotone operator in the sense that $I(X) \subseteq X$. Hence the sequence $\{S_k\}$ defined by Algorithm 1 is also monotonically decreasing with respect to

Algorithm 1 Safety

Input: $\mathcal{T} = (S, A, R, \Pi, L)$, $\varphi = \square\Omega$
Output: $Y = \mathbf{Win}_{\mathcal{T}}(\varphi)$, κ

```
1: S_0 ← Ω
2: k ← 0
3: while 1 do                                ▷ Compute the winning set
4:     S_{k+1} ← CPre(S_k) ∩ S_k
5:     if S_{k+1} = S_k then
6:         break
7:     end if
8:     k ← k + 1
9: end while
10: Y ← S_{k+1}
11: for s ∈ Y do                             ▷ Assign controls
12:     κ(s) ∈ {a ∈ A : Post(s, a) ⊆ Y}
13: end for
```

set inclusion. Since $S_0 = \Omega \subseteq S$ is finite, a fixed point is reached after a finite number of steps (in fact at most $|\Omega|$ steps).

We show that $Y = \mathbf{Win}_{\mathcal{T}}(\varphi)$. Since $\text{CPre}(Y) \cap Y = Y$, we have $Y \subseteq \text{CPre}(Y)$; that is, for any $y \in Y$, there exists $a \in A$ such that $\text{Post}(y, a) \subseteq Y$. We can define $\kappa(y) = a$ for any such a as in Line 12 of Algorithm 1. Then all κ-controlled paths starting from any $y \in Y$ remain in $Y \subseteq \Omega$. We have $Y \subseteq \mathbf{Win}_{\mathcal{T}}(\varphi)$.

Define a sequence $Z_0 = S \setminus \Omega$ and

$$Z_{k+1} = Z_k \cup \text{DPre}(Z_k),$$

where

$$\text{DPre}(X) = \{s' \in S \mid \forall a \in A\, \exists s \in X \text{ s.t. } (s', a, s) \in R\}.$$

Claim: $Z_k \cup S_k = S$ for all $k \geq 0$. We show this by induction on k. The statement clearly holds for $k = 0$. We assume it holds for $k = n$ and show it also holds for $k = n + 1$. For any $s \in S \setminus S_{k+1}$, we show that $s \in Z_{k+1}$. We only need to consider the case $s \in S_k$. Otherwise, we have $s \in Z_k \subseteq Z_{k+1}$. Since $s \in S_k$ and $s \notin S_{k+1} = \text{CPre}(S_k) \cap S_k$, we have $s \notin \text{CPre}(S_k)$. That is, for all $a \in A$, there exists $s' \notin S_k$ such that $(s, a, s') \in R$. Since $Z_k \cup S_k = S$ and $s' \in Z_k$, this shows $s \in \text{DPre}(Z_k) \subseteq Z_{k+1}$.

Clearly, when $S_{k+1} = S_k = Y$, we have $Z_{k+1} = Z_k = S \setminus Y$. For any $y \notin Y$, we have $y \in Z_k$. By the definition of Z_k, for all possible actions in A, there exists a path of $\mathcal{T}$ that starts in y and reaches $Z_0 = S \setminus \Omega$ within k steps. This shows $y \notin \mathbf{Win}_{\mathcal{T}}(\varphi)$. Hence $\mathbf{Win}_{\mathcal{T}}(\varphi) \subseteq Y$. ■

Complexity Analysis

Algorithm 1 can make at most $|\Omega|$ calls to compute $\text{CPre}(S_k) \cap S_k$, which requires checking for each $s \in S_k$ whether there exists $a \in A$ such that $\text{Post}(s, a)$ is contained entirely in S_k. Each is upper-bounded by checking at most all the m transitions in $\mathcal{T}$. Hence, the overall time complexity is $\mathcal{O}(|\Omega| \, m)$, which is linear in terms of the number of transitions. A slightly more careful examination shows that one can implement the algorithm by removing an action a for a state s (and all associated transitions), if $\text{Post}(s, a) \cap Z_k \neq \emptyset$. If all actions for a state s are removed, then s cannot belong to $\text{CPre}(S_k)$. Since a transition can only be removed once, the overall complexity of this implementation is $\mathcal{O}(m)$.

3.6.2 Reachability

We next consider reachability control problems for $\mathcal{T} = (S, A, R, \Pi, L)$. Similar to safety, the specification can be written as an LTL formula $\Diamond\Omega$, where Ω is the target set. It is not difficult to see that a reachability control problem is dual to a safety control problem in the following sense. While we aim to design a controller such that *all* controlled paths remain in a set Ω, i.e., for $\Box\Omega$, we are also trying to prevent (by choosing appropriate actions) the non-determinism from steering *any* path to the set $S \setminus \Omega$. This argument is already evident from the proof of Proposition 3.3 and the complexity analysis of Algorithm 1.

If we treat the controller as a player and the non-determinism as another player, this can be seen as a two-player game [92]. In a reachability game, roles of the controller and the non-determinism are reversed from that for a safety game. The controller aims to control all paths to reach a target set Ω, whereas the non-determinism tries to steer at least a path to remain in $S \setminus \Omega$. The duality of reachability will become even more clear after we present the algorithm for solving a reachability control problem.

Given $\mathcal{T} = (S, A, R, \Pi, L)$ and $\varphi = \Diamond\Omega$, Algorithm 2 aims to compute $Y = \mathbf{Win}_{\mathcal{T}}(\varphi)$ and output a memoryless control strategy $\kappa : S \to A$ such that all κ-controlled paths of $\mathcal{T}$ starting from Y satisfy φ.

Lines 2–8 of Algorithm 2 arbitrarily assign controls to states already in the target set. This is because paths starting in the target sets always satisfy $\Diamond\Omega$ by definition. One can also consider a variant of reachability where the target set has to be reached after applying some control. We shall not consider this variant here. Lines 10–19 of Algorithm 2 incrementally computes all the states that can reach the target set until a fixed point is reached, i.e., no more states can be added to the "attractor" of the target set. Controls assigned are the ones that can take the states to the target set in a minimum number of steps in the worst-case scenario.

Algorithm 2 Reachability

```
Input: T = (S, A, R, Π, L), φ = ◇Ω
Output: Y = Win_T(φ), κ
 1: S_0 ← Ω
 2: for s ∈ Y do                  ▷ Assign (arbitrary) controls for the target set
 3:     if s ∈ S_0 then
 4:         κ(s) ∈ A
 5:     else
 6:         κ(s) = ∅
 7:     end if
 8: end for
 9: k ← 0
10: while 1 do                    ▷ Compute the winning set and assign controls
11:     S_{k+1} ← CPre(S_k) ∪ S_k
12:     for s ∈ S_{k+1} \ S_k do
13:         κ(s) ∈ {a ∈ A | Post(s, a) ⊆ S_k}
14:     end for
15:     if S_{k+1} = S_k then
16:         break
17:     end if
18:     k ← k + 1
19: end while
20: Y ← S_{k+1}
```

Proposition 3.4 *Given* $\mathcal{T} = (S, A, R, \Pi, L)$ *and* $\varphi = \Diamond\Omega$, *let* Y *and* κ *be returned by Algorithm 2. Then* $Y = \mathbf{Win}_\mathcal{T}(\varphi)$ *and all* κ*-controlled paths of* $\mathcal{T}$ *starting from* Y *satisfy* φ.

Proof: Algorithm 2 terminates in a finite number of steps because $\{S_k\}$ is monotonically increasing with respect to set inclusion and the state space S is finite. Upon reaching a fixed-point, we have $S_{k+1} = S_k = Y$. We show that $Y = \mathbf{Win}_\mathcal{T}(\varphi)$ and κ is a winning controller.

For any $y \in Y$, there exists some $i \geq 0$ such that $y \in S_i$ and $y \notin S_j$ for any $j \leq i - 1$ (assuming $i \geq 1$). By the definition of S_i and the construction of κ in Algorithm 2, all κ-controlled paths starting from y will reach Ω within at most i steps (and 0 steps for $y \in S_0$). Hence $Y \subseteq \mathbf{Win}_\mathcal{T}(\varphi)$.

To show $\mathbf{Win}_\mathcal{T}(\varphi) \subseteq Y$. Define a sequence $Z_0 = S \setminus \Omega$ and

$$Z_{k+1} = Z_k \cap \mathrm{DPre}(Z_k),$$

where DPre$(\cdot)$ is as defined in the proof of Proposition 3.3. We can show that $Z_k \cup S_k = S$ for all $k \geq 0$ using a similar argument. A more direct (and abstract) argument can be obtained by noting $S \setminus \mathrm{DPre}(X) = \mathrm{CPre}(S \setminus X)$

and

$$\begin{aligned} S_{k+1} &= S_k \cup \mathrm{CPre}(S_k) = (S \setminus Z_k) \cup \mathrm{CPre}(S \setminus Z_k) \\ &= (S \setminus Z_k) \cup (S \setminus \mathrm{DPre}(Z_k)) \\ &= S \setminus (Z_k \cap \mathrm{DPre}(Z_k)) = S \setminus Z_{k+1}, \end{aligned}$$

where we used the inductive assumption $S_k = S \setminus Z_k$. Let $Z = S \setminus Y$. Then $Z_{k+1} = Z_k = Z$, which implies $\mathrm{DPre}(Z) = Z$. If $y \notin Y$, then $y \in Z$. Since $\mathrm{DPre}(Z) = Z$, for any $z \in Z$ and any action $a \in A$, there exists some $z' \in Z$ such that $(z, a, z') \in R$. It follows that no controller can steer all paths starting from y to escape Z. Note that $\Omega \subseteq Y$ and $Z \cap \Omega = \emptyset$. Hence $y \notin \mathbf{Win}_{\mathcal{T}}(\varphi)$, which implies $\mathbf{Win}_{\mathcal{T}}(\varphi) \subseteq Y$. ∎

Complexity Analysis

Algorithm 2 can be implemented in $\mathcal{O}(m)$ time complexity, where m is the number of transitions in $\mathcal{T}$. Starting from the target set $S_0 = \Omega$, one can mark whether a given transition (s, a, s') reaches S_k, i.e., $s' \in S_k$, where S_k can be seen as the current under-approximation of the winning set (or attractor of Ω). For any state s that is not already in S_k, if all transitions resulting from taking an action a from state s are marked as reaching S_k, s is added to S_{k+1}, with a recorded as the action to be taken at s. Of course, there may be multiple actions that can steer s to S_k. One may choose to record all or only one of them.

3.6.3 ω-Regular Properties

In this section, we discuss control of finite NTS to satisfy a general ω-regular property described by a deterministic Rabin automaton (DRA).

Let $\mathcal{A} = (Q, \Sigma, r, q_0, \Omega)$ be a DRA and $\mathcal{T} = (S, A, R, \Pi, L)$ be a finite transition system. Suppose $\Sigma = 2^{\Pi}$. Denote by φ the LT property that contains all infinite words accepted by $\mathcal{A}$. As defined in Problem 3.1, the objective of DRA control is to determine the winning set $\mathbf{Win}_{\mathcal{T}}(\varphi)$ and to obtain a winning control strategy.

To solve this problem, we define a product transition system

$$\mathcal{T} \otimes \mathcal{A} := (S \times Q, A, R_{\otimes}, Q, L_{\otimes}),$$

where

- $((s,q), a, (s',q')) \in R_{\otimes}$ if $(s, a, s') \in R$ and $q' = r(q, L(s))$;
- $L_{\otimes}(s,q) = \{q\}$.

The Rabin acceptance condition $\Omega = \{(E_i, F_i)\}_{i=1}^{k}$, where $E_i, F_i \subseteq Q$ for $1 \le i \le k$, can be written as an LTL formula

$$\psi = \bigvee_{1 \le i \le k} (\Box\Diamond F_i \wedge \Diamond\Box\neg E_i), \tag{3.4}$$

where $F_i = \bigvee_{q \in F_i} \{q\}$ and $E_i = \bigvee_{q \in E_i} \{q\}$. This is a slight abuse of notation, but the intuitive meaning is clear. The formula is satisfied by a path $\{(s_i, q_i)\}$ of $\mathcal{T} \otimes \mathcal{A}$ if and only if there exists a pair (F_i, E_i) such that F_i is visited by $\{q_i\}$ infinitely often and E_i is visited by $\{q_i\}$ finitely often.

The following lemma shows that there is a one-to-one correspondence between a control strategy for $\mathcal{T} \otimes \mathcal{A}$ to satisfy ψ defined above and a control strategy for $\mathcal{T}$ to satisfy the LT property φ defined by $\mathcal{A}$.

Lemma 3.1 *We have*

$$\mathbf{Win}_{\mathcal{T}}(\varphi) = \{s \mid (s, q_0) \in \mathbf{Win}_{\mathcal{T}\otimes\mathcal{A}}(\psi)\}. \tag{3.5}$$

Moreover, if $\mathcal{T} \otimes A$ has a memoryless winning strategy, then $\mathcal{T}$ has a finite memory winning strategy with $\mathcal{Q}$ as the memory set, $m(s, q) = r(q, L(s))$ as the memory update function, and $m_0 = q_0$ as the initial memory state.

Proof: We first show (3.5). There is a one-to-one correspondence between control strategies for $\mathcal{T}$ and control strategies for $\mathcal{T} \otimes \mathcal{A}$, because $\{q_i\}$ is determined completely by $\{s_i\}$ (due to determinism of $\mathcal{A}$). Moreover, controlled paths $\{s_i\}$ of $\mathcal{T}$ correspond to controlled paths $\{(s_i, q_i)\}$ of $\mathcal{T} \otimes \mathcal{A}$. The conclusion follows from the equivalence of the following: 1) $\{s_i\}$ satisfies φ, i.e., $\{L(s_i)\}$ is accepted by $\mathcal{A}$; 2) $\{q_i\}$ is the run of $\mathcal{A}$ for $\{L(s_i)\}$ and there exists a pair $(E_i, F_i) \in \Omega$ such that $\{q_i\}$ visits F_i infinitely often and E_i finitely often; 3) $\{(s_i, q_i)\}$ satisfies ψ.

Let $\kappa_\otimes$ be a memoryless control strategy for $\mathcal{T} \otimes \mathcal{A}$ such that all $\kappa_\otimes$-controlled paths of $\mathcal{T}\otimes\mathcal{A}$ starting from $\mathbf{Win}_{\mathcal{T}\otimes\mathcal{A}}(\psi)$ satisfy ψ. We construct a control strategy κ for $\mathcal{T}$ with a finite memory set $M = Q$ as follows. Suppose $s_0 \in \mathbf{Win}_{\mathcal{T}}(\varphi)$. Then $(s_0, q_0) \in \mathbf{Win}_{\mathcal{T}\otimes\mathcal{A}}(\psi)$. Set the initial state of the memory by $m_0 = q_0$. For each $i \geq 0$, κ generates a control action by

$$a_i = \kappa(s_i, m_i) = \kappa_\otimes(s_i, m_i). \tag{3.6}$$

Let the memory update function $m : S \times M \to M$ be defined as

$$m(s, q) = r(q, L(s)).$$

We show that each κ-controlled path $\{s_i\}$ of $\mathcal{T}$ corresponds to a $\kappa_\otimes$-controlled path $\{(s_i, q_i)\}$ of $\mathcal{T}\otimes\mathcal{A}$, where $q_{i+1} = r(q_i, L(s_i))$. The corresponding sequence of memory states $\{m_i\}$ is precisely $\{q_i\}$, because $m_0 = q_0$ and, by induction, we have

$$m_{i+1} = m(s_i, m_i) = r(q_i, L(s_i)) = q_{i+1}, \quad i \geq 0.$$

By (3.6), this shows that, if $\{s_i\}$ is a κ-controlled path of $\mathcal{T}$, then $\{(s_i, q_i)\}$ is a $\kappa_\otimes$-controlled path of $\mathcal{T} \otimes \mathcal{A}$. Since $(s_0, q_0) \in \mathbf{Win}_{\mathcal{T}\otimes\mathcal{A}}(\psi)$ and $\kappa_\otimes$ is winning, $\{(s_i, q_i)\}$ satisfies ψ. It follows that $\{L(s_i)\}$ is accepted by $\mathcal{A}$ and $\{s_i\}$ satisfies φ. ■

Defining the product system $\mathcal{T} \otimes A$ allows us to formulate a control problem on $\mathcal{T} \otimes A$ with the Rabin specification (3.4). With this setup, the Rabin specification can be intuitively understood as visiting a set F_i infinitely often and a set E_i finitely often, where (E_i, F_i) is some pair of sets from a family $\Omega = \{(E_i, F_i\}_{i=1}^{k}$ that defines the Rabin condition. Similar to the safety and reachability problems, the propositions E_i and F_i in the Rabin specification (3.4) can be interpreted as subsets of the state space $\mathcal{T} \otimes A$, i.e., E_i as $\{(s,q) \mid q \in E_i\}$ and F_i as $\{(s,q) \mid q \in F_i\}$.

To clear up the notation, we state a Rabin control problem for nondeterministic transition systems separately as follows, which is essentially a restatement of Problem 3.1 with the specification substantiated by a Rabin specification directly defined on the state space.

Problem 3.4 (Discrete Synthesis for NTS with a Rabin Specification) *Let $\mathcal{T} = (S, A, R, \Pi, L)$ be a transition system and*

$$\varphi = \bigvee_{1 \le i \le k} (\Box\Diamond F_i \wedge \Diamond\Box\neg E_i)$$

be a Rabin specification, where E_i and F_i are subsets of S. The objectives of discrete controller synthesis of $\mathcal{T}$ with respect to φ are to:

- *determine $\mathbf{Win}_{\mathcal{T}}(\varphi)$; and*
- *if $\mathbf{Win}_{\mathcal{T}}(\varphi)$ is nonempty, obtain a control strategy κ such that all κ-controlled paths of $\mathcal{T}$ starting from $\mathbf{Win}_{\mathcal{T}}(\varphi)$ satisfy φ.*

The negation of a Rabin specification defines a Streett specification, i.e., a specification of the form

$$\neg\varphi = \bigwedge_{1 \le i \le k} (\Diamond\Box\neg F_i \vee \Box\Diamond E_i).$$

Similar to how we analyze the safety and reachability control problems, we need to consider a dual problem that asks whether, from an initial condition and for any control strategy, there exists a way for the nondeterminism to be resolved to give a path satisfying $\neg\varphi$. The set of such initial conditions is clearly complementary to $\mathbf{Win}_{\mathcal{T}}(\varphi)$.

To introduce an algorithm for solving the Rabin control problem (Problem 3.4), we need to introduce subproblems for control of NTS. Consider $\mathcal{T} = (S, A, R, \Pi, L)$ and $S' \subseteq S$. We can restrict the transition system $\mathcal{T}$ to S' and obtain a new transition system $T|_{S'} = (S', A, R|_{S'}, \Pi, L|_{S'})$, where

- $R|_{S'}$ is the restriction of R on S', i.e.,

$$R|_{S'} = \{(s, a, s') : s \in S', s' \in S'\};$$

- $L|_{S'}$ is the restriction of L on S'.

The reachability control algorithm (Algorithm 2) plays a basic role in solving the Rabin control problem. We refer to the input-output relation of

Algorithm 2 as $(Y, \kappa) = \text{Reach}_{\exists\forall}(\mathcal{T}, \Omega)$, where Ω is the target set. We use this notation to differentiate it from a variant of the reachability algorithm, which is referred to as $Y = \text{Reach}_{\forall\exists}(\mathcal{T}, \Omega)$ and detailed in Algorithm 3. We can easily see from the proof of Proposition 3.3 that Algorithm 3 is dual to the safety control algorithm (Algorithm 1). In particular, we have

$$\mathbf{Win}_{\mathcal{T}}(\Box\Omega) = \text{Reach}_{\exists\forall}(\mathcal{T}, \Omega) = S \setminus \text{Reach}_{\forall\exists}(\mathcal{T}, \neg\Omega).$$

We only care about the "winning set" returned by Algorithm 3, because we are only interested in recording a strategy that can be used by the controller.

Algorithm 3 Dual Reachability

Input: $\mathcal{T} = (S, A, R, \Pi, L)$, $\varphi = \Diamond\Omega$
Output: Y
1: $S_0 \leftarrow \Omega$
2: $k \leftarrow 0$
3: **while** 1 **do**
4: $\quad S_{k+1} \leftarrow \text{DPre}(S_k) \cup S_k$
5: $\quad$ **if** $S_{k+1} = S_k$ **then**
6: $\quad\quad$ **break**
7: $\quad$ **end if**
8: $\quad k \leftarrow k + 1$
9: **end while**
10: $Y \leftarrow S_{k+1}$

We are now ready to present Algorithm 4 for solving a Rabin control problem (Problem 3.4). The algorithm relies on calls to $\text{Reach}_{\exists\forall}(\mathcal{T}, \Diamond\Omega)$ and $\text{Reach}_{\forall\exists}(\mathcal{T}, \Diamond\Omega)$ described above. The algorithm also contains two recursive calls to itself. For this, we refer to the function defined by Algorithm 4 as $(Y, \kappa) = \text{Rabin}(\mathcal{T}, \varphi)$.

Description of Algorithm 4 (Rabin)

We provide an intuitive description of Algorithm 4 as follows:

- Line 1 initializes the winning set and strategy to be empty.
- In the **for** loop (lines 2–18), each iteration begins with a call to $\text{Reach}_{\forall\exists}(\mathcal{T}, E_i)$, which computes the set from which there exists a path of $\mathcal{T}$ that reaches E_i no matter what choice of controls. This set is removed from the computation for this iteration.
- The **repeat** block (lines 6–11) iteratively defines a sequence of Rabin control problems of decreasing sizes. In line (8), states in $\text{Reach}_{\exists\forall}(\mathcal{T}|_{S_i^j}, F_i)$ are removed to form a subproblem, which does not contain states in E_i or F_i. Hence a recursive call to $\text{Rabin}(\mathcal{T}|_{Z_i^j}, \varphi)$ in line 9 solves a Rabin control problem with one less pairs in the Rabin

Algorithm 4 Rabin

Input: $\mathcal{T} = (S, A, R, \Pi, L)$, $\varphi = \bigvee_{1 \leq i \leq k}(\Box\Diamond F_i \wedge \Diamond\Box\neg E_i)$
Output: $Y = \mathbf{Win}_{\mathcal{T}}(\varphi)$, κ

1: $Y \leftarrow \emptyset$, $\kappa \leftarrow \emptyset$
2: **for** $i \leftarrow 1, k$ **do**
3: $\quad S_i \leftarrow S \setminus \text{Reach}_{\forall\exists}(\mathcal{T}, E_i)$
4: $\quad S_i^0 \leftarrow S_i$
5: $\quad j \leftarrow 0$
6: $\quad$ **repeat**
7: $\quad\quad (W, \kappa_1) = \text{Reach}_{\exists\forall}(\mathcal{T}|_{S_i^j}, F_i)$
8: $\quad\quad Z_i^j \leftarrow S_i^j \setminus W$
9: $\quad\quad (Z, \kappa_2) \leftarrow \text{Rabin}(\mathcal{T}|_{Z_i^j}, \varphi)$
10: $\quad\quad S_i^{j+1} \leftarrow S_i^j \setminus \text{Reach}_{\forall\exists}(\mathcal{T}|_{S_i^j}, Z_i^j \setminus Z)$
11: $\quad$ **until** $S_i^{j+1} = S_i^j$
12: $\quad$ **if** $S_i^j \neq \emptyset$ **then**
13: $\quad\quad Y \leftarrow Y \cup S_i^j$, $\kappa \leftarrow \kappa \cup \kappa_1 \cup \kappa_2$
14: $\quad\quad (Y', \kappa') \leftarrow \text{Rabin}(\mathcal{T}|_{S \setminus Y}, \varphi)$
15: $\quad\quad Y \leftarrow Y \cup Y'$, $\kappa \leftarrow \kappa \cup \kappa'$
16: $\quad\quad$ **break**
17: $\quad$ **end if**
18: **end for**

specification. The fixed point (i.e., the set $S_i^{j+1} = S_i^j$) of this **repeat** block is either a nonempty or empty set.

- If this fixed point is nonempty, it can be shown to be contained in the winning set. This is added to the winning set Y. Another recursive call to $\text{Rabin}(\mathcal{T}|_{S \setminus Y}, \varphi)$ on the remaining subproblem solves a Rabin control problem with at least one less state. The winning set and strategy of the subproblem are added to the overall winning set and strategy. The **for** loop is exited and the algorithm terminates.
- If the fixed point is empty, no winning set is found for this iteration and the algorithm proceeds to the next iteration in the **for** loop. If the **for** loop is not exited ever by line 16, the algorithm would terminate after the **for** loop ends.

Correctness of Algorithm 4 is formally stated in the following proposition.

Proposition 3.5 *Given* $\mathcal{T} = (S, A, R, \Pi, L)$ *and* $\varphi = \bigvee_{1 \leq i \leq k}(\Box\Diamond F_i \wedge \Diamond\Box\neg E_i)$*, where* E_i *and* F_i *are subsets of* S*. Let* Y *and* κ *be returned by Algorithm 4. Then* $Y = \mathbf{Win}_{\mathcal{T}}(\varphi)$ *and all* κ*-controlled paths of* $\mathcal{T}$ *starting from* Y *satisfy* φ*.*

Proof: The proof can be broken down into two parts. One is to show that $Y \subseteq \mathbf{Win}_{\mathcal{T}}(\varphi)$ and κ obtained is a winning strategy from Y, i.e., the computed winning set and strategy are sound. The other is to show $\mathbf{Win}_{\mathcal{T}}(\varphi) \subseteq Y$, i.e., the algorithm is complete in characterizing the entire winning set.

We first establish a fact about the sequence of subproblems defined on $\mathcal{T}_{S_i^j}$. By the definitions of S_i^j and $\mathcal{T}_{S_i^j}$, any path of $\mathcal{T}$ starting from a state in S_i^j and resulting from a strategy of $\mathcal{T}_{S_i^j}$ will remain in S_i^j. As a result, any such path will not reach E_i, because $S_i^{j+1} \subseteq S_i^j \subseteq S_i^0 = S_i = S \setminus \text{Reach}_{\forall\exists}(\mathcal{T}, E_i)$.

(**Soundness**) We then show that, if there exist some i and j such that S_i^j returned after line 11 of Algorithm 4 is nonempty, then $S_i^j \subseteq \mathbf{Win}_{\mathcal{T}}(\varphi)$. This is because for the **repeat** block to reach a nonempty fixed point, we must have $Z_i^j \setminus Z = \emptyset$, which shows that the winning set Z for $\text{Rabin}(\mathcal{T}|_{Z_i^j}, \varphi)$ is Z_i^j, the entire state space of this subproblem. We have $S_i^j = Z_i^j \cup W$ (a disjoint union). The winning strategy for states starting from W is κ_1 and from Z_i^j is the corresponding winning strategy for the subproblem $\text{Rabin}(\mathcal{T}|_{Z_i^j}, \varphi)$ is κ_2. We show that the strategy $\kappa_1 \cup \kappa_2$ indeed wins on $S_i^j = Z_i^j \cup W$. By the fact we proved earlier, starting from S_i^j and applying actions available in $\mathcal{T}_{S_i^j}$ lead to paths of $\mathcal{T}$ entirely contained in S_i^j. If the path remains in Z_i^j after exiting at most a finite number of times, the path will satisfy φ (with respect to the remaining pairs excluding (E_i, F_i)) by the definition of κ_2 and the subproblem $\text{Rabin}(\mathcal{T}|_{Z_i^j}, \varphi)$. If the path visits W infinitely often, it will reach F_i infinitely often under κ_1 (and not reaching E_i since $S_i^j \cap E_i = \emptyset$). Hence the path will satisfy φ. Note that Y is incrementally built by adding sets S_i^j (for subproblems). The resulting strategy κ is memoryless, because the nonrecursive part of the strategy (κ_1) is memoryless.

(**Completeness**) We show that, if $y \notin Y$, then, for all possible controls, there exists a path of $\mathcal{T}$ that violates φ. Note from the proof of soundness that the winning region Y is built incrementally by adding sets S_i^j that are fixed points of the **repeat** block. Once a set S_i^j is added, the entire current winning set Y is removed to form a subproblem. The algorithm terminates when no more such sets S_i^j are added. When this happens, from lines 12–17, we know that the last outermost call $\text{Rabin}(\mathcal{T}|_{S \setminus Y}, \varphi)$ does not further increase Y. It follows that for this call the fixed point of the **repeat** block is $S_i^j = \emptyset$ for each i.

Starting from $y \in S \setminus Y$, we construct a "strategy" for the nondeterminism to be resolved such that, for all possible controls, the corresponding path violates φ. To do so, consider the sequence of sets

$$\emptyset = S_i^j \subsetneq S_i^{j-1} \subsetneq \cdots \subsetneq S_i^0 = S_i = S \setminus \text{Reach}_{\forall\exists}(\mathcal{T}, E_i).$$

Note that

$$S_i^{j+1} = S_i^j \setminus \text{Reach}_{\forall\exists}(\mathcal{T}|_{S_i^j}, Z_i^j \setminus Z).$$

It follows that

$$S = \text{Reach}_{\forall\exists}(\mathcal{T}, E_i) \cup \bigcup_{k=0}^{j-1} \text{Reach}_{\forall\exists}(\mathcal{T}|_{S_i^k}, Z_i^k \setminus Z).$$

Consider a path starting from any $\text{Reach}_{\forall\exists}(\mathcal{T}|_{S_i^k}, Z_i^k \setminus Z)$. Assume first that the controller only takes actions are available in $\mathcal{T}|_{S_i^k}$. Then by the definition of $\text{Reach}_{\forall\exists}(\mathcal{T}|_{S_i^k}, Z_i^k \setminus Z)$, the path can be steered, by resolving the nondeterminism, to reach $Z_i^k \setminus Z$. From there, by the same assumption, the path can be steered by the nondeterminism to remain in Z_i^k. This is true by the definition of Z_i^k; a state that can be forced out of Z_i^k to reach $W = S_i^k \setminus Z_i^k$ should be in W and not Z_i^k in the first place. Within Z_i^k, this path would violate φ without the pair (E_i, F_i), by following a strategy to resolve nondeterminism for the subproblem $\text{Rabin}(\mathcal{T}|_{Z_i^k}, \varphi)$. Since the path is trapped in Z_i^k and thus cannot reach F_i, the path violates the original φ. Now suppose that at some state on the path the controller takes actions that are not available in $\mathcal{T}|_{S_i^k}$. Such actions are not available in $\mathcal{T}|_{S_i^k}$ because they would force the path to exit S_i^k no matter how the nondeterminism is resolved. Following such an action, the path we reach $\mathcal{T}|_{S_i^l}$ for some $l < k$ or $\text{Reach}_{\forall\exists}(\mathcal{T}, E_i)$. Repeating this argument, either the path can stay in one of S_i^k after a finite number of crossings, in which case the above argument shows that we can steer the path to violates φ, or the path will enter $\text{Reach}_{\forall\exists}(\mathcal{T}, E_i)$ infinitely often. From there, the nondeterminism can steer the path to visit E_i infinitely often. Since the above argument holds for all i, the nondeterminism can devise a strategy (with memory k) to visit each E_i infinitely often. As a result, the path violates φ. ■

Complexity Analysis

Algorithm 4 makes two recursive calls to itself. To analyze its time complexity, let $C(k, n)$ denote an upper bound on the time complexity of run $\text{Rabin}(\mathcal{T}, \varphi)$ with n states in $\mathcal{T}$ and k pairs in the Rabin specification φ. Apparently, the recursive calls are dominating the time complexity. By ignoring the time complexity of calls to Reach, which are linear in terms of the number of transitions, we have

$$
\begin{aligned}
C(k,n) &\leq k\,[C(k-1,n-1) + C(k-1,n-1) + \cdots + C(k-1,1)] \\
&\qquad + C(k,n-1) \\
&\leq k\,[C(k-1,n-1) + C(k-1,n-1) + \cdots + C(k-1,1)] \\
&\qquad + k\,[C(k-1,n-2) + C(k-1,n-1) + \cdots + C(k-1,1)] \\
&\qquad + C(k,n-2) \\
&\leq k\,[C(k-1,n-1) + C(k-1,n-1) + \cdots + C(k-1,1)] \\
&\qquad + k\,[C(k-1,n-2) + C(k-1,n-1) + \cdots + C(k-1,1)] \\
&\qquad + \cdots + C(k,1) \\
&\leq k[(n-1) + (n-2) + \cdots 1]C(k-1,n-1) \\
&\leq kn^2 C(k-1,n-1),
\end{aligned}
$$

which is clearly a conservative estimate. By this, we obtain that the overall time complexity is bounded by $\mathcal{O}(k!n^{2k})$, where k is the number of pairs in the specification and n is the number of states in $\mathcal{T}$.

Remark 3.2 *We now can remark on the time complexity of performing LTL control synthesis on finite NTS. Suppose that $\mathcal{T}$ is a finite NTS with n states and m transitions. Let φ be an LTL formula and $\mathcal{A}$ be a DRA translation of φ. According to Corollary 3.1, $\mathcal{A}$ can have at most $2^{2^{\mathcal{O}(|\varphi|)}}$ states and $2^{\mathcal{O}(|\varphi|)}$ pairs of states in the acceptance condition. By definition, $\mathcal{T} \otimes \mathcal{A}$ has $n \cdot 2^{2^{\mathcal{O}(|\varphi|)}}$ states. By the complexity analysis of Algorithm 4, one can solve a Rabin game on $\mathcal{T} \otimes \mathcal{A}$ with $n \cdot 2^{2^{\mathcal{O}(|\varphi|)}}$ states and $2^{\mathcal{O}(|\varphi|)}$ pairs, which results in a time complexity of $2^{\mathcal{O}(|\varphi|)}!(n \cdot 2^{2^{\mathcal{O}(|\varphi|)}})^{2^{\mathcal{O}(|\varphi|)}} = n^{2^{\mathcal{O}(|\varphi|)}} 2^{2^{\mathcal{O}(|\varphi|)}} = n^{2^{\mathcal{O}(|\varphi|)}}$, assuming $n \gg 2$. This double-exponential time complexity (in terms of the size of the formula φ) is unavoidable for general LTL synthesis [186].*

Remark 3.3 *A potentially different approach to LTL control synthesis is to translate an LTL formula φ to a deterministic ω-automaton with a parity acceptance condition [64]. Nonetheless, the double-exponential time complexity cannot be avoided. More specifically, the resulting deterministic parity automaton (DPA) has $2^{2^{\mathcal{O}(|\varphi|)}}$ states and $2^{\mathcal{O}(|\varphi|)}$ colors. The currently best algorithm for parity game has quasi-polynomial time complexity [44]. Even if parity game turns out to be in P (having polynomial time complexity), we still cannot avoid the double-exponential time complexity due to the $2^{2^{\mathcal{O}(|\varphi|)}}$ size of the DPA. There is, however, hope in improving the complexity in terms of the size of the NTS.*

Remark 3.4 *The special case of control of NTS with a one-pair deterministic Rabin specification has time complexity $\mathcal{O}(n^2)$, where n is the number of states in the Rabin automaton. Denote the pair of sets in the specification by (E, F). A further special case when $\neg E = \emptyset$ reduces to $\varphi = \Box\Diamond F$, which is known as a deterministic Büchi automaton (DBA) specification. Another special case is given by $F = \emptyset$, which reduces to $\varphi = \Box\Diamond\neg E$. This is known as a deterministic co-Büchi specification. As shown in Example 3.1, not every LTL formula can be translated to a DBA. Even if this exists, it has been shown that the translation from LTL formulas to DBAs is still doubly experiential [130]. While the doubly exponential time complexity in terms of the size of the LTL formula cannot be avoided, the complexity in terms of the size of the NTS is improved due to the quadratic complexity in solving a DBA control problem for NTS.*

3.7 Summary

This chapter introduces the basic concepts of formal methods, including transition system models and formal specifications, and formulates control problems for transition systems. We then extend such problem formulations to discrete-time and continuous-time dynamical systems. Finally, a few basic algorithms for solving control problems on finite nondeterministic transition systems (NTS) are introduced. This shows that, in principle, how control problems with respect to ω-regular properties can be solved for finite NTS.

The main references for writing this chapter include [23, 92]. All algorithms presented for discrete synthesis are standard, albeit they are usually presented in the form of solutions for two-player infinite games [92]. The algorithm for solving the Rabin control problem is from [106]. We choose to write the algorithms directly for nondeterministic transition systems, without translating the problem to a two-player game first, because this is closer to the discrete abstractions we usually compute for continuous systems (Chapter 5). The nondeterministic transition system representation also naturally agrees with the model of difference equations with inputs for discrete-time dynamical systems. Other references on formal verification and model checking include [49, 50].

Chapter 4

Interval Computation

In this chapter, we introduce *interval analysis* as a numerical framework for set-oriented computation. In this framework, the basis representation of sets is *intervals* in $\mathbb{R}^n$, which are Cartesian products of intervals in $\mathbb{R}$. A salient feature of interval analysis is to provide rigorous error bounds in such computation. Interval computation plays a fundamental role in connecting dynamical systems, which evolve in a continuous state space, with formal specifications (LTL or ω-regular properties), which are discrete in nature, for verification and control synthesis in later chapters.

4.1 Interval Analysis

We shall only consider compact (closed and bounded) intervals in $\mathbb{R}$ and $\mathbb{R}^n$.

Definition 4.1 *An **interval** in* $\mathbb{R}$ *is of the form*

$$[x] := [\underline{x}, \overline{x}] = \{x \mid \underline{x} \leq x \leq \overline{x}\},$$

where $\underline{x}, \overline{x} \in \mathbb{R}$ *are the lower and upper bounds of* $[x]$*, respectively. We denote the set of all intervals in* $\mathbb{R}$ *by* $\mathbb{IR}$.

Given an interval $[x] \in \mathbb{IR}$, we define its

- *width*: $\mathrm{w}([x]) := \overline{x} - \underline{x}$;
- *midpoint* (or *center*) : $\mathrm{mid}([x]) := (\underline{x} + \overline{x})/2$;
- *magnitude*: $|[x]| := \max\{|\underline{x}|, |\overline{x}|\}$.

Similar to the basic arithmetic operations for real numbers, we can define the binary arithmetic operations $\star \in \{+, -, \cdot, /\}$ for any intervals $[x]$ and $[y]$ as

$$[x] \star [y] := \{x \star y \mid x \in [x], y \in [y]\}.$$

DOI: 10.1201/9780429270253-4

It is easy to verify that

$$\begin{aligned}
[x] + [y] &= [\underline{x} + \underline{y}, \overline{x} + \overline{y}];\\
[x] - [y] &= [\underline{x} - \overline{y}; \overline{x} - \underline{y}];\\
[x] \cdot [y] &= [\min\left\{\underline{x}\underline{y}, \underline{x}\overline{y}, \overline{x}\underline{y}, \overline{x}\overline{y}\right\}, \max\left\{\underline{x}\underline{y}, \underline{x}\overline{y}, \overline{x}\underline{y}, \overline{x}\overline{y}\right\}];\\
[x]/[y] &= [\underline{x}, \overline{x}][1/\overline{y}, 1/\underline{y}],
\end{aligned}$$

where $[x]/[y]$ requires $0 \notin [y]$. Given $\alpha \in \mathbb{R}$ and $[x] \in \mathbb{IR}$, we can define

$$\alpha[x] := \{\alpha x \mid x \in [x]\} = \begin{cases} [\alpha\underline{x}, \alpha\overline{x}], & \alpha \geq 0,\\ [\alpha\overline{x}, \alpha\underline{x}], & \alpha < 0. \end{cases}$$

We can easily generalize these definitions to intervals in $\mathbb{R}^n$.

Definition 4.2 *An* ***interval*** *(also called a box or a hyperrectangle) in* $\mathbb{R}^n$ *is of the form*

$$[x] := [x_1] \times \cdots \times [x_n] \subseteq \mathbb{R}^n,$$

where $[x_i] = [\underline{x_i}, \overline{x_i}] \in \mathbb{IR}$ *for* $i = 1, \cdots, n$. *With* $\underline{x} = (\underline{x_1}, \cdots, \underline{x_n})$ *and* $\overline{x} = (\overline{x_1}, \cdots, \overline{x_n})$, *we can also write* $[x] = [\underline{x}, \overline{x}]$. *We denote the set of all intervals in* $\mathbb{R}^n$ *by* $\mathbb{IR}^n$.

The operations we defined for intervals in $\mathbb{IR}$ can be straightforwardly extended to intervals in $\mathbb{IR}^n$ (in the same way these operations are extended from real numbers in $\mathbb{R}$ to real vectors $\mathbb{R}^n$). We have

$$\begin{aligned}
\alpha[x] &= (\alpha[x_1]) \times \cdots \times (\alpha[x_n]),\\
[x] + [y] &= ([x_1] + [y_1]) \times \cdots \times ([x_n] + [y_n]),\\
[x] \cdot [y] &= [x_1] \cdot [y_1] + \cdots + [x_n] \cdot [y_n],
\end{aligned}$$

for any $[x], [y] \in \mathbb{IR}^n$ and $\alpha \in \mathbb{R}$. The width, magnitude, and midpoint of intervals in $\mathbb{IR}^n$ are naturally defined as

$$\begin{aligned}
\mathrm{w}([x]) &:= \max_{1 \leq i \leq n} \mathrm{w}([x_i]),\\
\|[x]\| &:= \max_{1 \leq i \leq n} \|[x_i]\|,\\
\mathrm{mid}([x]) &:= (\mathrm{mid}(x_1), \cdots, \mathrm{mid}(x_n)).
\end{aligned}$$

4.1.1 Inclusion Functions

For any function $f : \mathbb{R}^n \to \mathbb{R}^m$, we can extend it to a function $[f]^* : \mathbb{IR}^n \to \mathbb{IR}^m$ by

$$[f]^*([x]) := \bigcap \{[y] \in \mathbb{IR}^m \mid f([x]) \subseteq [y]\}, \tag{4.1}$$

where $f([x])$ is the image of $[x]$ under f, i.e., $f([x]) = \{f(x) \mid x \in [x]\}$. In other words, $[f]^*$ is the "smallest" interval such that $f([x]) \subseteq [f]^*([x])$ for any $[x] \in \mathbb{IR}$. For real-valued functions, we have the following result.

Proposition 4.1 *If $f : \mathbb{R}^n \to \mathbb{R}$ is continuous on some interval X_0, then $[f]^*([x]) = f([x])$ for any $[x] \subseteq X_0$.*

Proof: Since $[x] \subseteq X_0$ is connected and compact and f is continuous on X_0, the image $f([x])$ is also connected and compact. The only connected and compact sets in $\mathbb{R}$ are compact intervals of the form $[a, b]$. By definition, $[f]^*([x]) = f([x])$. ■

The above result clearly does not hold for vector-valued functions, because the image of a vector-valued function is not necessarily an interval. Nonetheless, we can show the following result.

Proposition 4.2 *If $f : \mathbb{R}^n \to \mathbb{R}^m$ is continuous on some interval X_0, then $[f]^*([x]) = f_1([x]) \times \cdots \times f_m([x])$ for any $[x] \subseteq X_0$.*

Proof: We can apply Proposition 4.1 to each f_i, $1 \le i \le m$. ■

Given an interval $[x]$, to compute the tightest interval $[f]^*([x])$ that contains the image $f([x])$ is not a trivial task. In principle, it can be cast as an optimization problem that seeks to compute the minimal and maximum value of a real-valued function over a compact interval. This optimization problem, however, can be difficult to solve for general nonlinear functions. We are often satisfied with an interval over-approximation of $f([x])$ that is sufficiently tight. Indeed, the main use of interval analysis in this book is to compute arbitrarily tight approximations of the (forward and backward) reachable sets of dynamical systems. For this purpose, we define convergent inclusion functions as follows.

Definition 4.3 *Consider a function $f : \mathbb{R}^n \to \mathbb{R}^m$ and an interval-valued function $[f] : \mathbb{IR}^n \to \mathbb{IR}^m$. The function $[f]$ is called an* ***inclusion function*** *of f if $f([x]) \subseteq [f]([x])$ for all $[x] \in \mathbb{IR}^n$. We say that an inclusion function of f is* ***convergent****, if $\lim_{\mathrm{w}([x]) \to 0} \mathrm{w}([f]([x])) = 0$. If f is only defined on some interval X_0 and $[f]$ defined for any $[x] \subseteq X_0$, we say that $[f]$ is an (convergent) inclusion function of f on X_0, if the above conditions hold for all $[x] \subseteq X_0$.*

The inclusion function $[f]^*$ defined in (4.1) is called the *minimal inclusion function* for f. By (4.1), $[f]^*$ is indeed minimal in the sense that, for any inclusion function $[f]$, we have

$$[f]^*([x]) = \bigcap \{[y] \in \mathbb{IR}^m \mid f([x]) \subseteq [y]\} \subseteq [f]([x]), \quad \forall [x] \in \mathbb{IR}^n.$$

The minimal inclusion function is convergent, provided that f is continuous. This is stated in the following proposition, which follows immediately from the fact that continuous functions on compact intervals are uniform continuous.

Proposition 4.3 *If $f : \mathbb{R}^n \to \mathbb{R}^m$ is continuous on some interval X_0, then $[f]^*$ is convergent on X_0.*

A particular class of convergent inclusion functions is given by Lipschitz inclusions.

Definition 4.4 *Let* $[f] : \mathbb{IR}^n \to \mathbb{IR}^m$ *be an interval inclusion of* $f : \mathbb{R}^n \to \mathbb{R}^m$*. Let* $X_0 \in \mathbb{IR}^n$*. We say that* $[f]$ *is* ***Lipschitz*** *on* X_0*, if there exists a constant* $L \geq 0$ *such that* $\mathrm{w}([f]([x])) \leq L\mathrm{w}([x])$*, for all* $[x] \subseteq X_0$*.*

We now consider how to construct inclusion functions. We restrict our attention to real-valued functions, because inclusions for a vector-valued function $f : \mathbb{R}^n \to \mathbb{R}^m$ can be easily obtained by constructing inclusion functions for the components of $f = (f_1, \cdots, f_m)$. Indeed, let $[f_i]$, $1 \leq i \leq m$, be inclusion functions for f_i. Then $[f]([x]) := [f_1]([x]) \times \cdots \times [f_m]([x])$ is an inclusion function for f. Furthermore, if each $[f_i]$ is minimal, convergent, or Lipschitz, then so is $[f]$.

Lipschitz inclusions can be easily obtained for a wide class of functions. If a function $f : \mathbb{R}^n \to \mathbb{R}$ is Lipschitz on some interval X_0, i.e.,

$$|f(x) - f(y)| \leq L \left\| x - y \right\|_\infty, \quad \forall x, y \in X_0,$$

then the following construction

$$[f]_L([x]) := f(\mathrm{mid}([x])) + \frac{L}{2}[-\mathrm{w}([x]), \mathrm{w}([x])] \tag{4.2}$$

is a Lipschitz inclusion of f. Furthermore, the minimal inclusion function $[f]^*$ is also Lipschitz. Another way to construct Lipschitz inclusions is through natural inclusion functions.

Definition 4.5 *Suppose that* $f : \mathbb{R}^n \to \mathbb{R}$ *is a function expressed using a finite composition of the arithmetic operators* $\{+, -, \cdot, /\}$ *and elementary functions* $\left\{\sin(x_i), \cos(x_i), \sqrt{x_i}, e^{x_i}, \ln(x_i), \cdots\right\}$*. A* ***natural inclusion function*** $[f]_n$ *of* f *is obtained by replacing each occurrence of real variables* x_i *by* $[x_i]$ *and each operator* $\{+, -, \cdot, /\}$ *by its interval counterpart.*

Example 4.1 *Consider* $f(x_1, x_2) = x_1^2 + \sin(x_2)$*. The natural inclusion function of* f *is*

$$[f]_{\mathrm{n}}([x_1], [x_2]) = [x_1]^2 + \sin([x_2]).$$

With $[x_1] = [-1, 1]$ *and* $[x_2] = [-\pi, \pi]$*, we have*

$$[f]_{\mathrm{n}}([-1, 1], [-\pi, \pi]) = [-1, 1]^2 + \sin([-\pi, \pi]) = [0, 1] + [-1, 1] = [-1, 2],$$

which is in fact the exact image $f([-1, 1], [-\pi, \pi])$*. Note that the construction of natural inclusion functions critically depends on the expression of* f *used for evaluation. For example, in this example, if we write* $f(x_1, x_2) = x_1 \cdot x_1 + \sin(x_2)$*, which is an equivalent expression for* f*, then the natural inclusion function obtained becomes*

$$[f]_{\mathrm{n}}([x_1], [x_2]) = [x_1] \cdot [x_1] + \sin([x_2]),$$

which gives

$$[f]_n([-1,1],[-\pi,\pi]) = [-1,1]\cdot[-1,1]+\sin([-\pi,\pi]) = [-1,1]+[-1,1] = [-2,2],$$

which is an over-approximation of the image set $f([-1,1],[-\pi,\pi])$.

Remark 4.1 *In the above example and in the evaluation of natural inclusion functions, it is implicitly assumed that one can readily access the (exact) image of a given interval under the basic unary functions* $\sin(x)$, $\cos(x)$, $\ln(x)$, e^x, $\sqrt{x}$, *etc. Of course, in practical implementations, we are not able to exactly compute such intervals. Nonetheless, we can assume that such images can be accurately over-approximated by outward rounding the computed values by an appropriate amount to compensate for the approximation errors for such functions. Indeed, this is how interval analysis tools have been implemented.*

The natural inclusion function enjoys the following properties.

Proposition 4.4 *Let f satisfy the assumption of Definition 4.5 and $[f]_n$ denote the natural inclusion function for $f : \mathbb{R}^n \to \mathbb{R}$. Suppose that the operators and elementary functions involved are continuous on the domain of interest. Then $[f]_n$ is convergent. Furthermore, if the expression used for evaluating $[f]_n$ has at most one occurrence of each variable x_1, x_2, …, x_n, then $[f]_n$ is also minimal.*

Proof: The fact that $[f]_n$ is convergent follows from that sums, differences, products, quotients, and compositions of convergent inclusion functions remain to be convergent, provided the operators and functions involved are continuous on the domain of interest. When each variable only appears once, $[f]_n$ is minimal because we can exactly evaluate the image of intervals under the unary functions involved (Remark 4.1) and the binary operators $\{+,-,\cdot,/\}$ only give intervals whose ranges are exact. Since each x_i only appears once, the minimal property holds for $[f]_n$ by induction on the sub-expressions of f. ■

Without additional assumptions, a natural inclusion function may fail to be Lipschitz, in the same way a continuous function can fail to be Lipschitz. For example, $f(x) = \sqrt{x}$ on the interval $[0, 1]$. If all the unary functions involved are Lipschitz continuous on the domain of interest, then $[f]_n$ is Lipschitz continuous.

Proposition 4.5 *Let f satisfy the assumption of Definition 4.5 and $[f]_n$ denote the natural inclusion function for $f : \mathbb{R}^n \to \mathbb{R}$. Suppose that all the unary elementary functions involved are Lipschitz on the domain of interest. Then $[f]_n$ is Lipschitz continuous.*

Proof: We show that Lipschitz continuity is preserved by linear combination, product, division, and composition. Consider any real numbers a and b and intervals $[x]$ and $[y]$ in $\mathbb{IR}$. By Lemma B.1 in Appendix B, we have

$$\mathrm{w}(a[x] + b[y]) \leq |a|\,\mathrm{w}([x]) + |b|\,\mathrm{w}([y]),$$

and

$$\mathrm{w}([x] \cdot [y]) \leq |a| \, \mathrm{w}([x]) + |b| \, \mathrm{w}([y]).$$

If $0 \notin [x]$, we also have

$$\mathrm{w}(1/[x]) = \frac{1}{|\underline{x}| \, |\overline{x}|} \mathrm{w}([x]).$$

For any unary function g that is Lipschitz continuous with constant L, we have

$$\mathrm{w}(g([x])) \leq L\mathrm{w}([x]). \tag{4.3}$$

It follows that a finite number of these operations and compositions would define a Lipschitz inclusion function. ■

Remark 4.2 *Following Remark 4.1, if we can compute interval over-approximations of images of any unary function g such that such over-approximations define a Lipschitz inclusion function [g], the conclusion of Proposition 4.5 still holds, in view of (4.3).*

It is sometimes important to characterize the convergence rate of the inclusion functions. We introduce distance between intervals for this purpose.

Definition 4.6 *The* ***distance*** *between two intervals* $[x] = [\underline{x}, \overline{x}]$ *and* $[y] = [\underline{y}, \overline{y}]$ *is given by*

$$d([x], [y]) = \max(\left|\underline{x} - \underline{y}\right|, |\overline{x} - \overline{y}|).$$

Definition 4.7 *Consider* $f : \mathbb{R}^n \to \mathbb{R}$ *and let* $[f] : \mathbb{IR}^n \to \mathbb{IR}$ *be an inclusion function of f. We say that [f]* ***converges in order*** *k on some interval* X_0, *if there exists a positive integer k and a constant c (possibly dependent on* X_0, *but not [x]) such that*

$$d([f]([x]), [f]^*([x])) \leq c\mathrm{w}([x])^k, \quad \forall [x] \subseteq X_0.$$

It is easy to verify that this definition of distance between intervals is exactly the Hausdorff distance induced by the metric on $\mathbb{R}$.

A Lipschitz inclusion function converges in order 1, because

$$d([f]([x]), [f]^*([x])) = \mathrm{w}([f]([x])) - \mathrm{w}([f]^*)([x]) \leq \mathrm{w}([f]([x])) \leq L\mathrm{w}([x]),$$

where $[f]$ is a Lipschitz interval inclusion of f and we used $[f]^*([x]) \subseteq [f]([x])$ and the fact that $A \subseteq B$ implies $d(A, B) = \mathrm{w}(B) - \mathrm{w}(A)$ for intervals A and B (see Lemma B.1 in Appendix B).

4.1.2 Mean-Value and Taylor Inclusion Functions

We introduce two commonly used forms of inclusion functions that can be shown to converge faster than natural/Lipschitz inclusion functions.

Definition 4.8 *Suppose that $f : \mathbb{R}^n \to \mathbb{R}$ is differentiable on an interval X_0. The* ***mean-value inclusion function*** *of f is given by*

$$[f]_{\mathrm{m}}([x]) := f(\mathrm{mid}([x])) + [f']([x]) \cdot ([x] - \mathrm{mid}([x])),$$

where $[f']$ is an inclusion function of the gradient fuction $f'(x) = (\frac{\partial f}{\partial x_1}, \dots, \frac{\partial f}{\partial x_n})^T$.

Note that, for a continuously differentiable function $f : \mathbb{R}^n \to \mathbb{R}$ on X_0, one can take the Lipschitz constant of f as

$$L := \max_{1 \le i \le n} \max_{x \in X_0} \left| \frac{\partial f}{\partial x_i}(x) \right|.$$

Since $|[x_i] - \mathrm{mid}([x_i])| \le \frac{1}{2}\mathrm{w}([x_i]) \le \frac{1}{2}\mathrm{w}([x])$, it follows that

$$[f']([x]) \cdot ([x] - \mathrm{mid}([x])) \subseteq \frac{L}{2}[-\mathrm{w}([x]), \mathrm{w}([x])],$$

provided that $|[f']([x])| \le L$ (clearly satisfied if $[f']$ is the minimal inclusion function). Hence we recover the Lipschitz form inclusion function in (4.2) as a special case of the mean-value form inclusion function.

Taking the mean-value inclusion function to a higher-order approximation naturally leads to Taylor inclusion functions. For easy of notation, we consider only functions of one variable.

Definition 4.9 *Suppose that $f : \mathbb{R} \to \mathbb{R}$ is k-times continuously differentiable on an interval X_0. The kth-order* ***Taylor inclusion function*** *of f is given by*

$$[f]_{\mathrm{t}}([x]) := f(\mathrm{mid}([x])) + f'(\mathrm{mid}([x])) \cdot ([x] - \mathrm{mid}([x])) + \cdots + \\ f^{(k-1)}(\mathrm{mid}([x])) \cdot \frac{([x] - \mathrm{mid}([x]))^{k-1}}{(k-1)!} + [f^{(k)}]([x]) \cdot \frac{([x] - \mathrm{mid}([x]))^k}{k!}.$$

where $[f^{(k)}]$ is an inclusion function of the kth derivative function $f^{(k)}(x)$.

Proposition 4.6 *Suppose that the inclusion function for $[f^{(k)}]$ is Lipschitz on X_0 ($k = 1$ for the mean-value inclusion). Then the mean-value and Taylor inclusion functions both converge in order 2 on X_0.*

We provide a proof of Proposition 4.6 for real-valued functions of one variable in Appendix B. Unfortunately, using a Taylor inclusion function (beyond order 1) does not necessarily improve the convergence rate as shown in the following example.

Example 4.2 *Consider* $f(x) = x^2$. *Let* $[x] = [1-\varepsilon, 1+\varepsilon]$, *where* $\varepsilon \in (0,1)$. *Then*

$$\begin{aligned}[f]_{\mathrm{t}}([x]) &= f(1) + f'(1)\cdot([x]-1) + \frac{1}{2}\cdot 2\cdot([x]-1)^2 \\ &= 1 + 2[-\varepsilon,\varepsilon] + [-\varepsilon,\varepsilon]^2 \\ &= [1-2\varepsilon,(1+\varepsilon)^2],\end{aligned}$$

whereas

$$[f]^*([x]) = f([x]) = [(1-\varepsilon)^2,(1+\varepsilon)^2].$$

Then

$$d([f]_{\mathrm{t}}([x]),[f]^*([x])) = \varepsilon^2 = \frac{1}{4}\mathrm{w}([x])^2, \quad \forall\varepsilon \in (0,1).$$

4.1.3 Inclusion Functions based on Mixed Monotonicity

When a function $f : \mathbb{R} \to \mathbb{R}$ is monotone on a certain interval X_0, one can easily construct the minimal inclusion $f^*([x])$ as $[f(\underline{x}), f(\overline{x})]$ or $[f(\overline{x}), f(\underline{x})]$ for $[x] \subseteq X_0$, depending on whether f is monotonically increasing or decreasing on X_0. For a function $f : \mathbb{R}^n \to \mathbb{R}$, if $f(x_1, \cdots, x_n)$ is monotone in each variable x_i, the minimal inclusion $f^*([x])$ can be constructed in a similar fashion. For vector-valued functions $f : \mathbb{R}^n \to \mathbb{R}^m$, we can apply the same idea to each of its component functions f_i.

Example 4.3 *Consider the function* $f : \mathbb{R}^2 \to \mathbb{R}^2$ *defined by*

$$\begin{aligned}f_1(x_1,x_2) &= \ln(e^{2x_1} + x_2^2), \\ f_2(x_1,x_2) &= \sin(x_1) - x_2^4,\end{aligned}$$

on the interval $X_0 = [-\frac{\pi}{2}, \frac{\pi}{2}] \times [0, c]$, *for any* $c > 0$. *Then* $f_1(x_1, x_2)$ *is monotonically increasing in both* x_1 *and* x_2, *while* $f_2(x_1, x_2)$ *is monotonically increasing in* x_1 *and monotonically decreasing in* x_2 *on* X_0. *Hence, for any* $[x] = [\underline{x}, \overline{x}] \subseteq X_0$, *we have*

$$\begin{aligned}[f]^*([x]) &= f([x]) = f_1([x]) \times f_2([x]) \\ &= [f_1(\underline{x_1},\underline{x_2}), f_1(\overline{x_1},\overline{x_2})] \times [f_2(\underline{x_1},\overline{x_2}), f_2(\overline{x_1},\underline{x_2})],\end{aligned}$$

where we also used Proposition 4.2 and the fact that f *is continuous in its domain.*

In many cases, however, the function f_i may not be monotone on a domain of interest X_0. The notion of *mixed monotonicity* generalizes monotonicity by extending the domain of the function such that the extended function can be readily decomposed into monotonically increasing and monotonically decreasing variables.

Definition 4.10 *Consider a function $f : X_0 \to \mathbb{R}^m$, where $X_0 \subseteq \mathbb{R}^n$. We say that f is* ***mixed monotone*** *on X_0 with respect to some function $g : X_0 \times X_0 \to \mathbb{R}^m$, if the following conditions hold:*

(i) $g(x,x) = f(x)$ for all $x \in X_0$;

(ii) $g(x,\hat{x}) \geq g(y,\hat{x})$ for all $x \geq y$ in X_0 and $\hat{x} \in X_0$; and

(iii) $g(x,\hat{x}) \leq g(x,\hat{y})$ for all $\hat{x} \geq \hat{y}$ in X_0 and $x \in X_0$,

where the inequalities $\geq$ and $\leq$ on vectors are interpreted component-wise, i.e., $x \geq y$ meaning $x_i \geq y_i$ for all $i \in \{1, 2, \cdots, n\}$. The function g is called a decomposition function for f.

The idea behind mixed monotonicity is simple. Item (i) in the definition above says that we can recover the value of f when g is evaluated at the diagonal of its domain. Items (ii) and (iii) indicate that $g(x,y)$ is monotonically increasing with respect to the argument x and monotonically decreasing with respect to the argument y. Based on this definition, we have the following result, which can be used to obtain an inclusion function based on mixed monotonicity.

Proposition 4.7 *If $f : X_0 \to \mathbb{R}^m$, where $X_0 \subseteq \mathbb{R}^n$, is mixed monotone with respect to some $g : X_0 \times X_0 \to \mathbb{R}^m$, then, for any $[x] \subseteq X_0$, we have*

$$f([x]) \subseteq [g(\underline{x}, \overline{x}), g(\overline{x}, \underline{x})].$$

In other words, the interval function $[f]_g$ defined by

$$[f]_g([x]) := [g(\underline{x}, \overline{x}), g(\overline{x}, \underline{x})]$$

is an inclusion function of f on X_0.

Proof: For any $x \in [x] \subseteq X_0$, we have $x \geq \underline{x}$ and $x \leq \overline{x}$. Hence, by the definition of mixed monotonicity, we have $f(x) = g(x,x) \leq g(\overline{x}, \underline{x})$, and $f(x) = g(x,x) \geq g(\underline{x}, \overline{x})$. ■

A natural question to ask about mixed monotonicity is how to construct the decomposition function g for f with respect to which mixed monotonicity of f is defined. For special cases where f has bounded partial derivatives, we can use the following result to construct a decomposition function for f.

Proposition 4.8 *Suppose that $f : X_0 \to \mathbb{R}^m$, where $X_0 \subseteq \mathbb{R}^n$, satisfies*

$$-\underline{L}_{ij} \leq \frac{\partial f_i}{\partial x_j}(x) \leq \overline{L}_{ij}, \quad 1 \leq i, j \leq n, \quad x \in X_0,$$

where $\underline{L}_{ij}, \overline{L}_{ij} \in [0, \infty]$ but at least one of them is finite. Then f is mixed monotone on X_0 with respect to the following decomposition function

$$g_i(x, \hat{x}) = f_i(\xi^i) + \alpha^i \cdot (x - \hat{x}),$$

where $\xi^i = (\xi^i_1, \cdots, \xi^i_n)$ *and* $\alpha^i = (\alpha^i_1, \cdots, \alpha^i_n)$ *are defined as*

$$\xi^i_j = \begin{cases} x_j, & \text{if } \overline{L}_{ij} \geq \underline{L}_{ij}, \\ \hat{x}_j, & \text{otherwise}, \end{cases}$$

and

$$\alpha^i_j = \begin{cases} \underline{L}_{ij}, & \text{if } \overline{L}_{ij} \geq \underline{L}_{ij}, \\ \overline{L}_{ij}, & \text{otherwise}. \end{cases}$$

Proof: We verify the conditions of mixed monotonicity. Clearly, $g_i(x, x) = f_i(x)$. We show that $g_i(x, \hat{x})$ is monotonically increasing on X_0 with respect to x by showing $\frac{\partial g_i}{\partial x_j}(x, \hat{x}) \geq 0$ for all $j \in \{1, 2, \cdots, n\}$ and all $x, \hat{x} \in X_0$. Indeed, when $\overline{L}_{ij} \geq \underline{L}_{ij}$, we have

$$\frac{\partial g_i}{\partial x_j}(x, \hat{x}) = \frac{\partial f_i}{\partial x_j}(\xi^i) + \alpha^i_j = \frac{\partial f_i}{\partial x_j}(\xi) + \underline{L}_{ij} \geq 0.$$

Otherwise, we have

$$\frac{\partial g_i}{\partial x_j}(x, \hat{x}) = \alpha^i_j = \overline{L}_{ij} \geq 0.$$

Similarly, we can verify that, when $\overline{L}_{ij} \geq \underline{L}_{ij}$, we have

$$\frac{\partial g_i}{\partial \hat{x}_j}(x, \hat{x}) = -\alpha_{ij} = -\underline{L}_{ij} \leq 0,$$

and, when $\overline{L}_{ij} < \underline{L}_{ij}$, we have

$$\frac{\partial g_i}{\partial \hat{x}_j}(x, \hat{x}) = \frac{\partial f_i}{\partial x_j}(\xi^i) - \alpha^i_j = \frac{\partial f_i}{\partial x_j}(\xi) - \overline{L}_{ij} \leq 0.$$

Hence $g_i(x, \hat{x})$ is monotonically decreasing on X_0 with respect to $\hat{x}$. ■

When each partial derivative $\frac{\partial f_i}{\partial x_j}(x)$ is sign-stable on X_0 in the sense that either $\frac{\partial f_i}{\partial x_j}(x) \geq 0$ for all $x \in X_0$ or $\frac{\partial f_i}{\partial x_j}(x) \leq 0$ for all $x \in X_0$, we obtain a special case of the proposition above by constructing the decomposition function

$$g_i(x, \hat{x}) = g_i(\xi^i)$$

with

$$\xi^i_j = \begin{cases} x_j, & \text{if } \dfrac{\partial f_i}{\partial x_j}(x) \geq 0, \quad \forall x \in X_0, \\ \hat{x}_j, & \text{if } \dfrac{\partial f_i}{\partial x_j}(x) \leq 0, \quad \forall x \in X_0. \end{cases}$$

Proposition 4.8 extends this fact by adding an additional linear term to ensure that the partial derivatives of decomposition function g are sign-stable on X_0 and that g possesses the required monotonicity properties.

Clearly, with bounded partial derivatives on X_0, f is Lipschitz continuous on X_0. A natural question to ask is whether $[f]_g$ obtained under Proposition 4.8 is a Lipschitz inclusion function. The following corollary gives an affirmative answer.

Corollary 4.1 *The inclusion function $[f]_g$ obtained from Proposition 4.8 under the assumptions there is a Lipschitz inclusion function.*

Proof: Considering the expression for $g_i(x, \hat{x})$ and the definition of $[f]_g$, it follows that

$$\begin{aligned}|g_i(\overline{x}, \underline{x}) - g_i(\underline{x}, \overline{x})| &= |f_i(\xi_1) - f_i(\xi_2) + 2\alpha_i \cdot (\overline{x} - \underline{x})| \\ &\leq L \|\overline{x} - \underline{x}\| + 2 \|\alpha_i\| \|\overline{x} - \underline{x}\| \\ &\leq (L\sqrt{n} + 2 \|\alpha_i\| \sqrt{n}) \mathrm{w}([x]),\end{aligned}$$

where L is a Lipschitz constant for f_i on X_0. ■

We use a simple example to illustrate how an inclusion function can be obtained by constructing a decomposition function for monotonicity.

Example 4.4 *Consider $f(x) = x^2$ on an interval of the form $X_0 = [-a, b]$, where $b \geq a \geq 0$. We have $-2a \leq \frac{df}{dx}(x) = f'(x) = 2x \leq 2b$. According to Proposition 4.8, we can construct a decomposition function $g(x, \hat{x}) = x^2 + 2a(x - \hat{x})$. For any $[x] = [\underline{x}, \overline{x}] \subseteq X_0$, using the monotonicity of g with respect to x and $\hat{x}$ and by Proposition 4.8, we have*

$$[f]_g([x]) = [\underline{x}^2 + 2a(\underline{x} - b), \overline{x}^2 + 2a(\overline{x} - \underline{x})].$$

More specifically, for $[x] = [-a, b]$, we have

$$[f]_g([-a, b]) = [a^2 + 2a(-a - \overline{x}), b^2 + 2a(b + a)] = [-a^2 - 2ab, 2a^2 + b^2 + 2ab].$$

As a comparison, consider the mean-value inclusion function, which gives

$$\begin{aligned}[f]_m([-a, b]) &= f(\frac{b-a}{2}) + [f']([-a, b]) \cdot ([-a, b] - \frac{b-a}{2}) \\ &= (\frac{b-a}{2})^2 + [-2a, 2b] \cdot [-\frac{a+b}{2}, \frac{a+b}{2}] \\ &= \frac{a^2 - 2ab + b^2}{4} + [-ab - b^2, ab + b^2] \\ &= [\frac{1}{4}a^2 - \frac{3}{4}b^2 - \frac{3}{2}ab, \frac{1}{4}a^2 + \frac{5}{4}b^2 + \frac{1}{2}ab],\end{aligned}$$

where we used the fact that $b \geq a \geq 0$. To draw a comparison, consider first the case $a \to 0$. Then the inclusion given by mixed monotonicity tends to $[f]_g([x]) = [0, b^2]$, which is the exact image $f([x])$, whereas $[f]_m([x])$ approaches $[-\frac{3}{4}b^2, \frac{5}{4}b^2]$, which a is a conservative over-approximation. However, when $a = b$, we have $[f]_g([-a, b]) = [-3a^2, 5a^2]$, while $[f]_m([-a, b]) =$

$[-2a^2, \frac{7}{2}a^2] \subseteq [f]_g([-a,b])$. *This is not surprising, because when* $a \to 0$, *the function is close to being monotone on* $[x]$ *and the inclusion* $[f]_g$ *is close to being exact. When the function is not monotone on* $[x]$, *it is unclear whether the inclusion given by mixed monotonicity (using Proposition 4.8) offers tighter bounds than the inclusion functions introduced earlier (e.g., mean-value and Taylor forms).*

To apply Proposition 4.8, one needs to obtain bounds on the partial derivatives of f, which can either rely on analytically or numerically solving an optimization problem. Nonetheless, solving an optimization to obtain such bounds would defeat the purpose of finding efficient inclusion functions because one can directly optimize over the original function f and obtain inclusion functions directly.

4.2 Interval Over-Approximations of One-Step Forward Reachable Sets of Discrete-Time Systems

In this section, we apply convergent interval inclusion functions discussed in the previous section to compute arbitrarily precise over-approximations of the one-step reachable sets of discrete-time dynamical systems.

Consider the discrete-time system

$$x(t+1) = f(x(t), w(t)), \tag{4.4}$$

where $x(t) \in X \subseteq \mathbb{R}^n$ is the state, $w(t) \in W \subseteq \mathbb{R}^p$ is a disturbance input, and $f : X \times W \to X$. We assume that X and W are compact sets and f is Lipschitz continuous with respect to both variables on $X \times W$. For the sake of computation, we further assume that X, and W are intervals.

Suppose that we have access to an inclusion function $[f]$ with the following property:

$$\mathrm{w}([f]([x],[w])) \le L_1\mathrm{w}([x]) + L_2\mathrm{w}([w]), \quad \forall [x] \subseteq X, \forall [w] \subseteq W. \tag{4.5}$$

This is always possible, given the Lipschitz continuity assumption on f. One may use any inclusion function that satisfies this property (or stronger convergence properties), as discussed in the previous section.

The following result shows that we can use $[f]$ to compute arbitrarily precise over-approximation of the following set

$$f(X_0, W) = \{f(x,w) \mid x \in X_0,\ w \in W\},$$

which is the one-step reachable set of (4.4) with initial conditions in $X_0 \subseteq X$ and disturbance values in W.

Proposition 4.9 *For any $\varepsilon > 0$, there exists a union of intervals Y such that*

$$f(X_0, W) \subseteq Y \subseteq f(X_0, W) + \mathbb{B}_\varepsilon, \tag{4.6}$$

where $\mathbb{B}_\varepsilon$ denotes the ball of radius ε with respect to the infinity norm in $\mathbb{R}^n$.

Proof: The proof uses the simple idea of subdivision. Dividing X_0 uniformly into N_1 intervals $\{[x_i]\}_{i=1}^{N_1}$ such that $\mathrm{w}([x_i]) \leq \varepsilon_1$ for all $1 \leq i \leq N_1$. Similarly, divide W into N_2 intervals $\{[w_j]\}_{j=1}^{N_2}$ such that $\mathrm{w}([w_j]) \leq \varepsilon_2$ for all $1 \leq j \leq N_2$. We assume that ε_1 and ε_2 are chosen such that

$$L_1\varepsilon_1 + L_2\varepsilon_2 \leq \varepsilon.$$

Define

$$Y = \bigcup_{i=1}^{N_1}\bigcup_{j=1}^{N_2}[f]([x_i],[w_j]).$$

We show that Y satisfies (4.6). Clearly, $f(X_0, W) \subseteq Y$ because $[f]$ is an inclusion function. For each $y \in Y$, there exist $[x_i]$ and $[w_j]$ such that $y \in [f]([x_i],[w_j])$. Pick any $x \in [x_i]$ and $w \in [w_j]$. We have

$$\|y - f(x, w)\|_\infty \leq \mathrm{w}([f]([x_i],[w_j])) \leq L_1\mathrm{w}([x_i]) + L_2\mathrm{w}([w_j]) \leq L_1\varepsilon_1 + L_2\varepsilon_2 \leq \varepsilon,$$

which implies $Y \subseteq f(X_0, W) + \mathbb{B}_\varepsilon$. ■

We now add a control input to (4.4) and consider

$$x(t + 1) = f(x(t), u(t), w(t)), \tag{4.7}$$

where $x(t) \in X \subseteq \mathbb{R}^n$ is the state, $w(t) \in W \subseteq \mathbb{R}^p$ is a disturbance input, $u(t) \in U \subseteq \mathbb{R}^m$ is a control input, and $f : X \times U \times W \to X$.

For each $u \in U$, we can denote $f_u(x, w) = f(x, u, w)$ and assume that we have access to an inclusion function $[f_u]$ with the following property:

$$\mathrm{w}([f_u]([x],[w])) \leq L_1\mathrm{w}([x]) + L_2\mathrm{w}([w]), \quad \forall [x] \subseteq X, \forall [w] \subseteq W.$$

Define

$$f_u(X_0, W) = \{f_u(x, w) \mid x \in X_0,\ w \in W\},$$

which is the one-step controlled reachable set of (4.7) with input $u \in U$, initial conditions in $X_0 \subseteq X$, and disturbance values in W. Following Proposition 4.9, the following result is immediate.

Proposition 4.10 *For any $\varepsilon > 0$ and each $u \in U$, there exists a union of intervals Y such that*

$$f_u(X_0, W) \subseteq Y \subseteq f_u(X_0, W) + \mathbb{B}_\varepsilon, \tag{4.8}$$

where $\mathbb{B}_\varepsilon$ denotes the ball of radius ε with respect to the infinity norm in $\mathbb{R}^n$.

4.3 Interval Over-Approximations of One-Step Forward Reachable Sets of Continuous-Time Systems

We now consider the continuous-time control system[1]

$$x'(t) = f(x(t), u(t), w(t)), \tag{4.9}$$

where $x(t) \in X \subseteq \mathbb{R}^n$ is the state, $w(t) \in W \subseteq \mathbb{R}^p$ is a disturbance input, $u(t) \in U \subseteq \mathbb{R}^m$ is a control input, and $f : X \times U \times W \to \mathbb{R}^n$. We assume that X, U, and W are intervals of their respective dimensions. The vector field f is assumed to be Lipschitz with respect to each of its variables on $X \times U \times W$. It follows by the existence and uniqueness result (Proposition 1.1) in Chapter 1 that, for each $u(\cdot)$ and $w(\cdot)$ and initial condition $x(0) = x_0 \in X$, there exists a unique solution to (4.9) that can be extended to the boundary of X.

To implement sampled-data control, we are interested in computing the one-step reachable set of system (4.9) with a fixed constant input $u \in U$ over a time interval $[0, \tau]$.

Definition 4.11 (τ-Forward Reachable Set) *The* ***τ-forward reachable set*** *for system (4.9) after a sampling time τ from a set of initial states $X_0 \subseteq X$ under a constant control input $u \in U$ over the interval $[0, \tau]$ is defined by*

$$\mathcal{R}_\tau^{(4.9)}(X_0, u) = \{x(\tau) \mid x'(t) = f(x(t), u, w(t)) \wedge x(0) \in X_0 \wedge w(t) \in W, \\ \textit{for almost all } t \in [0, \tau]\}.$$

If clear from the context, we simply write it as $\mathcal{R}_\tau(X_0, u)$.

4.3.1 A Priori Enclosure

To estimate solutions of (4.9), we shall use the fact that the right-hand side f is locally Lipschitz in x. Lipschitz continuity implies that solutions start from close initial conditions remain close on a bounded time interval (cf. Proposition A.4 on continuous dependence). To access the Lipschitz constant, however, one needs to ensure that solutions remain in a bounded set over this time interval. If such a bounded enclosure is available, we can take it to be compact without loss of generality. One way to estimate the Lipschitz constant is to compute bounds on the partial derivatives of f with respect to x on this compact set.

[1] While we have presented the basic theory of ordinary differential equations (ODEs) for a general ODE model involving an explicit dependence on t, we omit t for the models considered in subsequent chapters. This in general is not a serious limitation, since we can always define t as an extra state and modify the right-hand side accordingly, or modify the technical conditions such that they hold uniformly in t.

For this purpose, we are interested in obtaining an interval *a priori enclosure* of solutions of system (1.1), starting from an interval $[x_0]$, defined as follows.

Definition 4.12 (τ-Forward Reachable Tube) *The* ***τ-forward reachable tube*** *for system (4.9) over an interval $[0,\tau]$ from a set of initial states $X_0 \subseteq X$ under a constant control signal $u(t) = u \in U$ over $[0,\tau]$ is defined by*

$$\begin{aligned}\mathcal{R}^{(4.9)}_{[0,\tau]}(X_0,u) &= \bigcup_{t\in[0,\tau]} \mathcal{R}^{(4.9)}_t(X_0,u)\\ &= \bigcup_{t\in[0,\tau]} \{x(t) \mid x'(s) = f(x(s),u,w(s)) \wedge x(0) \in X_0 \wedge w(s) \in W,\\ &\qquad\qquad \textit{for almost all } s \in [0,t]\}.\end{aligned}$$

If clear from the context, we simply write it as $\mathcal{R}_{[0,\tau]}(X_0,u)$.

Definition 4.13 *An interval $\widehat{[x_0]}$ is called an* ***a priori enclosure*** *for the solutions of system (4.9) on $[0,\tau]$ starting from an interval $[x_0] \subseteq \mathbb{R}^n$ if $\mathcal{R}^{(4.9)}_{[0,\tau]}([x_0],u) \subseteq \widehat{[x_0]}$.*

An easy way to find an a priori enclosure is through the following lemma.

Lemma 4.1 *Suppose that W is an interval. If $\widehat{[x_0]} \subseteq X$ satisfies*

$$[x_0] + [0,\tau][f](\widehat{[x_0]},u,W) \subseteq \widehat{[x_0]}, \tag{4.10}$$

then $\widehat{[x_0]}$ is an a priori enclosure for system (4.9) on $[0,\tau]$.

Proof: Note that $x(t)$ solves the IVP for (4.9) with the initial condition $x(0) = x_0$ with a constant input $u \in U$ on $[0,\tau]$ if and only if

$$x(t) = x_0 + \int_0^t f(x(s),u,w(s))ds, \quad \forall t \in [0,\tau].$$

Suppose that (4.10) holds. Since $[x_0] + [0,t][f](\widehat{[x_0]},u,W)$ is contained in the interior of $\widehat{[x_0]}$ for every $t \in [0,\tau)$, a solution $x(t)$ starting from $[x_0]$ cannot reach the boundary of $\widehat{[x_0]}$ for any $t < \tau$. Hence, $\mathcal{R}_{[0,\tau]}([x_0],u) \subseteq \widehat{[x_0]}$. ■

A more sophisticated way for finding an a priori enclosure involving interval Taylor expansions will be introduced in Section 4.5 (Lemma 4.3).

4.3.2 Over-Approximations by Lipschitz Growth Bound

Pick any $w_0 \in W$ and write (4.9) as

$$x'(t) = f(x,u,w_0) + f(x,u,w) - f(x,u,w_0).$$

Let $[w] = [\underline{w}, \overline{w}] \subseteq \mathbb{R}^p$ be an interval such that $f(x, u, w) - f(x, u, w_0) \in [w]$ for all $x \in X$, $u \in U$, and $w \in W$. Clearly, $\underline{w} \leq 0$. Denote $f(x, u) = f(x, u, w_0)$ and consider a continuous-time control system of the form

$$x'(t) = f(x(t), u(t)) + w(t), \tag{4.11}$$

where $w(t) \in [w]$. We further assume that $[w] = [-\overline{w}, \overline{w}]$ for some $\overline{w} \geq 0$[2]. Clearly, reachable sets of (4.9) can be over-approximated by reachable sets of (4.11).

The following result provides an interval over-approximation of the reachable set of (4.11) starting from an interval $[x_0]$ in terms of a growth bounded using a one-sided Lipschitz condition on the system dynamics.

Proposition 4.11 *Let* $[x_0] = [\underline{x_0}, \overline{x_0}] \subseteq X$ *be an interval and* $u \in U$. *Let* $\widehat{[x_0]}$ *be an a priori enclosure of solutions of (4.11) starting from* $[x_0]$ *with the constant input* u. *Suppose that* $f(x, u)$ *satisfies the following one-sided Lipschitz condition*

$$f_i(x, u) - f_i(y, u) \leq \sum_{j=1}^{n} L_{ij} \left| x_i - y_i \right|, \tag{4.12}$$

for all $i \in \{1, 2, \cdots, n\}$ *and* $x, y \in \widehat{[x_0]}$ *such that* $x_i \geq y_i$, *where* $L := (L_{ij}) \in \mathbb{R}^{n \times n}$ *is a matrix whose off-diagonal entries are all nonnegative. Let* $y(t)$ *be the solution of the IVP for (4.11) with* $w = 0$ *and* $y(0) = y_0 = \frac{\underline{x_0} + \overline{x_0}}{2} \in [x_0]$. *Let* $\rho : [0, \tau] \to \mathbb{R}^n$ *denote the solution to the IVP*

$$\rho'(t) = L\rho(t) + \overline{w}, \quad \rho(0) = \frac{\overline{x_0} - \underline{x_0}}{2}. \tag{4.13}$$

Let $x(t)$ *be any solution of (4.11) satisfying* $x(0) = x_0 \in [x_0]$ *with the constant control input* u *and an arbitrary disturbance signal* $w(\cdot)$ *taking values in* $[w]$. *Then*

$$x(t) \in y(t) + [-\rho(t), \rho(t)], \quad t \in [0, \tau],$$

and ρ *is a* ***growth bound*** *for (4.11) on* $X \times U$.

It follows that

$$\mathcal{R}_t^{(4.11)}([x_0], u) \subseteq [y(t) - \rho(t), y(t) + \rho(t)], \quad t \in [0, \tau].$$

Proof: We first show that $\rho(t) \geq 0$. Since L has only nonnegative off-

[2]This is always achievable by modifying $\overline{w} := \overline{w} - \frac{\overline{w} + \underline{w}}{2} = \frac{\overline{w} - \underline{w}}{2}$, $\underline{w} := \underline{w} - \frac{\overline{w} + \underline{w}}{2} = -\frac{\overline{w} - \underline{w}}{2}$, and $f(x, u) := f(x, u) + \frac{\overline{w} + \underline{w}}{2}$.

diagonal entries, all entries of e^{Lt} $(t \geq 0)$ are nonnegative[3]. It follows that

$$\rho(t) = e^{Lt}\rho(0) + \int_0^t e^{L(t-s)}\overline{w}ds \geq 0.$$

Let $r(t) = x(t) - y(t)$. We show that $|r(t)| \leq \rho(t)$ for $t \in [0, \tau]$. To do so, we construct $m(t)$ by $m_i(t) := \rho_i(t) + \varepsilon e^{Mt}$ for each i, where $\varepsilon > 0$ and $M \geq \sum_{j=1}^n L_{ij}$. We show that $|r(t)| \leq m(t)$ for $t \in [0, \tau]$. Then the conclusion follows by letting $\varepsilon \to 0$.

Suppose that this is not the case. Since

$$-m(0) < -\rho(0) = \underline{x_0} - y_0 \leq r(0) = x_0 - y_0 \leq \overline{x_0} - y_0 = \rho(0) < m(0),$$

by continuity of r and m, there exists some $\bar{t} \in (0, \tau)$ such that $|r(t)| < m(t)$ for all $t \in [0, \bar{t})$ and some $i \in \{1, 2, \cdots, n\}$ such that $r_i(\bar{t}) = m_i(\bar{t})$ or $r_i(\bar{t}) = -m_i(\bar{t})$. In fact, we can let

$$\bar{t} := \inf\{t \in [0, \tau] : |r_i(t)| \geq m_i(t) \text{ for some } i\}.$$

It follows by continuity of r and m that $r_i(\bar{t}) = m_i(\bar{t})$ or $r_i(\bar{t}) = -m_i(\bar{t})$. Consider the former case. Then $x_i(\bar{t}) - y_i(\bar{t}) = r_i(\bar{t}) = m(\bar{t}) > 0$. There exists an interval $[\bar{t} - \theta, \bar{t}]$ such that $x_i(t) - y_i(t) > 0$ for all $t \in [\bar{t} - \theta, \bar{t}]$. It follows that, for almost all $t \in [\bar{t} - \theta, \bar{t}]$,

$$\begin{aligned}
r_i'(t) = x_i'(t) - y_i'(t) &= f_i(x(t), u) - f_i(y(t), u) + w_i \\
&\leq \sum_{j=1}^n L_{ij}|x_i(t) - y_i(t)| + \overline{w}_i \\
&= L_{ii}(x_i(t) - y_i(t)) + \sum_{j \neq i} L_{ij}|x_i - y_i| + \overline{w}_i \\
&\leq L_{ii}(r_i(t) - m_i(t)) + \sum_{j=1}^n L_{ij}m_j(t) + \overline{w}_i \\
&= L_{ii}(r_i(t) - m_i(t)) + m_i'(t) + \varepsilon e^{Mt}(\sum_{j=1}^n L_{ij} - M) \\
&\leq L_{ii}(r_i(t) - m_i(t)) + m_i'(t),
\end{aligned} \tag{4.14}$$

where we used $m_i = \rho_i(t) + \varepsilon e^{Mt}$, $\rho_i'(t) = \sum_{j=1}^n L_{ij}\rho_j(t) + \overline{w}_i$, and $M \geq \sum_{j=1}^n L_{ij}$.

Let

$$v(t) = -(r_i(\bar{t} - t) - m_i(\bar{t} - t))$$

[3]Here is a short proof of this fact. Clearly Lt has only nonnegative off-diagonal entries. Let B be a matrix of only nonnegative off-diagonal entries. We can add a constant multiple of the identity matrix kI such that $B + kI$ has only nonnegative entries. By the definition of matrix exponential, e^{B+kI} has only nonnegative entries. We have $e^B = e^{B+kI}e^{-k}$ with only nonnegative entries as well.

on $[0,\theta]$. Then $v(t)$ is nonnegative on $[0,\theta]$ and, by (4.14),

$$v'(t) = r_i'(\bar{t}-t) - m_i'(\bar{t}-t) \leq L_{ii}(r_i'(\bar{t}-t) - m_i'(\bar{t}-t)) = -L_{ii}v(t),$$

for almost all $t \in [0,\theta]$. Note also that $v(0) = 0$. By Gronwall's inequality (Lemma A.2), $v(t) \equiv 0$ on $[0,\theta]$. In particular, $v(\theta) = 0$ implies $r_i(\bar{t}-\theta) = m_i(\bar{t}-\theta)$, which contradicts the definition of $\bar{t}$.

The case for $r_i(\bar{t}) = -m_i(\bar{t})$ is identical by redefining $r(t) := y(t) - x(t)$. ■

Remark 4.3 (One-sided Lipschitz condition) *Fix $u \in U$. Suppose that $f(x,u)$ above is differentiable with respect to x. By the mean value theorem, we can write*

$$f_i(x,u) - f_i(y,u) = \frac{\partial f_i}{\partial x}(\xi_i,u)\cdot(x-y) = \sum_{j=1}^{n}\frac{\partial f_i}{\partial x_j}(\xi_i,u)\cdot(x_j-y_j),$$

where $\xi_i = (1-c)x + cy$ for some $c \in (0,1)$. Let $\widehat{[x_0]}$ be the interval enclosure from Proposition 4.11. If the partial derivatives of f with respect to x on $\widehat{[x_0]}$ satisfy the following bounds:

$$\frac{\partial f_i}{\partial x_i}(x,u) \leq L_{ii}, \qquad \left|\frac{\partial f_i}{\partial x_j}(x,u)\right| \leq L_{ij}, \tag{4.15}$$

for all $i,j \in \{1,2,\cdots,n\}$, $j \neq i$ and all $x \in \widehat{[x_0]}$, then we have

$$\begin{aligned} f_i(x,u) - f_i(y,u) &= \sum_{j=1}^{n}\frac{\partial f_i}{\partial x_j}(\xi_i,u)\cdot(x_j-y_j) \\ &\leq \sum_{j=1}^{n} L_{ij}\,|x_i - y_i|, \end{aligned}$$

or all $i \in \{1,2,\cdots,n\}$ and $x,y \in \widehat{[x_0]}$ such that $x_i \geq y_i$. In other words, bounded derivatives imply Lipschitz continuity. The one-sided Lipschitz continuity is slightly more general than the usual Lipschitz continuity by allowing L_{ii} to be negative. The analysis in this remark remains valid when replacing $\widehat{[x_0]}$ with any convex set K. Furthermore, the bounds L_{ij} always exist if the partial derivatives are continuous in x and K is compact.

4.3.3 Over-Approximations by Mixed Monotonicity

Recall the continuous-time control system (4.9):

$$x'(t) = f(x(t), u(t), w(t)),$$

where $x(t) \in X \subseteq \mathbb{R}^n$, $u(t) \in U$, and $w(t) \in W$. For convenience of computation, we assume that $W = [w] = [\underline{w}, \overline{w}] \subseteq \mathbb{R}^n$.

In this book, we mostly consider piecewise constant control inputs. For each $u \in U$, we can conveniently write $f_u(x,w) = f(x,u,w)$ and focus instead on the system

$$x'(t) = f_u(x(t), w(t)). \tag{4.16}$$

The mixed monotonicity property for discrete-time dynamical systems can be extended to continuous-time dynamical systems as follows.

Definition 4.14 *Consider the dynamical system*

$$x' = f(x, w),$$

*where $f : D \times W \to \mathbb{R}^n$, $D \subseteq \mathbb{R}^n$, and $W \subseteq \mathbb{R}^p$. We say that it is **mixed monotone** on $X \times W$, where $X \subseteq D$, with respect to some function $g : X \times W \times X \times W \to \mathbb{R}^n$, if the following conditions hold:*

(i) $g(x,w,x,w) = f(x,w)$ for all $x \in X$ and all $w \in W$;

(ii) $g_i(x,w,\hat{x},\hat{w}) \leq g(y,v,\hat{x},\hat{w})$ for all $x,y,\hat{x} \in X$ such that $x \leq y$ and $x_i = y_i$ and all $w,v,\hat{w} \in W$ such that $w \leq v$; and

(iii) $g(x,w,\hat{x},\hat{w}) \leq g(x,w,\hat{y},\hat{v})$ for all $x,\hat{x},\hat{y} \in X$ such that $\hat{x} \geq \hat{y}$ and all $w,\hat{w},\hat{v} \in W$ such that $\hat{w} \geq \hat{v}$.

The function g is called a decomposition function for f.

Similar to mixed monotonicity for discrete-time dynamical systems, the condition above intuitively says that the function $g(x,w,\hat{x},\hat{w})$ is monotonically increasing with respect to (x,w) and decreasing with respect to $(\hat{x},\hat{w})$. Furthermore, the function $g(x,w,\hat{x},\hat{w})$ agrees with the right-hand side of (4.16) when $\hat{x} = x$ and $\hat{w} = w$. The only difference is that the monotonicity with respect to the component x is *quasi-monotonicity* (see Definition A.1 in Section A.4). It is well known in the context of comparison theorems and monotone dynamical systems that quasi-monotonicity is sufficient for ensuring the monotonicity of continuous dynamical systems. In fact, the property above is termed *mixed quasi-monotonicity* in the literature of differential inequalities [41, 133].

If we have access to a decomposition function for the right-hand side of system (4.16), we can over-approximate the one-step forward reachable set $\mathcal{R}_{\tau}^{(4.9)}(X_0,u)$ for the system (4.9) using the following result.

Suppose that $g : X \times W \times X \times W \to \mathbb{R}^n$ is a decomposition function for f_u. Consider the dynamical system

$$\begin{bmatrix} x'(t) \\ y'(t) \end{bmatrix} = \begin{bmatrix} g(x(t), \underline{w}, \hat{x}(t), \overline{w}) \\ g(\hat{x}(t), \overline{w}, x(t), \underline{w}) \end{bmatrix}. \tag{4.17}$$

Proposition 4.12 *Consider $[x_0] = [\underline{x_0}, \overline{x_0}] \subseteq X$. Suppose that $g : X \times W \times X \times W \to \mathbb{R}^n$ is a decomposition function for f_u. We assume that g is locally Lipschitz continuous on $X \times W \times X \times W$. Let $(x(t), \hat{x}(t))$ denote a solution to (4.17) with initial condition $(x(0), \hat{x}(0)) = (\underline{x_0}, \overline{x_0})$ such that $(x(t), \hat{x}(t)) \in int(X) \times int(X)$[4] for all $t \in [0, \tau]$. Fix any $u \in U$. Then*

$$\mathcal{R}_t^{(4.9)}(X_0, u) \subseteq [x(t), \hat{x}(t)], \quad t \in [0, \tau].$$

Proof: Let $\varphi(t)$ denote any solution of (4.9) with an initial condition $\varphi(0) = \varphi_0 \in [x_0]$ and $u(t) \equiv u \in U$; that is,

$$\varphi'(t) = f(\varphi(t), u, w(t)) = f_u(\varphi(t), w(t)), \quad t \in [0, \tau],$$

where $w(\cdot)$ is any disturbance signal such that $w(t) \in W$ for all $t \in [0, \tau]$.

Consider an auxiliary system

$$\begin{bmatrix} x'(t) \\ \hat{x}'(t) \end{bmatrix} = \begin{bmatrix} g(x(t), w(t), \hat{x}(t), w(t)) \\ g(\hat{x}(t), w(t), x(t), w(t)) \end{bmatrix}. \tag{4.18}$$

If we set the initial condition $(x(0), \hat{x}(0)) = (\varphi_0, \varphi_0)$, then $(\varphi(t), \varphi(t))$ is the unique solution of (4.18) on $[0, \tau]$. In fact, it is a solution because $g(\varphi(t), w(t), \varphi(t), w(t)) = f_u(\varphi(t), w(t))$ and uniqueness follows from Lipschitz continuity.

Define

$$r(t) = \begin{bmatrix} x(t) \\ -\hat{x}(t) \end{bmatrix}, \quad m(t) = \begin{bmatrix} \varphi(t) \\ -\varphi(t) \end{bmatrix},$$

For convenience, write $r_1(t) = x(t)$, $r_2(t) = -\hat{x}(t)$, $m_1(t) = \varphi(t)$, and $m_2(t) = -\varphi(t)$. We have

$$r'(t) = \begin{bmatrix} r_1'(t) \\ r_2'(t) \end{bmatrix} = \begin{bmatrix} g(r_1(t), \underline{w}, -r_2(t), \overline{w}) \\ g(-r_2(t), \overline{w}, r_1(t), \underline{w}) \end{bmatrix}, \tag{4.19}$$

and

$$m'(t) = \begin{bmatrix} \varphi'(t) \\ -\varphi'(t) \end{bmatrix} = \begin{bmatrix} g(m_1(t), w(t), -m_2(t), w(t)) \\ g(-m_2(t), w(t), m_1(t), w(t)) \end{bmatrix}. \tag{4.20}$$

Define

$$F(t, x, \hat{x}) := \begin{bmatrix} g(x, w(t), -\hat{x}, w(t)) \\ g(-\hat{x}, w(t), x, w(t)) \end{bmatrix}.$$

By the definition of F and g, it can be easily verified that, for each t, $(x, \hat{x}) \mapsto F(t, x, \hat{x})$ as a mapping from $\mathbb{R}^{2n}$ to $\mathbb{R}^{2n}$ is quasi-monotone. Furthermore,

[4]This is a technical requirement from the comparison theorem (Proposition A.5) that is used to prove this proposition. We need g to have the monotonicity property on an open set that contains an interval enclosure of the images of the functions involved.

in view of (4.18), (4.19), and (4.20), r and m satisfy the conditions in the comparison theorem (Proposition A.5) with respect to F on $[0,\tau]$. It follows that $r(t) \leq m(t)$ for all $t \in [0,\tau]$. Hence $\varphi(t) \in [x(t), \hat{x}(t)]$ for $t \in [0,\tau]$. The conclusion follows. ■

Similar to Proposition 4.8, one can use bounded partial derivatives of the function $f(x,w)$ to construct a decomposition function g for f. A slight modification is needed in light of the quasi-monotonicity requirement (instead of monotonicity). This is stated in the next result.

Proposition 4.13 *Suppose that $f : X \times W \to \mathbb{R}^n$ satisfies*

$$-\underline{L}_{ij} \leq \frac{\partial f_i}{\partial x_j}(x,w) \leq \overline{L}_{ij}, \quad 1 \leq i,j \leq n, \quad \forall x \in X, \forall w \in W,$$

where $\underline{L}_{ij}, \overline{L}_{ij} \in [0,\infty]$ but at least one of them is finite. Furthermore,

$$-\underline{M}_{ij} \leq \frac{\partial f_i}{\partial w_j}(x,w) \leq \overline{M}_{ij}, \quad 1 \leq i,j \leq p, \quad \forall x \in X, \forall w \in W,$$

where $\underline{M}_{ij}, \overline{M}_{ij} \in [0,\infty]$ but at least one of them is finite. Then f is mixed monotone on $X \times W$ with respect to the following decomposition function

$$g_i(x,\hat{x},w,\hat{w}) = f_i(\xi^i,\zeta^i) + \alpha^i \cdot (x - \hat{x}) + \beta^i \cdot (w - \hat{w}),$$

where $\xi^i = (\xi^i_1,\cdots,\xi^i_n)$, $\zeta = (\zeta^i_1,\cdots,\zeta^i_n)$, $\alpha^i = (\alpha^i_1,\cdots,\alpha^i_n)$, and $\beta^i = (\beta^i_1,\cdots,\beta^i_n)$ are defined as

$$\xi^i_j = \begin{cases} x_j, & \text{if } \overline{L}_{ij} \geq \underline{L}_{ij} \text{ or } i = j, \\ \hat{x}_j, & \text{otherwise,} \end{cases}$$

$$\zeta^i_j = \begin{cases} w_j, & \text{if } \overline{M}_{ij} \geq \underline{M}_{ij}, \\ \hat{w}_j, & \text{otherwise,} \end{cases}$$

$$\alpha^i_j = \begin{cases} 0, & \text{if } j = i, \\ \underline{L}_{ij}, & \text{if } \overline{L}_{ij} \geq \underline{L}_{ij} \text{ and } j \neq i, \\ \overline{L}_{ij}, & \text{otherwise,} \end{cases}$$

and

$$\beta^i_j = \begin{cases} \underline{M}_{ij}, & \text{if } \overline{M}_{ij} \geq \underline{M}_{ij}, \\ \overline{M}_{ij}, & \text{otherwise,} \end{cases}$$

Proof: We verify the conditions of mixed monotonicity. Clearly, $g_i(x,w,x,w) = f_i(x,w)$.

We show that $g_i(x, w, \hat{x}, \hat{w})$ is quasi-monotonically increasing on $X \times W \times X \times W$ with respect to x by showing[5] $\frac{\partial g_i}{\partial x_j}(x, w, \hat{x}, \hat{w}) \geq 0$ for all $j \in \{1, 2, \cdots, n\} \setminus \{i\}$, all $x, \hat{x} \in X$, and all $w, \hat{w} \in W$. Indeed, when $\overline{L}_{ij} \geq \underline{L}_{ij}$ and $i \neq j$, we have

$$\frac{\partial g_i}{\partial x_j}(x, w, \hat{x}, \hat{w}) = \frac{\partial f_i}{\partial x_j}(\xi^i, \zeta^i) + \alpha^i_j = \frac{\partial f_i}{\partial x_j}(\xi^i, \zeta^i) + \underline{L}_{ij} \geq 0.$$

Otherwise, we have

$$\frac{\partial g_i}{\partial x_j}(x, w, \hat{x}, \hat{w}) = \alpha^i_j = \overline{L}_{ij} \geq 0.$$

Similarly, we can verify that, when $\overline{L}_{ij} \geq \underline{L}_{ij}$, we have

$$\frac{\partial g_i}{\partial \hat{x}_j}(x, w, \hat{x}, \hat{w}) = -\alpha^i_j = -\underline{L}_{ij} \leq 0,$$

and, when $\overline{L}_{ij} < \underline{L}_{ij}$, we have

$$\frac{\partial g_i}{\partial \hat{x}_j}(x, w, \hat{x}, \hat{w}) = \frac{\partial f_i}{\partial x_j}(\xi^i, \zeta^i) - \alpha^i_j = \frac{\partial f_i}{\partial x_j}(\xi^i, \zeta^i) - \overline{L}_{ij} \leq 0.$$

Hence, $g_i(x, w, \hat{x}, \hat{w})$ is monotonically decreasing on $X \times W \times X \times W$ with respect to $\hat{x}$.

The verification of monotonicity with respect to w and $\hat{w}$ proceeds in the same way. We have

$$\frac{\partial g_i}{\partial w_j}(x, w, \hat{x}, \hat{w}) = \frac{\partial f_i}{\partial w_j}(\xi^i, \zeta^i) + \beta^i_j = \frac{\partial f_i}{\partial w_j}(\xi^i, \zeta^i) + \underline{M}_{ij} \leq 0,$$

and

$$\frac{\partial g_i}{\partial \hat{w}_j}(x, w, \hat{x}, \hat{w}) = -\beta^i_j = -\underline{M}_{ij} \geq 0,$$

when $\overline{M}_{ij} \geq \underline{M}_{ij}$, and

$$\frac{\partial g_i}{\partial w_j}(x, w, \hat{x}, \hat{w}) = \beta^i_j = \underline{M}_{ij} \geq 0,$$

and

$$\frac{\partial g_i}{\partial \hat{w}_j}(x, w, \hat{x}, \hat{w}) = \frac{\partial f_i}{\partial w_j}(\xi^i, \zeta^i) - \beta^i_j = \frac{\partial f_i}{\partial w_j}(\xi^i, \zeta^i) - \overline{M}_{ij} \leq 0,$$

when $\overline{M}_{ij} < \underline{M}_{ij}$. ∎

[5]By the mean value theorem, the monotonicity conditions in Definition 4.14 can be reduced to verifying these conditions on partial derivatives.

4.3.4 Over-Approximations by Validated Integration

Consider the dynamical system

$$x'(t) = f(x(t), u), \quad x(0) \in [x_0], \tag{4.21}$$

where $f : D \times U \to \mathbb{R}^n$ and $D \subseteq \mathbb{R}^n$ and $u \in U$. This can correspond to, e.g., the case (4.9) without the disturbance w and with a given constant input $u \in U$. We assume that the right-hand side f is sufficiently regular that the solution $x(t)$ admits a Taylor series expansion of any order.

Given a set of initial states $X_0 \subseteq D$ and a time $\tau > 0$, we denote the reachable set at $t = \tau$ by $\mathcal{R}_\tau(X_0, u)$. A standard method for over-approximating the reachable set from an initial interval $[x_0]$ builds on high-order interval Taylor expansions around the system solutions during the concerned time horizon that are confined in an a priori enclosure.

Let $\widehat{[x_0]}$ be an a priori enclosure on $[0, \tau]$ for solutions of (4.21) starting from $[x_0]$. By Taylor's theorem, we have

$$\mathcal{R}_\tau([x_0], u) \subseteq \sum_{i=0}^{k} f^{[i]}([x_0], u)\frac{\tau^i}{i!} + f^{[k+1]}(\widehat{[x_0]}, u)\frac{\tau^{k+1}}{(k+1)!}, \tag{4.22}$$

where the sequence of functions $f^{[i]}(x, u)$ $(i \geq 0)$ are defined by

$$\begin{aligned} f^{[0]}(x, u) &= x, \\ f^{[i]}(x, u) &= \frac{\partial f^{[i-1]}(x, u)}{\partial x} f(x, u),\ i \geq 1. \end{aligned}$$

We can over-approximate the function $f^{[i]}(\cdot, u)$ in (4.22) using convergent inclusion functions $[f]^{[i]}(\cdot, u)$. By doing so, we obtain a Taylor over-approximation of $\mathcal{R}_\tau([x_0], u)$ as follows:

$$\begin{aligned} \widehat{\mathcal{R}}_\tau^k([x_0], u) &:= \sum_{i=0}^{k} [f]^{[i]}([x_0], u)\frac{\tau^i}{i!} + [f]^{[k+1]}(\widehat{[x_0]}, u)\frac{\tau^{k+1}}{(k+1)!} \\ &\supseteq \mathcal{R}_\tau([x_0], u). \end{aligned} \tag{4.23}$$

4.3.5 Examples

In this section, we present numerical examples to illustrate the different approaches discussed for computing one-step forward reachable sets for continuous-time systems.

Example 4.5 *Consider the nonlinear system*

$$x' = \begin{bmatrix} x_1' \\ x_2' \end{bmatrix} = f(x, u) = \begin{bmatrix} x_2^2 + 2 \\ x_1 + u \end{bmatrix}. \tag{4.24}$$

Consider a region $X = [-5,5] \times [-5,5] \subseteq \mathbb{R}^2$ and an initial set $[x_0] = [-0.5, 0.5]$. We aim to approximate $\mathcal{R}_\tau([x_0], u)$ for $\tau = 0.25$ and $u = 0$.

We can first verify that $\widehat{[x_0]} = [-0.5, 1.1556] \times [-0.625, 0.7889]$ is a valid a priori enclosure. Indeed, this follows from Lemma 4.1 and the fact that

$$[-0.5, 0.5] + [0, 0.25] \cdot ([-0.625, 0.7889]^2 + 2) \subseteq [-0.5, 1.1556],$$

and

$$[-0.5, 0.5] + [0, 0.25] \cdot [-0.5, 1.1556] \subseteq [-0.625, 0.7889],$$

where we used the natural inclusion function to evaluate $[f]$.

Using this priori enclosure and by Remark 4.3 and (4.15), we can compute a matrix of one-sided Lipschitz constants as

$$L_{11} = 0, \quad L_{12} = 2 \cdot 0.7889 = 1.5778, \quad L_{21} = 1, \quad L_{22} = 0.$$

By Proposition 4.11, we can integrate the above differential equation from the initial condition $x(0) = (0, 0)$ to obtain $x(\tau) = (0.500195..., 0.062508...)$. In practice, this cannot be computed exactly and we may need to use a validated numerical solver to over-approximate $x(\tau)$ (see Section 4.3.4). For this example, we assume the numerical solution is the exact solution. We also do not have any disturbance. Hence, by Proposition 4.11, we obtain an over-approximation of $\mathcal{R}_\tau([x_0], u)$ using the growth bound as

$$\begin{aligned}\mathcal{R}_\tau([x_0], u) &\subseteq [x(\tau) - e^{L\tau}\rho_0, x(\tau) - e^{L\tau}\rho_0] \\ &\subseteq [-0.225144, 1.225535] \times [-0.589413, 0.714430],\end{aligned}$$

where $\rho_0 = (0.5, 0.5)$ and we used conservative rounding of the endpoints. We also assumed that the matrix exponential can be computed exactly.

To use the bound by mixed monotonicity, we can construct a decomposition function g using Proposition 4.13 as

$$g_1(x, \hat{x}) = x_2^2 + 2 \cdot 0.625(x_2 - \hat{x}_2),$$

and

$$g_2(x, \hat{x}) = x_1,$$

Integrating the corresponding system (4.17) and using Proposition 4.12, we obtain another over-approximation of $\mathcal{R}_\tau([x_0], u)$ as

$$\mathcal{R}_\tau([x_0], u) \subseteq [-0.283768, 1.451651] \times [-0.596338, 0.740255],$$

where we also assumed that numerical integration gives the exact solution and used conservative rounding of the endpoints. It can be seen that this bound is more conservative than the growth bound obtain above.

Finally, we used validated integration to obtain an over-approximation. For $u = 0$, we compute

$$f^{[0]}(x,u) = x,$$

$$f^{[1]}(x,u) = f(x,u) = \begin{bmatrix} x_2^2 + 2 \\ x_1 \end{bmatrix},$$

$$f^{[2]}(x,u) = \frac{\partial f^{[1]}(x,u)}{\partial x} f(x,u) = \begin{bmatrix} 0 & 2x_2 \\ 1 & 0 \end{bmatrix} \begin{bmatrix} x_2^2 + 2 \\ x_1 \end{bmatrix} = \begin{bmatrix} 2x_1x_2 \\ x_2^2 + 2 \end{bmatrix},$$

$$f^{[3]}(x,u) = \frac{\partial f^{[2]}(x,u)}{\partial x} f(x,u) = \begin{bmatrix} 2x_2 & 2x_1 \\ 0 & 2x_2 \end{bmatrix} \begin{bmatrix} x_2^2 + 2 \\ x_1 \end{bmatrix} = \begin{bmatrix} 2x_2^3 + 4x_2 + 2x_1^2 \\ 2x_1x_2 \end{bmatrix}.$$

By (4.23) with $k = 2$, we obtain

$$\begin{aligned} \mathcal{R}_\tau([x_0],u) &\subseteq \widehat{\mathcal{R}}_\tau^k([x_0],u) \\ &= \sum_{i=0}^{k} [f]^{[i]}([x_0],u)\frac{\tau^i}{i!} + [f]^{[k+1]}(\widehat{[x_0]},u)\frac{\tau^{k+1}}{(k+1)!} \\ &\subseteq [-0.023407, 1.094919] \times [-0.566262, 0.700061], \end{aligned}$$

we used the natural inclusion function to evaluate $[f]$ and conservative rounding of the endpoints with the number of digits shown above. It can be seen that this bound obtain by validated integration using interval Taylor expansion is the tightest among the three bounds computed for this example.

Example 4.6 *Consider the nonlinear system*

$$x' = \begin{bmatrix} x_1' \\ x_2' \end{bmatrix} = f(x,u) = \begin{bmatrix} x_1^2 + 10x_2 + u_1 \\ -x_1 - x_2^3 + u_2 \end{bmatrix}. \tag{4.25}$$

Consider a region $X = [-5,5] \times [-5,5] \subseteq \mathbb{R}^2$ and an initial set $[x_0] = [-0.5, 0.5]$. We aim to approximate $\mathcal{R}_\tau([x_0],u)$ for $\tau = 0.1$ and $u = (0,0)$.

Following the same procedure as the previous example, we obtain an a priori enclosure as $\widehat{[x_0]} = [-1.2, 1.4] \times [-0.7, 0.7]$, which can be verified using Lemma 4.1. Accordingly, we can choose

$$L = \begin{bmatrix} 2.8 & 10 \\ 1 & 0 \end{bmatrix}.$$

The over-approximations obtained the three approaches are shown in the following table.

methods	*computed over-approximations*
Prop. 4.11	$[-1.278679, 1.278679] \times [-0.586408, 0.586408]$
Props. 4.12 and 4.13	$[-0.971270, 1.090092] \times [-0.563525, 0.559015]$
(4.23), $k = 1$	$[-1.202430, 1.227590] \times [-0.618642, 0.610312]$

It can be seen that for this example, the bound given by mixed monotonicity (Proposition 4.12 using the decomposition function given by Proposition 4.13) is the sharpest.

Remark 4.4 *From the examples above, it can be seen that it is in general difficult to state which one of the approaches would provide a tighter over-approximation of the one-step forward reachable set of a continuous-time dynamical system.*

4.4 Interval Under-approximations of Controlled Predecessors of Discrete-Time Systems

We now consider the nominal discrete-time control system

$$x(t+1) = f(x(t), u(t)), \tag{4.26}$$

where $x(t) \in X \subseteq \mathbb{R}^n$ is the system state space, $u(t) \in U$ is the control input. In the presence of additive disturbances, system (4.26) becomes

$$x(t+1) = f(x(t), u(t)) + w(t), \tag{4.27}$$

where $w(t) \in W$ is the additive disturbance input, and we assume that

$$W := \{w \in \mathbb{R}^n \mid \|w\|_\infty \leq \delta,\, \delta \geq 0\}. \tag{4.28}$$

Nonlinear control design for control system (4.27) is often built on the evaluation of the controlled predecessor of an target set of states and the set of valid control inputs defined as follows.

Definition 4.15 *Given a set $B \subseteq X$, the* ***controlled predecessors*** *of B for system (4.27) is a set of states defined by*

$$Pre^\delta(B) = \{x \in X \mid \exists u \in U, \forall w \in W \text{ s.t. } f(x,u) + w \in B\}. \tag{4.29}$$

The set of ***valid control inputs*** *that lead to one-step transition to B for an $x \in Pre^\delta(B)$ is*

$$K_B^\delta(x) = \{u \in U \mid f(x,u) + w \in B,\, \forall w \in W\}. \tag{4.30}$$

For system (4.26), i.e., the special case of system (4.27) with $\delta = 0$, we simplify the notation of the controlled predecessor (4.29) to $\text{Pre}(B)$. For $A, B \subseteq X$, the controlled predecessor of B that resides in a set A is the set $A \cap \text{Pre}^\delta(B)$. To simplify the notation, we let

$$\text{Pre}^\delta(B|A) := A \cap \text{Pre}^\delta(B). \tag{4.31}$$

Recall that, in Section 3.5, a discrete-time control system in the form of (3.1) can be treated as a transition system and the controlled predecessor of a set is defined in (3.3). The definition (4.29) is nothing but a general controlled predecessor for a system with additive and bounded disturbances.

The computation of controlled predecessors is essentially the computation of set inversions under a nonlinear function with a given set of control inputs and a bounded disturbance set. Without further assumptions, such set inversions are not guaranteed to be any specific shape, and exact computation is nontrivial. Therefore, we often aim at finding their approximations that are good enough to serve our purposes (e.g., control). Intervals are of simple geometric shape and easy to manipulate. Any compact set can be arbitrarily approximated by a finite union of intervals because of the Borel-Lebesgue finite covering theorem (see Appendix C.2). An example of interval approximations of set inversions is the SIVIA algorithm described in [108, Section 3.4], which, by using a set of non-overlapping intervals in $\mathbb{IR}^n$, approximates the set $\{x \in \mathbb{R}^n \mid f(x) \in A\}$ where f is a nonlinear function from $\mathbb{R}^n$ to $\mathbb{R}^m$. Following the same line, we can approximate the controlled predecessor $\mathrm{Pre}(B|A)$ by using Algorithm 5.

Algorithm 5 $[\underline{A}, \Delta A, A_c] = \textsc{Pre}([f], B, A, U, \varepsilon)$

1: $\underline{A} \leftarrow \emptyset, \Delta A \leftarrow \emptyset, A_c \leftarrow \emptyset$
2: $List \leftarrow A$
3: **while** $List \neq \emptyset$ **do**
4: $\quad [x] \leftarrow List.first$
5: $\quad$ **if** $[f]([x], u) \cap B = \emptyset$ for all $u \in U$ **then**
6: $\quad\quad A_c \leftarrow A_c \cup [x]$
7: $\quad$ **else if** $[f]([x], u) \subseteq B$ for some $u \in U$ **then**
8: $\quad\quad \underline{A} \leftarrow \underline{A} \cup [x]$
9: $\quad$ **else**
10: $\quad\quad$ **if** $\mathrm{w}([x]) < \varepsilon$ **then**
11: $\quad\quad\quad \Delta A \leftarrow \Delta A \cup [x]$
12: $\quad\quad$ **else**
13: $\quad\quad\quad Left[x], Right[x] = Bisect([x])$ ▷ Perform bisection to $[x]$.
14: $\quad\quad\quad List.add(Left[x]), List.add(Right[x])$
15: $\quad\quad$ **end if**
16: $\quad$ **end if**
17: **end while**

The inputs of Algorithm 5 are: a convergent interval inclusion function $[f]$ of f in (4.26), the sets $A, B \subseteq X$, the control input set U, and a precision parameter ε. During each iteration, Algorithm 5 checks if the image $[f]([x], u)$ of a particular interval $[x]$ is contained in B, or completely outside of Y. If neither, and the width of the interval is greater than ε, then $[x]$ is considered undetermined and divided into two sub-intervals $Left[x]$ and $Right[x]$ by bisection, which are given by

$$Left[x] = [\underline{x}_1, \overline{x}_1] \times \cdots \times [\underline{x}_j, (\underline{x}_j + \overline{x}_j)/2] \times \cdots \times [\underline{x}_n, \overline{x}_n],$$
$$Right[x] = [\underline{x}_1, \overline{x}_1] \times \cdots \times [(\underline{x}_j + \overline{x}_j)/2, \overline{x}_j] \times \cdots \times [\underline{x}_n, \overline{x}_n],$$

where j is the dimension in which the interval $[x]$ attains its width. The intervals $Left[x]$ and $Right[x]$ are then pushed into a queue *List* to be determined later. An interval will go through subdivision if necessary until its width is less than the precision parameter ε.

In the outputs of Algorithm 5, $\underline{A}$ denotes the set of intervals that absolutely belong to $\mathrm{Pre}(B|A)$, A_c is the set of intervals that do not, and those intervals that intersect with but not fully inside of $\mathrm{Pre}(B|A)$, i.e., undetermined intervals, are collected in ΔA. The parameter ε controls the minimum width of intervals for approximating $\mathrm{Pre}(B|A)$.

Clearly, any intervals in $\underline{A}$ is a subset of $\mathrm{Pre}(B|A)$ and $\mathrm{Pre}(B|A)$ is fully covered by the union of intervals contained in $\underline{A}$ and ΔA. Let

$$[\underline{\mathrm{Pre}}]^{\varepsilon}(B|A) := \bigcup_{[x]\in\underline{A}} [x], \tag{4.32}$$

$$[\overline{\mathrm{Pre}}]^{\varepsilon}(B|A) := \bigcup_{[x]\in(\underline{A}\cup\Delta A)} [x], \tag{4.33}$$

where $\underline{A}$ and ΔA are returned by Algorithm 5. Then $[\underline{\mathrm{Pre}}]^{\varepsilon}(B|A)$ and $[\overline{\mathrm{Pre}}]^{\varepsilon}(B|A)$ represent an under and over approximations of $\mathrm{Pre}(B|A)$, respectively, i.e.,

$$[\underline{\mathrm{Pre}}]^{\varepsilon}(B|A) \subseteq \mathrm{Pre}(B|A), \quad \mathrm{Pre}(B|A) \subseteq [\overline{\mathrm{Pre}}]^{\varepsilon}(B|A).$$

The control inputs that satisfy line 7 in Algorithm 5 are often of particular interest, since they are the atomic components in the construction of a control strategy. To be consistent with (4.30), we define the set of valid control inputs found by Algorithm 5 as follows:

$$K_B^{\varepsilon}([x]) := \{u \in U \mid [f]([x], u) \subseteq B\}, \; [x] \subseteq X \tag{4.34}$$
$$[K]_B^{\varepsilon}(x) := K_B([x]), \; x \in [x]. \tag{4.35}$$

The monotonicity of the under-approximation $[\underline{\mathrm{Pre}}]^{\varepsilon}(B|A)$ can be preserved by using Algorithm 5.

Proposition 4.14 *The interval under-approximation* $[\underline{Pre}]^{\varepsilon}(B|A)$ *obtained by Algorithm 5 for system (4.26) satisfies* $[\underline{Pre}]^{\varepsilon}(B_1|A) \subseteq [\underline{Pre}]^{\varepsilon}(B_2|A)$ *for* $B_1 \subseteq B_2 \subseteq X$.

Proof: Algorithm 5 can be considered as a breadth-first search algorithm. Denote the i-the iteration by the operations of line 4-14 on the intervals at depth $i-1$. For example, the 1st iteration is performed to the intervals representing the original input set A, which are at depth 0 of the binary tree. Let the sets $\underline{A}$, ΔA and A_c in Algorithm 5 after iteration i for set B_1 and B_2 be $\underline{A}_1^i$, ΔA_1^i, A_{c1}^i and $\underline{A}_2^i$, ΔA_2^i, A_{c2}^i, respectively. Given $B_1 \subseteq B_2$, for an interval $[x] \subseteq A$,

$$\begin{aligned} [x] \in \underline{A}_1^i &\implies [x] \in \underline{A}_2^i, \\ [x] \in A_{c1}^i &\implies [x] \in A_{c2}^i \text{ or } \underline{A}_2^i \text{ or } \Delta A_2^i, \\ [x] \in \Delta A_1^i &\implies [x] \in \underline{A}_2^i \text{ or } \Delta A_2^i. \end{aligned} \tag{4.36}$$

Initially, $List_1^0 = List_2^0$. By (4.36), $\underline{A}_1^1 \subseteq \underline{A}_2^1$, and for any $[x] \in List_1^1$, $[x] \in List_2^1$ or $[x] \in \underline{A}_2^1$, i.e., $List_1^1 \subseteq \left(List_2^1 \cup \underline{A}_2^1\right)$. Suppose that $\underline{A}_1^j \subseteq \underline{A}_2^j$ and $List_1^j \subseteq \left(List_2^j \cup \underline{A}_2^j\right)$. For $[x] \in \left(List_1^j \cap List_2^j\right)$, if $[x] \in A_1^{j+1}$ then $[x] \in A_2^{j+1}$. Since $[x] \in \left(List_1^j \cap \underline{A}_2^j\right)$ implies that $[x] \in \underline{A}_2^{j+1}$, we have $\underline{A}_1^{j+1} \subseteq \underline{A}_2^{j+1}$. Similarly, if $[x] \in List_1^{j+1}$, then $[x]$ is the result of bisection from an interval $[x]' \in \left(List_1^j \cap \underline{A}_2^j\right)$ or $\left(List_1^j \cap List_2^j\right)$. For the former, $[x] \subseteq [x]' \in \underline{A}_2^j \subseteq \underline{A}_2^{j+1}$. For the latter, we can have the same conclusion as for the base case, i.e., for any $[x] \in List_1^{j+1}$, $[x] \in List_2^{j+1}$ or $[x] \in \underline{A}_2^{j+1}$. Hence, $List_1^{j+1} \subseteq \left(List_2^{j+1} \cup \underline{A}_2^{j+1}\right)$. Through induction, it then straightforward to see that $\underline{A}_1^i \subseteq \underline{A}_2^i$ for all $i \in \mathbb{N}$. ■

Remark 4.5 *The input sets A, B of Algorithm 5 are usually assumed to be intervals or unions of a finite number of intervals, since Algorithm 5 follows interval operations. However, more generally, the input set B can also be defined by equations or inequalities, i.e.,*

$$B := \{y \in X \mid g(y) \leq 0\}, \quad g : \mathbb{R}^n \to \mathbb{R}^l.$$

In this case, the condition $[f]([x], u) \cap B = \emptyset$ *and* $[f]([x], u) \subseteq B$ *can be respectively tested by*

$$\begin{aligned} &[g \circ f]([x], u) \subseteq [0, \infty]^l, \\ &[g \circ f]([x], u) \subseteq [-\infty, 0]^l, \end{aligned}$$

respectively, where $[g \circ f]([x], u)$ *denotes the convergent inclusion function of the composite function* $g(f(x, u))$.

The control input set U is required to be finite to be enumerated in lines 5 and 7. Systems with finite sets of control inputs, however, are not uncommon. For electrical power converters [78] and DISC engines [205], the system state is controlled by switching between different operating modes, and system evolution under each mode may be determined by different functions. Control synthesis for systems with complex dynamics or specifications, e.g., robot motion planning [125] and flight management [74], is usually simplified to switching control between different operating modes and motion primitives. Such systems can be described by the form

$$x(t+1) = f_{u(t)}(x(t)), \tag{4.37}$$

where $u(t) \in U$ indicates the active mode at time $t \in \mathbb{N}$, and the input space U is finite.

For control systems with infinite sets of control inputs, it will be impractical to run Algorithm 5 directly. To under-approximate $\text{Pre}(B|A)$, a straightforward solution is to use an under-sampled set of controls, e.g., a set of uniformly sampled points within U with a granularity $\mu > 0$ defined as

$$[U]_\mu := \mu\mathbb{Z}^m \cap U, \tag{4.38}$$

where $\mu\mathbb{Z}^m := \{\mu z \mid z \in \mathbb{Z}^m, \mu > 0\}$.

By using Algorithm 5 to approximate the exact controlled predecessor $\text{Pre}(B|A)$ $(A, B \subseteq X)$, it is often of great interest to know how close the returned approximations are to the real one and in which way the precision parameter ε affects the approximations.

Let us first consider the situation where the the control space is finite.

Assumption 4.1 (Lipschitz in X) *There exists a constant $L > 0$ for the function $f : \mathbb{R}^n \times \mathbb{R}^m \to \mathbb{R}^n$ in (4.26) such that for all $u \in U$*

$$\|f(x,u) - f(y,u)\|_\infty \leq L \|x - y\|_\infty, \quad \forall x, y \in X \subseteq \mathbb{R}^n. \tag{4.39}$$

The following lemma gives the error bounds of the under and over approximations of $\text{Pre}(B|A)$ under Assumption 4.1.

Lemma 4.2 *Consider system (4.26). Let $B, A \subseteq X$ be compact. If Assumption 4.1 holds in an neighborhood of A with the Lipschitz constant ρ_1 and a Lipschitz inclusion function $[f]$ is used in Algorithm 5, then*

$$Pre(B|A) \subseteq [\overline{Pre}]^\varepsilon(B|A) \subseteq Pre(B + \mathbb{B}_{\rho_1\varepsilon}|A), \tag{4.40}$$

$$Pre(B - \mathbb{B}_{\rho_1\varepsilon}|A) \subseteq [\underline{Pre}]^\varepsilon(B|A) \subseteq Pre(B|A), \tag{4.41}$$

where $[\overline{Pre}^\varepsilon(B|A)]$ and $[\underline{Pre}]^\varepsilon(B|A)$ are given in (4.33) and (4.32), respectively.

Proof: It follows straightforwardly from Algorithm 5 that $\text{w}([x]) < \varepsilon$ for all $[x] \in \Delta A$, and

$$[\underline{\text{Pre}}]^\varepsilon(B|A) \subseteq \text{Pre}(B|A) \subseteq [\overline{\text{Pre}}]^\varepsilon(B|A) \subseteq A.$$

By (4.39) and using a Lipschitz inclusion function $[f]$, we have $\text{w}([f]([x], u)) \leq \rho_1 \text{w}([x]) < \rho_1\varepsilon$ for all $[x] \in \Delta A$ and $u \in U$. Then for any

$[x] \in \Delta A$, there exists a $u \in U$ such that $[f]([x], u) \cap B \neq \emptyset$ and $[f]([x], u) \subseteq B + \mathbb{B}_{\rho_1 \varepsilon}$ by the definition of the Minkowski sum. Also, $\underline{A} \subseteq \text{Pre}(B|A)$. Hence,

$$\overline{A} = (\underline{A} \cup \Delta A) \subseteq \text{Pre}(B + \mathbb{B}_{\rho_1 \varepsilon}|A) \subseteq \text{Pre}(B + \mathbb{B}_{\rho_1 \varepsilon}|A),$$

which shows (4.40).

We now show that $\text{Pre}(B - \mathbb{B}_{\rho_1 \varepsilon}|A) \subseteq \underline{[\text{Pre}]}^{\varepsilon}(B|A)$. If not, there exists an $x \in \text{Pre}(B - \mathbb{B}_{\rho_1 \varepsilon}|A)$, but $x \notin \underline{[\text{Pre}]}^{\varepsilon}(B|A)$. Then x has to be in $\bigcup_{[x] \in \Delta A}[x]$, since $x \in \bigcup_{[x] \in A_c}[x]$ implies that $x \notin \text{Pre}(B - \mathbb{B}_{\rho_1 \varepsilon}|A)$, which is contradictory to the fact that $x \in \text{Pre}(B - \mathbb{B}_{\rho_1 \varepsilon}|A)$. Let $x \in [x] \in \Delta A$. By Proposition C.2 (ii), there exists $u \in U$ such that

$$f(x, u) \in [f]([x], u) \subseteq B - \mathbb{B}_{\rho_1 \varepsilon} + \mathbb{B}_{\rho_1 \varepsilon} \subseteq B.$$

It indicates that $[x] \subseteq \underline{[\text{Pre}]}^{\varepsilon}(B|A)$, which is a contradiction. Hence, (4.41) holds. ■

As we have discussed before, an under-sampled set of controls $[U]_\mu$ can be used as a replacement of the real set U of control inputs in Algorithm 5 when U is infinite. By using this "cheaper" control space, we can surprisingly achieve a similar result to Lemma 4.2, but the following assumption is additionally required.

Assumption 4.2 (Lipschitz in U) *There exists a Lipschitz constant $L > 0$ for the function $f : \mathbb{R}^n \times \mathbb{R}^m \to \mathbb{R}^n$ in (4.26) such that for all $x \in A \subseteq \mathbb{R}^n$*

$$\|f(x, u) - f(x, v)\|_\infty \leq L \|u - v\|_\infty, \quad \forall u, v \in U. \tag{4.42}$$

Lemma 4.3 *Let $B, A \subseteq X$ be compact and μ be a parameter given in (4.38). If Assumption 4.1 and 4.2 hold in a neighborhood of A with Lipschitz constant $\rho_1 > 0$ and $\rho_2 > 0$, respectively, then*

$$\textit{Pre}(B - \mathbb{B}_{\rho_1 \varepsilon + \rho_2 \mu}|A) \subseteq \underline{[\textit{Pre}_\mu]}^{\varepsilon}(B|A) \subseteq \textit{Pre}(B|A), \tag{4.43}$$

where $\underline{[\textit{Pre}_\mu]}^{\varepsilon}(B|A)$ is the set of intervals given in (4.32) with $\underline{A}$ returned by PRE*$([f]_{\rho_1}, B, A, [U]_\mu, \varepsilon)$ and $[f]_{\rho_1}$ is a Lipschitz inclusion function that satisfies $\mathrm{w}([f]_{\rho_1}([x], u)) \leq \rho_1 \mathrm{w}([x])$ for all $u \in U$.*

Proof: First of all, we define a new map

$$\text{Pre}_\mu(B) := \{x \in X \mid \exists u \in [U]_\mu, \text{ s.t. } f(x, u) + w \in B, \forall w \in W\},$$

and let $Z = \text{Pre}(B|A)$, $Z_\mu = \text{Pre}_\mu(B|A)$, $\underline{Z} = \underline{[\text{Pre}_\mu]}^{\varepsilon}(B|A)$, and $\widetilde{B} = B - \mathbb{B}_{\rho_2 \frac{\mu}{2}}$.

Claim: $\text{Pre}(\widetilde{B}|A) \subseteq Z_\mu \subseteq Z$. Trivially $Z_\mu \subseteq Z$ because $[U]_\mu$ is a subset of U. By Definition 4.15, for all $z \in \text{Pre}(\widetilde{B}|A)$, there exists a $u \in U$ such that $f(z, u) + w \in \widetilde{B}$ for all $w \in W$. With Assumption 4.2, for all $u \in U$, there exists a $v \in [U]_\mu$ such that $f(z,v) \in f(z,u) + \mathbb{B}_{\rho_2 \frac{\mu}{2}}$. Then by Proposition C.2 (ii),

$$\begin{aligned} f(z,v) + d \in f(z,u) + \mathbb{B}_{\rho_2 \frac{\mu}{2}} + d &= (f(z,u) + d) + \mathbb{B}_{\rho_2 \frac{\mu}{2}} \\ \in \widetilde{B} + \mathbb{B}_{\rho_2 \frac{\mu}{2}} &= B - \mathbb{B}_{\rho_2 \frac{\mu}{2}} + \mathbb{B}_{\rho_2 \frac{\mu}{2}} \subseteq B, \end{aligned}$$

which means that $z \in Z_\mu$. Hence the claim holds.

By (4.41) in Lemma 4.2, $\mathrm{Pre}_\mu(\widetilde{B} - \mathbb{B}_{\rho_1 \varepsilon} | A) \subseteq \underline{Z} \subseteq Z_\mu$. Applying the claim above, we have

$$\mathrm{Pre}(B - \mathbb{B}_{\rho_1 \varepsilon + \rho_2 \mu} | A) = \mathrm{Pre}(\widetilde{B} - \mathbb{B}_{\rho_1 \varepsilon} - \mathbb{B}_{\rho_2 \frac{\mu}{2}} | A) \subseteq \mathrm{Pre}_\mu(\widetilde{B} - \mathbb{B}_{\rho_1 \varepsilon} | A).$$

Therefore, $\mathrm{Pre}(B - \mathbb{B}_{\rho_1 \varepsilon + \rho_2 \mu} | A) \subseteq \underline{Z} \subseteq Z_\mu \subseteq \mathrm{Pre}(B|A)$. ■

The outer approximation obtained from $\mathrm{PRE}([f]_{\rho_1}, B, A, \varepsilon)$ does not necessarily satisfy a relationship similar to (4.40) in Lemma 4.2, because the set of control values $[U]_\mu$ is only a finite subset of U. Any evaluation of $[f]_{\rho_1}$ is only an under-approximation in terms of control input.

Because of the bisection operation in Algorithm 5, an input set $A \subseteq \mathbb{R}^n$ of states will be partitioned into a set of intervals of various widths, which, by the following definition, is a partition of A, and each interval is a cell.

Definition 4.16 *Given a set $\Omega \subseteq \mathbb{R}^n$ and a positive integer N, a finite collection of sets $\mathcal{P} = \{P_1, P_2, \cdots, P_N\}$ is said to be a* ***partition*** *of Ω if*

(i) $P_i \subseteq \Omega$, for all $i \in \{1, \cdots, N\}$;

(ii) $int(P_i) \cap int(P_j) = \varnothing$ for all $i, j \in \{1, \cdots, N\}$ where $int(P_i)$ denotes the interior of set P_i;

(iii) $\Omega \subseteq \bigcup_{i=1}^N P_i$.

Each element P_i of the partition $\mathcal{P}$ is called a ***cell.***

4.5 Interval Under-approximations of Controlled Predecessors of Continuous-Time Systems

Consider the continuous-time nonlinear control system of the form:

$$x'(t) = f(x(t), u(t)) + w(t), \tag{4.44}$$

where $x(t) \in X \subseteq \mathbb{R}^n$ is the system state, $u(t) \in U \subseteq \mathbb{R}^m$ is the control input, and $w(t) \in W$ is the disturbance input. More specifically, $w(t) \in W$ given in (4.28), i.e., $W = \mathbb{B}_\delta$ where $\delta > 0$. We assume that the function f satisfies the existence and uniqueness conditions (assumptions in Proposition 1.1) on $D \times U \times W$ (D is an open set containing X) and any solution of system (4.44) is essentially bounded.

Unlike discrete-time systems, controlled predecessors of a set for continuous-time systems are considered with a fixed constant input $u \in U$ over a sampling time interval $[0, \tau]$.

Definition 4.17 *Given a set $Y \subseteq X$ and a sampling time $\tau > 0$, the τ-* ***controlled predecessor*** *from Y after for system (4.44) is a set of states defined by*

$$\mathit{Pre}_\tau^\delta(Y) := \{x_0 \in X \mid \exists u \in U, \;\; s.t. \; (x(0) = x_0) \wedge (x(\tau) \in Y) \\ \wedge \, x'(t) \in f(x(t), u(t)) + \mathbb{B}_\delta, u(t) = u \, \forall t \in [0, \tau]\}.$$

The set of ***valid control inputs*** *that lead to the transition to Y for an $x_0 \in \mathit{Pre}_\tau^\delta(Y)$ after time τ is*

$$K_{\tau,Y}(x_0) := \{u \in U \mid x'(t) \in f(x(t), u(t)) + \mathbb{B}_\delta, u(t) = u, \forall t \in [0, \tau] \\ \wedge \, (x(0) = x_0) \wedge (x(\tau) \in Y)\}.$$

For the sake of simplicity, we denote by $\mathrm{Pre}_\tau(Y)$ the τ-controlled predecessor for the nominal system ($\delta = 0$)

$$x'(t) = f(x(t), u(t)). \tag{4.45}$$

We also let $\mathrm{Pre}_\tau^\delta(B|A) = \mathrm{Pre}_\tau^\delta(B) \cap A$, where $A, B \subseteq X$.

Here, we focus on the under-approximation of τ-controlled predecessors by using interval arithmetic tools. In a nutshell, instead of direct under-approximation, we adopt the same framework in Algorithm 5 and over-approximate the τ-forward reachable sets based on validated interval computation. While any validated over-approximations suffice, we consider the over-approximation by Taylor method obtained in (4.23), recalled as

$$\widehat{\mathcal{R}}_\tau^k([x_0], u) = \sum_{i=0}^{k} [f]^{[i]}([x_0], u)\frac{\tau^i}{i!} + [f]^{[k+1]}(\widehat{[x_0]}, u)\frac{\tau^{k+1}}{(k+1)!}.$$

To apply this, we need an a priori enclosure for solutions starting from $[x_0]$. An interval $\widehat{[x_0]}$ can function as such an a priori enclosure if there exists some $\bar{k}$ that

$$[x_0] + \sum_{i=1}^{\bar{k}-1} [f]^{[i]}([x_0], u)\frac{[0, \tau^i]}{i!} + [f]^{[\bar{k}]}(\widehat{[x_0]}, u)\frac{[0, \tau^{\bar{k}}]}{\bar{k}!} \subseteq \widehat{[x_0]}. \tag{4.46}$$

This extends Lemma 4.1 using Taylor method (however, without any disturbance). Such an *a priori* enclosure is usually unknown beforehand, and therefore, the standard interval Taylor method used in [177] and the related works such as [90] involves an extra "predictor-corrector" procedure in which the a priori enclosure is guessed and verified against (4.46) repeatedly until the condition is satisfied.

Such an a priori enclosure is usually unknown beforehand, and therefore the standard interval Taylor method used in [177] and the related works such as [90] involves an extra "predictor-corrector" procedure in which the a priori enclosure is guessed and verified against (4.46) repeatedly until the condition is satisfied.

The computation of over-approximations of the reachable sets after one sampling time plays a fundamental role in abstraction-based and specification-guided controller synthesis to be introduced in the next chapters. Such computation has to be done frequently and iteratively verifying (4.46) can be expensive. To reduce the computational complexity, we provide in Lemma 4.4 a method to parametrically construct the a priori enclosure if the following Assumption 4.3 is satisfied. In this way, rather than checking the interval inclusion condition (4.46), we can simply choose the parameters that meet certain conditions.

Assumption 4.3 *Let X, U be compact and $[X]$ be an interval containing X. For a given order $k_{\max} \geq 1$, there exists a constant $L > 0$ and inclusion functions $[f]^{[i]}(\cdot, u)$ of $f^{[i]}(\cdot, u)$ such that for all $1 \leq i \leq k_{\max}$,*

$$\mathrm{w}([f]^{[i]}([x], u)) \leq L\mathrm{w}([x]), \ \forall [x] \subseteq [X], u \in U.$$

The above assumption can be guaranteed by $f(\cdot, u)$ being smooth for any $u \in U$, which implies bounded partial derivatives of $f^{[i]}(\cdot, u)$ on any compact set.

Lemma 4.4 *Suppose that there exists an order $k_{\max} \geq 1$ for system (4.44) such that Assumption 4.3 holds. Let*

$$M_u = \sup_{\substack{x \in X, \\ 1 \leq i \leq k_{\max}}} \left\| f^{[i]}(x, u) \right\|_\infty, \quad \bar{\mathrm{w}} = \sup_{[x] \subseteq X} \{\mathrm{w}([x])\}.$$

For an interval $[x_0] \subseteq X$, if γ, τ, and the order $\bar{k} \in [1, k_{\max}]$ are chosen such that

$$[x_0] + [-1, 1] \left((M_u + L\mathrm{w}([x_0])) (e^\tau - 1) + 2\gamma \right) \mathbf{1}^n \subseteq [X], \tag{4.47}$$

$$\frac{\tau^{\bar{k}}}{\bar{k}!} < \frac{2\gamma}{M_u + L\bar{\mathrm{w}}}, \tag{4.48}$$

then

$$\widehat{[x_0]} \triangleq [x_0] + \sum_{i=1}^{\bar{k}-1} [f]^{[i]}([x_0], u) \frac{[0, \tau^i]}{i!} + [-2\gamma, 2\gamma]\mathbf{1}^n \tag{4.49}$$

is an a priori enclosure.

Proof: For a given $k_{\max} \geq 1$, there exist $\tau \in (0,1)$, $\gamma > 0$, and $\bar{k} \in [1, k_{\max}]$ such that $\tau^{\bar{k}}/\bar{k}! < 2\gamma/(M_u + L\bar{\mathrm{w}})$. Under Assumption 4.3, we can construct a Lipschitz inclusion function for any interval $[x] \subseteq [X]$: $[f]^{[i]}([x], u) = f^{[i]}(\bar{x}, u) + L([x] - \bar{x})$ for $1 \leq i \leq k_{\max}$, where $\bar{x}$ is the center point of the interval $[x]$. Then, by (4.48),

$$\begin{aligned}
\mathrm{w}([f]^{[\bar{k}]}([x], u)) &= \mathrm{w}(L([x] - \bar{x})) \leq L\bar{\mathrm{w}} \Rightarrow \\
[f]^{[\bar{k}]}([x], u) &\subseteq [-1, 1](M_u + L\bar{\mathrm{w}})\mathbf{1}^n \Rightarrow \\
[f]^{[\bar{k}]}([x], u)\frac{[0, \tau^{\bar{k}}]}{\bar{k}!} &\subseteq [-1, 1](M_u + L\bar{\mathrm{w}})\frac{\tau^{\bar{k}}}{\bar{k}!}\mathbf{1}^n \subseteq [-2\gamma, 2\gamma]\mathbf{1}^n.
\end{aligned}$$

Let x_0 be the center point of $[x_0]$. Similarly, we have

$$\begin{aligned}
&\sum_{i=1}^{\bar{k}-1} \left(f^{[i]}(x_0, u) + L([x_0] - x_0) \right) \frac{[0, \tau^i]}{i!} \\
&\subseteq [-1, 1](M_u + L\mathrm{w}([x_0])) \left(\sum_{i=1}^{\infty} \frac{\tau^i}{i!} - \sum_{i=\bar{k}}^{\infty} \frac{\tau^i}{i!} \right) \mathbf{1}^n \\
&\subseteq [-1, 1](M_u + L\mathrm{w}([x_0]))\,(e^{\tau} - 1)\,\mathbf{1}^n.
\end{aligned}$$

We can see that $\widehat{[x_0]} \subseteq [X]$ by (4.47). Furthermore,

$$\begin{aligned}
&[x_0] + \sum_{i=1}^{\bar{k}-1} [f]^{[i]}([x_0], u)\frac{[0, \tau^i]}{i!} + [f]^{[\bar{k}]}(\widehat{[x_0]}, u)\frac{[0, \tau^{\bar{k}}]}{\bar{k}!} \subseteq \\
&\qquad [x_0] + \sum_{i=1}^{\bar{k}-1} [f]^{[i]}([x_0], u)\frac{[0, \tau^i]}{i!} + [-2\gamma, 2\gamma] = \widehat{[x_0]},
\end{aligned}$$

which means that the $\widehat{[x_0]}$ defined above satisfies (4.46). ∎

The sampling time τ, the Taylor expansion order $\bar{k}$ and γ are the parameters used in the construction (4.49) of the a priori enclosure. A sampling time $\tau \in (0, 1)$ is necessary to satisfy (4.48), which is also the case for real-world digital control systems. One would like to choose a small sampling time as $\bar{k}$ and γ can be relatively small by conditions (4.47) and (4.48), but a high order $\bar{k}$ (e.g., $\bar{k} \geq 3$) is often required. On the other hand, we will see in Chapter 6 that a smaller sampling time requires a higher partition precision of the system state space for control synthesis.

For system (4.44) that satisfies the assumptions of Lemma 4.4, we can always find an a priori enclosure on which (4.23) is an over-approximation of the reachable set $\mathcal{R}_\tau([x_0], u)$. Let

$$[\mathrm{Pre}_\tau]^\varepsilon(B|A) \triangleq \bigcup_{[x] \in \underline{A}} [x], \tag{4.50}$$

Algorithm 6 $[\underline{A}, \Delta A, A_c] = \text{PRE}(\tau, k, [f], B, A, U, \varepsilon)$

```
1: A ← ∅, ΔA ← ∅, A_c ← ∅, List ← A
2: while List ≠ ∅ do
3:     [x] ← List.first
4:     if R̂_τ^k([x], u) ∩ B = ∅ for all u ∈ U then
5:         A_c ← A_c ∪ [x]                          ▷ R̂_τ^k([x], u) is computed by (4.23)
6:     else if R̂_τ^k([x], u) ⊆ B for some u ∈ U then
7:         A ← A ∪ [x]
8:     else
9:         if w([x]) < ε then
10:            ΔA ← ΔA ∪ [x]
11:        else
12:            {Left[x], Right[x]} = Bisect([x])
13:            List.add({Left[x], Right[x]})
14:        end if
15:    end if
16: end while
```

where $\underline{A}$ is a set of intervals returned by Algorithm 6 that can be controlled into set B after a sampling time period. Thus, the τ-predecessor $\text{Pre}_\tau(B|A)$ of the set $B \subseteq X$ can be under-approximated by $[\text{Pre}_\tau]^\varepsilon(B|A)$. Algorithm 6 is basically Algorithm 5 with $\widehat{\mathcal{R}}_\tau^k([x], u)$ instead of $[f]$.

4.6 Summary

Interval analysis was developed by Moore in the 1960s [173, 174] to rigorously quantify the effects of rounding and propagated errors in numerical analysis. The main idea is to replace numerical computation with numbers with interval computation. Over the past decades, interval analysis has found applications in various areas [108]. In this chapter, we introduced the basic machinery of interval analysis for computation of one-step forward and backward reachable sets for discrete-time and continuous-time dynamical systems. The main references for writing the Section 6.3 on interval analysis include [1, 108, 173, 174]. One-step forward reachable set sets for discrete-time dynamical systems can be easily handled by inclusion functions. This was discussed in Section 4.2. The case for continuous-time dynamical systems is slightly more challenging because integration of a differential equation is involved. Section 4.3 summarizes different approaches for computing the one-step forward reachable sets for continuous-time dynamical systems. We provided self-contained proofs for all the results. Computing one-step reachable sets using a Lipschitz growth bound was used in [204] for computing discrete abstractions for controller synthesis. Mixed monotonicity was used in reachability analysis and controller synthesis in [55, 56] (see also [166]). It is evident from the proofs that differential inequalities and comparison techniques are behind both the growth bound and mixed monotonicity approaches (Propositions 4.11 and 4.12). The readers are referred to [133, 221, 237] for classical books on differential inequalities. In particular, the proofs for Propositions 4.11 and 4.12, as well as Proposition A.5 in Appendix A.4, are formulated following [237], but simplified for Lipschitz flows. The construction of decomposition functions for maps using Lipschitz constants (Proposition 4.8) was due to [246] (see [55] for the case with flows). Sections 4.3.4, 4.4, and 4.5 are based on classical interval analysis. These were used in [139, 142] and [143] for controller synthesis.

Chapter 5

Controller Synthesis via Finite Abstractions

For a dynamical system that evolves over a continuous state space under continuous inputs, the computation of controllers is highly nontrivial. In fact, even the reachability problem for low-dimensional dynamical systems without controls is undecidable [119]. On the other hand, for discrete transition systems with ω-regular objectives, one can solve the discrete controller synthesis problem effectively (albeit with high computational complexity). Discrete abstractions serve as a bridge between continuous control and discrete control synthesis problems. In this chapter, we introduce a notion of such abstractions and investigate theoretical guarantees of abstraction-based controller synthesis.

5.1 Control Abstractions

Consider the discrete-time control system

$$x(t+1) = f(x(t), u(t), w(t)), \tag{5.1}$$

where $x(t) \in X \subseteq \mathbb{R}^n$ is the state, $u(t) \in U \subseteq \mathbb{R}^m$ is the control input, $w(t) \in W \subseteq \mathbb{R}^p$ is a disturbance input, and $f : D \times U \times W \to \mathbb{R}^n$, where $D \subseteq \mathbb{R}^n$. We assume that $X \subseteq D$, U, and W are compact sets and f is Lipschitz continuous with respect to all of its variables on $X \times U \times W$. For the sake of computation, we further assume that X, U, and W are intervals[1].

We recall the controller synthesis problem (Problem 3.2) for system (5.1). Let $\mathcal{T} = (S, A, R, \Pi, L)$ denote the transition system for (5.1) and φ be an LT property over Π. We would like to determine the winning set $\mathbf{Win}_{\mathcal{T}}(\varphi)$ and obtain a control strategy κ such that all controlled paths of $\mathcal{T}$ starting from $\mathbf{Win}_{\mathcal{T}}(\varphi)$ satisfy φ.

A control abstraction of system (5.1) is a discrete (often finite for the purpose of computation) transition system that preserves the dynamical properties of the original dynamical system (5.1) relevant for the purpose of controller synthesis. This is made precise by the following definition.

[1]This is not a severe restriction in practice, as we can always under- or over-approximate these sets by intervals based on the problem we would like to address.

DOI: 10.1201/9780429270253-5

Definition 5.1 (Control abstraction) *For two transition systems*

$$\mathcal{T}_1 = (S_1, A_1, R_1, \Pi, L_1)$$

and

$$\mathcal{T}_2 = (S_2, A_2, R_2, \Pi, L_2),$$

a relation $\alpha \subseteq S_1 \times S_2$ is said to be a ***control abstraction*** *(or simply an* ***abstraction****) from $\mathcal{T}_1$ to $\mathcal{T}_2$, if the following conditions are satisfied:*

(i) for all $s_1 \in S_1$, there exists $s_2 \in S_2$ such that $(s_1, s_2) \in \alpha$ (i.e., $\alpha(s_1) \neq \emptyset$);

(ii) for all $s_2 \in S_2$ and $a_2 \in A_{\mathcal{T}_2}(s_2)$, there exists $a_1 \in A_1$ such that, for all $s \in \alpha^{-1}(s_2)$, we have $a_1 \in A_{\mathcal{T}_1}(s)$ and

$$\alpha(Post_{\mathcal{T}_1}(s, a_1)) \subseteq Post_{\mathcal{T}_2}(s_2, a_2); \tag{5.2}$$

(iii) for all $(s_1, s_2) \in \alpha$, $L_2(s_2) \subseteq L_1(s_1)$.

If such a relation α exists, we say that $\mathcal{T}_2$ ***abstracts*** *$\mathcal{T}_1$ and write $\mathcal{T}_1 \preceq_\alpha \mathcal{T}_2$ or simply $\mathcal{T}_1 \preceq \mathcal{T}_2$.*

Condition (i) states that each "concrete state" in S_1 is associated with an "abstract state" in S_2. Condition (ii) ensures that for each "abstract state" $s_2 \in S_2$ and each available action for $\mathcal{T}_2$ at s_2, there is a choice of action a_1 for $\mathcal{T}_1$, that possibly depends on s_2 and a_2 (but not the particular choice of "concrete states" in $\alpha^{-1}(s_2)$), such that if this action is implemented for any concrete state $s \in \alpha^{-1}(s_2)$, the successors of s under a_1 in $\mathcal{T}_1$ are related to the successors of s_2 under a_2 in $\mathcal{T}_2$ by the relation α. Condition (iii) implies that traces of $\mathcal{T}_2$ can satisfy fewer properties than corresponding traces of $\mathcal{T}_1$. All the conditions are important for ensuring soundness of control abstractions, which will be made precise in the next section.

5.2 Soundness

We show that, under certain semantic requirements, control abstractions are *sound* for controller synthesis in the sense that, when $\mathcal{T}_1 \preceq \mathcal{T}_2$, a controller found for $\mathcal{T}_2$ can be used to construct a controller for $\mathcal{T}_1$ with respect to the same objective.

We are interested in satisfying linear-time (LT) properties (e.g., ω-regular). Recall that an LT property over a set of atomic propositions Π is a subset of

$$(2^\Pi)^\omega := \left\{ \{\Pi_i\}_{i=0}^\infty \mid \Pi_i \in 2^\Pi, \forall i \geq 0 \right\}.$$

We introduce a technical requirement for proving soundness of control abstractions. In Definition 3.6, we say that a path ρ for $\mathcal{T}$ or its trace $L(\rho) := \{L(s_i)\}_{i=0}^{\infty}$ satisfies an LT property φ if $L(\rho) \in \varphi$. We assume that all LT properties are in positive normal form defined below.

Definition 5.2 *An LT property over* Π *is said to be in positive normal form if* $\sigma \vDash \varphi$ *implies* $\rho \vDash \varphi$ *for any* $\sigma = \{\sigma\}_{i=0}^{\infty} \in (2^{\Pi})^{\omega}$ *and* $\rho = \{\rho\}_{i=0}^{\infty} \in (2^{\Pi})^{\omega}$ *such that* $\sigma_i \subseteq \rho_i$ *for all* $i \geq 0$.

Intuitively, putting φ in positive normal form requires that all properties in Π are understood to be "good" properties. This can be achieved by adding negations of "bad" atomic propositions as new atomic propositions. This requirement is not stringent. For example, for LT properties described by linear temporal logic formulas, they can be transformed into positive normal form [23, Chapter 5], where all negations appear only in front of the atomic propositions. By doing so, all negations of atomic propositions can be replaced by new atomic propositions.

In the following, we always assume that LT properties are in positive normal form.

Definition 5.3 *Given an abstraction relation* α *from* $\mathcal{T}_1$ *to* $\mathcal{T}_2$ *and a control strategy* μ_i *for* $\mathcal{T}_i$ $(i = 1, 2)$, μ_1 *is called* ***α-implementation*** *of* μ_2, *if, for each* $n \geq 0$,

$$u_n = \mu_1(x_0, x_1, x_2, \cdots, x_n)$$

is chosen according to

$$a_n = \mu_2(s_0, s_1, s_2, \cdots, s_n)$$

in such a way (as guaranteed by Definition 5.1 for $\mathcal{T}_1 \preceq_{\alpha} \mathcal{T}_2$*) that*

$$\alpha(Post_{\mathcal{T}_1}(x, u_n)) \subseteq Post_{\mathcal{T}_2}(s_n, a_n)$$

for all $x \in \alpha^{-1}(s_n)$, *where* $q_n \in \alpha(x_n)$.

We end this section by stating a soundness result for abstractions.

Theorem 5.1 *Consider two transition systems*

$$\mathcal{T}_1 = (S_1, A_1, R_1, \Pi, L_1)$$

and

$$\mathcal{T}_2 = (S_2, A_2, R_2, \Pi, L_2).$$

Suppose that $\alpha \subseteq S_1 \times S_2$ *is an abstraction from* $\mathcal{T}_1$ *to* $\mathcal{T}_2$, *i.e.,* $\mathcal{T}_1 \preceq_{\alpha} \mathcal{T}_2$ *and let* φ *be an LT property. If there exists a control strategy* μ_2 *for* $\mathcal{T}_2$ *such that all* μ_2*-controlled paths of* $\mathcal{T}_2$ *starting from a set* $W_2 \subseteq S_2$ *satisfy* φ, *then there exists a control strategy* μ_1, *which is an* α*-implementation of* μ_2, *for* $\mathcal{T}_1$ *such that all* μ_1*-controlled paths of* $\mathcal{T}_1$ *starting from the set* $W_1 = \alpha^{-1}(W_2)$ *satisfy* φ. *Furthermore, if* μ_2 *is of finite memory (or memoryless), then* μ_1 *is also of finite memory (or memoryless).*

Proof: We show that, by Definitions 5.1 and 5.3, a μ_1-controlled path of $\mathcal{T}_1$ always leads to a μ_2-controlled path of $\mathcal{T}_2$. Suppose we start with $x_k \in S_1$ and let s_k be arbitrarily chosen from $\alpha(x_k)$, where $k \geq 0$. For any initial state $x_0 \in W_1$, we have $s_0 \in W_2$. Suppose $a_k = \mu_2(s_0, s_1, s_2, \cdots, s_k)$ and $u_k = \mu_1(x_0, x_1, x_2, \cdots, x_k)$. Since $\alpha(\text{Post}_{\mathcal{T}_1}(x_k, u_k)) \subseteq \text{Post}_{\mathcal{T}_2}(s_k, a_k)$, we know that for any $s_{k+1} \in \alpha(x_{k+1})$ and $x_{k+1} \in \text{Post}_{\mathcal{T}_1}(x_k, u_k)$, we have $s_{k+1} \in \text{Post}_{\mathcal{T}_2}(s_k, a_k)$. This implies that (s_k, a_k, s_{k+1}) is a valid transition in $\mathcal{T}_2$ and therefore, by induction, $s_0 s_1 s_2 \cdots$ is a μ_2-controlled path of $\mathcal{T}_2$, if $x_0 x_1 x_2 \cdots$ is a μ_1-controlled path of $\mathcal{T}_1$. Furthermore, by Definitions 5.1, we have $L_2(s_k) \subseteq L_1(x_k)$ for all $k \geq 0$. Since the trace of $s_0 s_1 s_2 \cdots$ satisfies φ, we know that the trace of $x_0, x_1, x_2 \cdots$ also satisfies φ, because φ is in positive normal form.

The last statement follows from the definition of α-implementation. ■

5.3 Completeness via Robustness

In this section, we discuss the construction of finite abstractions for system (5.1). We show that, under mild technical assumptions, one can always construct finite abstractions for system (5.1) that are not only sound but also complete via a robustness argument, which will be made precise later in the section. Such completeness results offer theoretical guarantees for abstraction-based controller synthesis in the sense that, by continually refining the abstractions, we will either be successful in finding a controller using discrete abstractions, or we can prove via such computation that certain robust controllers do not exist.

We first introduce the transition system associated with system (5.1). The definition is a slight modification of Definition 3.21 to explicitly deal with out-of-domain behaviors.

Definition 5.4 *The discrete-time control system (5.1) can be written as a transition system*

$$\mathcal{S} = (S, A, R, \Pi, L),$$

where

- $S = X \cup \{X^c\}$ *is the set of states;*
- $A = U$ *is the set of actions;*
- $R \subseteq S \times A \times S$ *is defined by* $(x, u, \hat{x}) \in R$ *if and only if one of the following holds: (i) there exists* $w \in W$ *such that* $f(x, u, w) = \hat{x}$ *and* $x, \hat{x} \in X$*; (ii)* $x \in X$, $\hat{x} = X^c$, *and* $f(x, u, w) \notin X$ *for some* $w \in W$*; (iii)* $x = \hat{x} = X^c$*;*

- Π *is the set of atomic propositions;*
- $L : S \to 2^{\Pi}$ *is the labeling function.*

The state X^c is introduced to precisely encode if an out-of-domain transition takes place.

We also consider a perturbed version of the discrete-time control system (3.1) given by

$$x(t+1) = f(x(t), u(t), w(t)) + d(t), \tag{5.3}$$

where $d(t)$ is an additional disturbance signal with infinity norm bounded by some $\delta > 0$, i.e., $d(t) \in \mathbb{B}_\delta$ for all t, where $\mathbb{B}_\delta$ is the ball of radius δ with respect to the infinity norm in $\mathbb{R}^n$. Similarly, we formulate this δ-perturbed system as a transition system.

Definition 5.5 *The δ-perturbed discrete-time control system (5.3) can be written as a transition system*

$$\mathcal{S}_\delta = (S, A, R_\delta, \Pi, L_\delta),$$

where

- $S = X \cup \{X^c\}$ *is the set of states;*
- $A = U$ *is the set of actions;*
- $R_\delta \subseteq S \times A \times S$ *is defined by $(x, u, \hat{x}) \in R_\delta$ if and only if one of the following holds: (i) there exists $w \in W$ and $d \in \mathbb{B}_\delta$ such that $f(x, u, w) + d = \hat{x}$ and $x, \hat{x} \in X$; (ii) $x \in X$, $\hat{x} = X^c$, and $f(x, u, w) + d \notin X$ for some $w \in W$ and $d \in \mathbb{B}_\delta$; (iii) $x = \hat{x} = X^c$;*
- Π *is the set of atomic propositions;*
- $L_\delta : S \to 2^{\Pi}$ *is the labeling function (to be chosen).*

We construct a finite transition system

$$\mathcal{T} = (S_\mathcal{T}, A_\mathcal{T}, R_\mathcal{T}, \Pi, L_\mathcal{T})$$

as follows.

For a positive integer k, let $\mathbb{Z}^k$ denote the k-dimensional integer lattice, i.e., the set of all k-tuples of integers. For parameters $\eta > 0$ and $\mu > 0$ (to be chosen later), define

$$[\mathbb{R}^n]_\eta := \eta\mathbb{Z}^n, \quad [\mathbb{R}^m]_\mu := \mu\mathbb{Z}^m,$$

where $\mu\mathbb{Z}^k = \{\mu z : z \in \mathbb{Z}^k\}$ (for $k = n, m$). Define a relation α over $[\mathbb{R}^n]_\eta \cup \{X^c\}$ and S by

$$\left\{(x, s) \mid s = \eta\lfloor\frac{x}{\eta}\rfloor, x \in X\right\} \cup \{(X^c, X^c)\},$$

where $\lfloor\cdot\rfloor$ is the floor function (i.e., $\lfloor x\rfloor = (\lfloor x_1\rfloor, \cdots, \lfloor x_n\rfloor)$ and $\lfloor x_i\rfloor$ gives the largest integer less than or equal to x_i). Now we restrict α to S and let $S_{\mathcal{T}} = \alpha(S)$ and $A_{\mathcal{T}} = \left\{a \mid \exists u \in A \text{ s.t. } a = \mu\lfloor\frac{u}{\mu}\rfloor\right\}$ (which are both non-empty by definition and finite because X and U are compact). Note that this gives a deterministic relation in the sense that $\alpha(x)$ is single-valued for all $x \in S$.

We next construct $R_{\mathcal{T}}$. For each $s \in S_{\mathcal{T}}$ and $a \in A_{\mathcal{T}}$, from Proposition 4.10, we can compute Y such that

$$f(X_0, u, W) \subseteq Y \subseteq f(X_0, u, W) + \mathbb{B}_\delta,$$

where $X_0 = \overline{\alpha^{-1}(s)}$, $u = a$, and $\theta > 0$ can be arbitrarily chosen.

Define $\hat{Y} = Y$ if $Y \subseteq X$ and $\hat{Y} = Y \cup \{X^c\}$ otherwise. Clearly, $\hat{Y}$ over-approximate the one-step reachable set of system (5.1) from X_0 under control u, with the out-of-domain encoded by X^c. Hence

$$\bigcup_{x\in\alpha^{-1}(s)} \text{Post}_{\mathcal{S}}(x, a) \subseteq \hat{Y}. \tag{5.4}$$

We let (s, a, s') be included in $R_{\mathcal{T}}$ if and only if $s' \in \alpha(\hat{Y})$, i.e.,

$$\text{Post}_{\mathcal{T}}(s, a) = \alpha(\hat{Y}). \tag{5.5}$$

Finally, we construct $L_{\mathcal{T}}$. The construction is based on the labeling functions L for $\mathcal{S}$ and L_δ for $\mathcal{S}_\delta$. We assume that L_δ is an *ε-strengthening* of L in the sense that if (1) $\pi \in L_\delta(x)$ and $x \in X$, then $\pi \in L(y)$ for all $y \in (x + \mathbb{B}_\varepsilon) \cap X$; and (2) $L(X^c) = L_\delta(X^c)$. We define $L_{\mathcal{T}}$ to be an η-strengthening of L.

Theorem 5.2 *Let $\delta > 0$ and $\varepsilon > 0$ be arbitrarily given. Suppose that L_δ is an ε-strengthening of L. If η, μ, and θ above are chosen such that $\eta \leq \frac{\varepsilon}{2}$ and*

$$L(\eta + \mu) + \eta + \theta \leq \delta, \tag{5.6}$$

where L is the uniform Lipschitz constant of $f(x, u, w)$ with respect to (x, u) on $X \times U$ for $w \in W$, then $\mathcal{T}$ and α constructed above satisfy

$$\mathcal{S} \preceq_\alpha \mathcal{T} \preceq_{\alpha^{-1}} \mathcal{S}_\delta. \tag{5.7}$$

Proof: We verify that conditions (i)–(iii) of Definition 5.1 hold for both $\mathcal{S} \preceq_\alpha \mathcal{T}$ and $\mathcal{T} \preceq_{\alpha^{-1}} \mathcal{S}_\delta$.

Condition (i) obviously holds.

To verify condition (ii) for $\mathcal{S} \preceq_\alpha \mathcal{T}$, we choose $u = a$ and note that

$$\alpha\left(\bigcup_{x\in\alpha^{-1}(s)} \text{Post}_{\mathcal{S}}(x, u)\right) \subseteq \alpha(\hat{Y}) = \text{Post}_{\mathcal{T}}(s, a),$$

where we used (5.4) and (5.5). This verifies condition (ii) of Definition 5.1 for $\mathcal{S}_1 \preceq_\alpha \mathcal{T}$. Now consider α^{-1} over $S_\mathcal{T}$ and S. For each $x \in S$ and $u \in A$, denote $s = \alpha(x)$ and pick $a = \mu\lfloor\frac{u}{\mu}\rfloor \in A_\mathcal{T}$ such that

$$\alpha^{-1}(\bigcup_{s\in\alpha(x)} \text{Post}_\mathcal{T}(s,a)) = \alpha^{-1}(\text{Post}_\mathcal{T}(s,a)) = \alpha^{-1}(\alpha(\hat{Y})), \tag{5.8}$$

where we used (5.5).

By the definition of $\hat{Y}$ and α, if $Y \subseteq X$, we have

$$\begin{aligned}\alpha^{-1}(\alpha(\hat{Y})) &\subseteq Y + \eta\mathbb{B}\\ &\subseteq f(X_0, a, W) + (\eta+\varepsilon)\mathbb{B}\\ &\subseteq f(u, x, W) + [L(\mu+\eta) + (\eta+\varepsilon)]\mathbb{B}\\ &\subseteq f(u, x, W) + \mathbb{B}_\delta\\ &\subseteq \text{Post}_{\mathcal{S}_\delta}(x,u),\end{aligned} \tag{5.9}$$

where we used (5.6). If $Y \nsubseteq X$, we have $X^c \in \hat{Y}$, which implies

$$\alpha^{-1}(\alpha(\hat{Y})) \subseteq (Y + \eta\mathbb{B}) \cup \{X^c\}. \tag{5.10}$$

It follows that $f(X_0, a, W) + \mathbb{B}_\varepsilon \nsubseteq X$. Similarly, by the Lipschitz continuity of f, we have

$$f(X_0, a, W) + \mathbb{B}_\varepsilon \subseteq f(u, x, W) + [L(\mu+\eta) + \varepsilon]\mathbb{B} \subseteq f(u, x, W) + \mathbb{B}_\delta,$$

which implies $f(u, x, W) + \mathbb{B}_\delta \nsubseteq X$ and $X^c \in \text{Post}_{\mathcal{S}_\delta}(x,u)$. Hence

$$\alpha^{-1}(\bigcup_{s\in\alpha(x)} \text{Post}_\mathcal{T}(s,a)) \subseteq \text{Post}_{\mathcal{S}_\delta}(x,u), \tag{5.11}$$

which verifies condition (ii) of Definition 5.1 for $\mathcal{T} \preceq_{\alpha^{-1}} \mathcal{S}_\delta$.

To verify condition (iii) of Definition 5.1 for $\mathcal{S} \preceq_\alpha \mathcal{T}$ and $\mathcal{T} \preceq_{\alpha^{-1}} \mathcal{S}_2$, we need to check that

$$L_\delta(x) \subseteq L_\mathcal{T}(q) \tag{5.12}$$

and

$$L_\mathcal{T}(s) \subseteq L(x) \tag{5.13}$$

for all $(x, s) \in \alpha$. Both clearly hold for $x = s = X^c$ by the definitions of L_δ and $L_\mathcal{T}$. Fix any $(x, s) \in \alpha$ with $x, s \neq X^c$. If $\pi \in L_\delta(x)$, then $\pi \in L(y)$ for all $y \in x + \varepsilon\mathbb{B}$. Since $s + \eta\mathbb{B} \subseteq x + 2\eta\mathbb{B} \subseteq x + \varepsilon\mathbb{B}$, we have $\pi \in L(y)$ for all $y \in s + \eta\mathbb{B}$ and $\pi \in L_\mathcal{T}(s)$. Hence, (5.12) holds. If $\pi \in L_\mathcal{T}(s)$, then $\pi \in L(y)$ for all $y \in s + \eta\mathbb{B}$ by the definition of $L_\mathcal{T}$. Since $x \in s + \eta\mathbb{B}$, we have $\pi \in L(x)$. Hence, (5.13) holds. ∎

5.4 Extension to Continuous-Time Dynamical Systems

Consider the continuous-time control system

$$x'(t) = f(x(t), u(t)) + w(t), \tag{5.14}$$

where $x(t) \in X \subseteq \mathbb{R}^n$ is the state, $u(t) \in U \subseteq \mathbb{R}^m$ is the control input, $w(t) \in \mathbb{B}_\delta \subseteq \mathbb{R}^n$ is a disturbance input, and $f : D \times U \to \mathbb{R}^n$ for some open set $D \subseteq \mathbb{R}^n$. We assume that $X \subseteq D$ and U are compact sets and f is Lipschitz continuous with respect to both of its variables on $X \times U$. To simplify the notation, system (5.14) can be equivalently written as a differential inclusion

$$x' \in f(x, u) + \mathbb{B}_\delta, \tag{5.15}$$

We denote this system by $\mathcal{S}_\delta$.

A *(sampled-data) control strategy* with sampling period $\tau > 0$ for (5.14) is a partial function of the form:

$$\sigma(x_0, \cdots, x_i) = u_i \in U, \ \forall i = 0, 1, 2, \cdots, \tag{5.16}$$

where $x_0, \cdots, x_i$ is a finite sequence of sampled states taken at sampling times $t_0 = 0, \cdots, t_i$ and u_i is a constant control input. The sampling times t_0, t_1, t_2, $\cdots$ satisfy $t_{i+1} - t_i = \tau$ for all $i \geq 0$, where $\tau > 0$ is the sampling period that represents the duration for which the constant u_i is applied to the system.

A *σ-controlled trajectory* is a pair of functions (x, u) resulting from executing the control strategy σ, where u is defined by $u(t) = u_i$ for $t \in [t_i, t_{i+1})$, where $t_i = i\tau$ and u_i is determined by (5.16).

With a fixed sampling period $\tau > 0$, we define the transition system representation of $\mathcal{S}_\delta$ as follows.

Definition 5.6 *The system $\mathcal{S}_\delta$ with a sampling period $\tau > 0$ can be interpreted as a transition system*

$$\mathcal{T}_{\delta,\tau} = (S, A, R, , \Pi, L_\delta)$$

by defining

- *$S = X$ is the set of states[2];*
- *$A = U$ is the set of actions;*
- *$(x_0, u, x_1) \in R$ if and only if there exists a trajectory $x : [0, \tau] \to X$ such that $x(0) = x_0$, $x_1 = x(\tau)$, and $x'(t) \in f(x(t), u) + \mathbb{B}_\delta$ for almost all $t \in [0, \tau]$;*
- *Π is the set of atomic propositions;*
- *$L_\delta : S \to 2^\Pi$ is the labeling function.*

[2]To simplify the presentation, we shall not treat the out-of-domain behaviors explicitly. They can be handled in a similar way as done in the discrete-time case.

5.4.1 Soundness and Robust Completeness

Suppose that there exists a transition system

$$\mathcal{T} = (S_{\mathcal{T}}, A_{\mathcal{T}}, R_{\mathcal{T}}, \Pi, L_{\mathcal{T}})$$

and a relation $\alpha \subseteq (S, S_{\mathcal{T}})$ such that $\mathcal{T}_{\delta,\tau} \preceq_{\alpha} \mathcal{T}$. By Theorem 5.1, if $\mathcal{T}$ has a winning strategy with respect to a linear-time property φ, then we can construct a winning strategy for $\mathcal{T}_{\delta,\tau}$ with respect to φ. A winning strategy for $\mathcal{T}_{\delta,\tau}$ is by definition a sampled-data control strategy for $\mathcal{S}_{\delta}$ with sampling period τ.

We next show that we can similarly construct complete abstractions for $\mathcal{T}_{\delta,\tau}$.

Theorem 5.3 (Robust completeness) *Fix any $\delta_2 > \delta_1 \geq 0$ and $\varepsilon > 0$. Suppose that L_{δ_2} is an ε-strengthening of L_{δ_1}. We can choose $\tau > 0$ and compute a finite transition system $\mathcal{T}$ such that*

$$\mathcal{T}_{\delta_1,\tau} \preceq \mathcal{T} \preceq \mathcal{T}_{\delta_2,\tau}.$$

Proof: We construct $\mathcal{T} = (S_{\mathcal{T}}, A_{\mathcal{T}}, R_{\mathcal{T}}, \Pi, L_{\mathcal{T}})$ as follows. Let $\eta > 0$ and $\mu > 0$ be parameters to be chosen. Let $S_{\mathcal{T}}$ consist of the centres of the grid cells in $[\mathbb{R}^n]_{\eta}$ that have a non-empty intersection with X. Let $A_{\mathcal{T}}$ consist of the centres of the grid cells in $[\mathbb{R}^m]_{\mu}$ that have a non-empty intersection with U. Because U and X are compact sets, $S_{\mathcal{T}}$ and $A_{\mathcal{T}}$ are both finite. Let L denote the Lipschitz constant of f with respect to (x, u) on $X \times U$. Denote $M := \max_{(x,u)\in X\times U} \|f(x, u)\|_{\infty}$.

We define a relation $\alpha \subseteq X \times S_{\mathcal{T}}$ by $(x, s) \in \alpha$ if and only if $\|x - s\|_{\infty} \leq \frac{\eta}{2}$. Clearly, α^{-1} is a relation on $S_{\mathcal{T}} \times X$. Define $R_{\mathcal{T}} \subseteq S \times A \times S$ by $(s, a, s_1) \in R_{\mathcal{T}}$ if and only if

$$|s_1 - (s + \tau f(s, a))| \leq \frac{\eta}{2} + \frac{\eta}{2}e^{L\tau} + (\frac{\delta_1}{L} + \frac{\mu}{2})(e^{L\tau} - 1) + \frac{M(e^{L\tau} - L\tau - 1)}{L}. \tag{5.17}$$

We show that, if η, μ, and τ are chosen sufficiently small, we have

$$\mathcal{T}_{\delta_1,\tau} \preceq_{\alpha} \mathcal{T} \preceq_{\alpha^{-1}} \mathcal{T}_{\delta_2,\tau}.$$

Condition (i) in Definition 5.1 is clearly satisfied by both α and α^{-1}.

We verify that condition (ii) in Definition 5.1 holds for $\mathcal{T}_{\delta_1,\tau} \preceq_{\alpha} \mathcal{T}$; that is, for $s \in S_{\mathcal{T}}$ and $a \in A_{\mathcal{T}}$, there exists $u \in U$ such that

$$\alpha(\mathrm{Post}_{\mathcal{T}_{\delta_1,\tau}}(x, u)) \subseteq \mathrm{Post}_{\mathcal{T}}(s, a), \tag{5.18}$$

for all $x \in \alpha^{-1}(s)$. Pick $u \in U$ with $|u - a| \leq \frac{\mu}{2}$. Given $x_1 \in \mathrm{Post}_{\mathcal{T}_{\delta_1,\tau}}(x, u)$, there exists a trajectory $x : [0, \tau] \to X$ such that $x(0) = x$, $x(\tau) = x_1$, and

$x'(s) \in f(x(s), u) + \mathbb{B}_{\delta_1}$ for all $s \in [0, \tau]$. Define $x_\tau(t) = q + tf(q, a)$ for $t \in [0, \tau]$. We have

$$\begin{aligned}
&\|x'(t) - x'_\tau(t)\|_\infty \le \|f(x(t), u) - f(s, a)\|_\infty + \delta_1 \\
&\le \|f(x(t), u) - f(x_\tau(t), u)\|_\infty + \|f(x_\tau(t), u) - f(q, u)\|_\infty \\
&\qquad + \|f(q, u) - f(q, a)\|_\infty + \delta_1 \\
&\le L\|x(t) - x_\tau(t)\|_\infty + L\|x_\tau(t) - q\|_\infty + L\|u - a\|_\infty + \delta_1 \\
&\le L\|x(t) - x_\tau(t)\|_\infty + LMt + \frac{L\mu}{2} + \delta_1, \quad t \in [0, \tau].
\end{aligned} \tag{5.19}$$

By Gronwall's inequality (Lemma A.2), we have

$$\begin{aligned}
&\|x_1 - (s + \tau f(s, u))\|_\infty = \|x(\tau) - x_\tau(\tau)\|_\infty \\
&\le \|x - s\|_\infty e^{L\tau} + \int_0^\tau (LMs + \frac{L\mu}{2} + \delta_1)e^{L(\tau - s)} ds \\
&\le \frac{\eta}{2}e^{L\tau} + (\frac{\delta_1}{L} + \frac{\mu}{2})(e^{L\tau} - 1) + \frac{M(e^{L\tau} - L\tau - 1)}{L}.
\end{aligned}$$

By (5.17), this shows $\alpha(x_1) \subseteq \mathrm{Post}_\mathcal{T}(q, a)$. Hence (5.18) holds.

We next verify that condition (ii) holds for $\mathcal{T} \preceq_{\alpha^{-1}} \mathcal{T}_{\delta_2,\tau}$; that is, for $x \in X$ and $u \in U$, there exists $a \in A_\mathcal{T}$ such that

$$\alpha^{-1}(\mathrm{Post}_\mathcal{T}(s, a)) \subseteq \mathrm{Post}_{\mathcal{T}_{\delta_2,\tau}}(x, u); \tag{5.20}$$

for all $s \in \alpha(x)$. Pick a be the center of the grid cell in $[\mathbb{R}^m]_\mu$ that contains u. Given $y_1 \in \alpha^{-1}(\mathrm{Post}_\mathcal{T}(s, a))$, there exists $s_1 \in \mathrm{Post}_\mathcal{T}(s, a)$ such that $\|y_1 - s_1\|_\infty \le \frac{\eta}{2}$. By the definition of $\mathrm{Post}_\mathcal{T}(s, a)$, we have

$$\|s_1 - (s + \tau f(s, a))\|_\infty \le \frac{\eta}{2} + \frac{\eta}{2}e^{L\tau} + (\frac{\delta_1}{L} + \frac{\mu}{2})(e^{L\tau} - 1) + \frac{M(e^{L\tau} - L\tau - 1)}{L}.$$

Consider the trajectory $x : [0, \tau] \to X$ such that $x(0) = x$, $x(\tau) = x_1$, and $x'(s) \in f(x(s), u)$. By a similar argument as in (5.19), we can show

$$\|x_1 - (s + \tau f(s, a))\|_\infty \le \frac{\eta}{2}e^{L\tau} + \frac{\mu}{2}(e^{L\tau} - 1) + \frac{M(e^{L\tau} - L\tau - 1)}{L}.$$

Hence, by the triangle inequality,

$$\|y_1 - x_1\|_\infty \le \eta + \eta e^{L\tau} + (\frac{\delta_1}{L} + \mu)(e^{L\tau} - 1) + \frac{2M(e^{L\tau} - L\tau - 1)}{L} \tag{5.21}$$

Define

$$z(\theta) = x(\theta) + \frac{\theta}{\tau}(y_1 - x_1), \quad \theta \in [0, \tau].$$

Then $z(0) = x(0) = x$ and $z(\tau) = y_1$, and

$$z'(\theta) \in f(x(\theta), u) + \frac{1}{\tau}(y_1 - x_1). \tag{5.22}$$

Note that

$$\|z(\theta) - x(\theta)\|_\infty = \left\|\frac{\theta}{\tau}[y_1 - x_1]\right\|_\infty \le \|y_1 - x_1\|_\infty, \quad \theta \in [0, \tau]. \tag{5.23}$$

Since $0 \le \delta_1 < \delta_2$, we can choose τ, μ, η sufficiently small such that

$$[\eta + \eta e^{L\tau} + (\frac{\delta_1}{L} + \mu)(e^{L\tau} - 1) + \frac{2M(e^{L\tau} - L\tau - 1)}{L}][L + \frac{1}{\tau}] < \delta_2. \tag{5.24}$$

To see this is possible, choose, e.g., $\eta = \tau^2$ and $\mu = \tau$, and note that the limit of the left-hand side as $\tau \to 0$ is given by $\lim_{\tau\to 0} \delta_1 \frac{e^{L\tau}-1}{L\tau} = \delta_1$. It follows from (5.21)–(5.24) and Lipschitz continuity of f that $z'(\theta) \in f(z(\theta), u) + \mathbb{B}_{\delta_2}$. Hence $y_1 \in \text{Post}_{\mathcal{T}_{\delta_2,\tau}}(x, u)$ and (5.20) holds.

The construction of $L_\mathcal{T}$ and verification of condition (ii) in Definition 5.1 are similar to that in the proof of Theorem 5.2. ■

Remark 5.1 *In the proof, we choose the simplest possible validated bounds on a one-step reachable set, i.e., a forward Euler scheme with an error bound. This suffices to prove the required convergence to show approximate completeness. With the template provided by the proof of Theorem 5.3, one can in fact use any accurate over-approximation of the one-step reachable set for* $\mathcal{S}_{\delta_1}$ *to replace (5.17) for defining the transitions in* $\mathcal{T}$ *and then show that this over-approximation is contained in the actual one-step reachable set of* $\mathcal{S}_{\delta_2}$.

Examples and case studies for abstraction-based controller synthesis will be presented in Chapter 7.

5.5 Summary

In this chapter, we discussed abstraction-based controller synthesis. Our focus is on the construction of control abstractions with theoretical guarantees of soundness and completeness. The underlying computation engine is interval analysis (or reachability analysis with convergence guarantees). We show that, under the assumption of Lipschitz continuity, we can construct arbitrarily precise abstractions in the sense of Theorems 5.2 and 5.3 for discrete-time and continuous-time systems, respectively. Such abstractions can be viewed as approximately complete, where the completeness is achieved via a robustness argument—we are either guaranteed to find a controller using these discrete abstractions or certain robust controllers do not exist. Section 5.3 is based on the results in [152] and Section 5.4 is based on [153]. We only handle linear-time properties with a discrete-time semantics. For extensions to continuous-time semantics, the readers are referred to [69] and its use in construction of sound and approximately complete abstractions [153–155]. One limitation of the abstraction-based approaches for controller synthesis discussed in this chapter is that we often rely on uniform discretization for constructing abstractions. In the next chapter, we shall consider specification-guided controller synthesis that can overcome this limitation by adaptively refining the state spaces according to given specifications.

5.5.1 A Brief Account of Formal Methods for Control

Formal methods for controller synthesis can be dated back to the late 1980s. Seminal work was done by Pnueli and Rosner [186, 207] in the context of *reactive synthesis*, which showed that synthesis from a linear temporal logic (LTL) formula unavoidably gives rise to double-exponential time complexity. This seminal result led to the impression that LTL synthesis is hopelessly intractable. More recent work, however, has shown that synthesis of interesting fragments of LTL [33, 183] can be achieved in polynomial time. The success of formal verification and synthesis in computer science sparked a series of work in the robotics and control communities on using temporal logic for synthesizing planners and controllers (see, e.g., [66–68, 116–118, 123–125, 127, 159, 169–171, 196, 239–242]), which extended previous work on planning as model checking [82, 88]. Separately in the control community, seminal work was done by Ramadge and Wonham [200, 243] for the control of *discrete-event systems* [45]. This also led to the development of control with temporal logic and ω-regular specifications for discrete-event systems [109, 149, 225–227]. The use of discrete-event models and temporal logic for automatic synthesis of locomotion controllers, in fact, can be traced back to as early as [11]. See [63] for a comparative survey of supervisory control and reactive synthesis.

Naturally, the use of discrete logic on continuous dynamical systems leads to what are called *hybrid systems* [7, 12, 37, 89, 93, 148, 151, 161, 222, 235]—a field at the confluence of computer science and control theory. Algorithmic analysis and controller synthesis for hybrid systems is a central topic that received considerable attention since the 1990s (see, e.g., [3–5, 15, 16, 18, 26, 35, 39, 40, 58, 60, 76, 77, 94, 95, 97, 101–103, 110, 122, 132, 136–138, 140, 142, 145, 156–158, 162, 163, 168, 179, 181, 189, 190, 192, 208, 219, 220, 230–232]). To work with formal specifications—and at the same time to leverage the rich set of tools for verification and synthesis for discrete transition systems—there has been a considerable literature on constructing discrete abstractions of continuous and hybrid systems (see, e.g., [6, 42, 43, 56, 57, 83–86, 131, 144, 152–156, 179, 181, 188, 191, 197, 198, 202, 204, 224, 247]). Notably, seminal work by Caines and Wei proposed the notion of dynamically consistent abstractions for hierarchical hybrid control [42, 43].

In a nutshell, abstraction serves as a bridge for connecting control theory and formal methods in the sense that hybrid control design for dynamical systems and high-level specifications can be done using finite abstractions of these systems. Early work focused on constructing abstractions that are equivalent to the original system (termed as bisimilar symbolic models). Seminal work by Tabuada and Pappas [223] showed that bisimilar symbolic models exist for controllable linear systems. For nonlinear and special classes of hybrid systems that are incrementally stable [10], approximately bisimilar models can be constructed [83, 86, 188]. For nonlinear systems without stability assumptions, the work by Zamani et al. [247] proposed symbolic models that are sound for controller synthesis. This led to the later development of sound control abstractions [154, 155, 203, 204] for nonlinear systems with temporal logic and linear-time specifications.

A central theoretical question around abstraction-based controller synthesis is: *to what extent and in what sense are such discrete abstractions faithful to the original systems with general nonlinear dynamics*? In its full generality, unfortunately, we cannot hope to use discrete abstractions to answer questions pertaining to the exact dynamical properties of the original system. Indeed, even reachability is undecidable for low-dimensional continuous or simple hybrid system models [17, 104, 119, 172, 215]. In a foundational work [52, 53], Collins proved that arbitrarily precise outer-approximations of reachable sets can be computed if and only if the reachable set is equal to the *chain reachable set*—a set containing all points which can be reached by the system perturbed with an arbitrarily small amount of noise. This agrees with early fundamental work by Conley [54] on isolated invariant sets and his belief that only such "structurally stable" objects could be detected in nature. Through the lens of computation, this means we can only hope to verify robust properties. In the context of reachability analysis, the work by Fränzle [72, 73] and Asarin and Bouajjani [14] showed that "*robustness implies decidability*" for reachability analysis of hybrid systems. Kong et al. [122] developed δ-decision procedure [80] for bounded reachability analysis of nonlinear hybrid systems, where δ is an arbitrary used-defined parameter indicting error tolerance. In this sense, the results of this chapter (Sections 5.3 and 5.4) provide a δ-decision procedure for nonlinear controller synthesis with ω-regular specifications. In the next chapter, we provide an alternative and more memory-efficient approach that also provides a δ-decision procedure using specification-guided synthesis.

This brief account is by no means comprehensive. Indeed, formal methods captured considerable attention of researchers in the robotics and control communities over the past two decades. The readers are referred to recent surveys and monographs [25, 126, 150, 160, 184, 187, 222] and references therein for a more complete overview of this research trend.

Chapter 6

Specification-Guided Controller Synthesis via Direct Interval Computation

In this chapter, we introduce methods that directly perform control synthesis on the continuous-state dynamical system without first constructing finite abstractions. The main idea behind these methods is to under-approximate the controlled predecessors of specification-related sets for dynamical systems to a desired precision. For this reason, we often call this class of methods *specification-guided.* Interval analysis is the key tool to perform the direct set approximations for dynamical systems. While most of the specification-guided methods are developed for discrete-time systems, they can be extended to continuous-time dynamical systems.

6.1 Discrete-Time Dynamical Systems

We recall the nominal discrete-time control system

$$x(t+1) = f(x(t), u(t)), \tag{6.1}$$

where $x(t) \in X \subseteq \mathbb{R}^n$ is the system state, $u(t) \in U$ is the control input. In the presence of additive disturbances, system (6.1) above becomes

$$x(t+1) = f(x(t), u(t)) + w(t), \tag{6.2}$$

where $w(t) \in W$ is the additive disturbance input, and we assume that

$$W = \mathbb{B}_\delta := \{w \in \mathbb{R}^n \mid \|w\|_\infty \leq \delta\} \tag{6.3}$$

for some $\delta \geq 0$.

DOI: 10.1201/9780429270253-6

6.2 Properties of Controlled Predecessors

Similar to transition systems, computation of controlled predecessors is also a building block for the control synthesis via direct interval computation. In Section 4.4, we have introduced controlled predecessors for discrete-time system (6.2), i.e.,

$$\mathrm{Pre}^{\delta}(B) = \{x \in X \mid \exists u \in U, \forall w \in W \text{ s.t. } f(x,u) + w \in B\}, \tag{6.4}$$

where $B \subseteq X$ is a target set and $W = \mathbb{B}_{\delta}$ ($\delta \geq 0$) is a set of bounded disturbances. A controlled predecessor can also be interpreted as a map between sets of system states, i.e., the map Pre^{δ} projects the set B to a set of states that can be controlled into B within one time step.

Without further assumptions on system (6.2), we can additionally derive the following properties of the map Pre^{δ}. Recall that Pre^{δ} reduces to Pre when $\delta = 0$.

Proposition 6.1 *Let $A, B \subseteq X$ and $\delta \geq 0$. Then*

(i) $\mathit{Pre}^{\delta}(A) \subseteq \mathit{Pre}^{\delta}(B)$ *if* $A \subseteq B$,

(ii) $\mathit{Pre}^{\delta_2}(A) \subseteq \mathit{Pre}^{\delta_1}(A)$ *if* $0 \leq \delta_1 \leq \delta_2$,

(iii) $\mathit{Pre}^{\delta}(A) = \mathit{Pre}(A - \mathbb{B}_{\delta})$.

Proof: The first two properties are straightforward by (6.4). For (iii),

$$\begin{aligned} \mathrm{Pre}(A - \mathbb{B}_{\delta}) &= \{x \in X \mid f(x,u) \in A - \mathbb{B}_{\delta}\} \\ &= \{x \in X \mid f(x,u) + y \in A, \forall y \in \mathbb{B}_{\delta}\} \qquad \text{Expand } A - \mathbb{B}_{\delta} \\ &= \mathrm{Pre}^{\delta}(A). \end{aligned}$$

Hence, (iii) is proved. ■

Proposition 6.1 (i) indicates that the map Pre^{δ} is increasing, and (iii) implies that controlled predecessors for the perturbed system (6.2) with $\delta > 0$ can be constructed by using the Minkowski difference and by computing controlled predecessors for the nominal system (6.2) with $\delta = 0$.

Additionally, the generic properties in the following proposition also hold for the map Pre^{δ} in view of Proposition 6.1(i).

Proposition 6.2 *Let $A, B \subseteq Y \subseteq X$. If a function $h : 2^Y \to 2^Y$ satisfies $h(A) \subseteq h(B)$ for any $A \subseteq B$, then*

(i) $h(A \cap B) \subseteq h(A) \cap h(B)$;

(ii) $h(A) \cup h(B) \subseteq h(A \cup B)$.

Proof: See Appendix C.4 ■

If continuity is imposed to (6.2), then $\mathrm{Pre}^{\delta}(\cdot)$ will have more favorable properties for LTL control synthesis.

Assumption 6.1 *The state space $X \subseteq \mathbb{R}^n$ and the input space $U \in \mathbb{R}^m$ of system (6.2) are compact, and the function $f : X \times U \to X$ is continuous with respect to both arguments.*

When u plays the role of a switching parameter, as in (4.37), that takes a finite number of values, we assume that f in (6.2) satisfies the following assumption.

Assumption 6.2 *The state space $X \subseteq \mathbb{R}^n$ is compact and the input space U is finite for system (6.2), and the function $f : X \times U \to X$ is continuous with respect to the first argument.*

Under Assumption 6.1 or 6.2, the first noticeable property is that the map Pre^{δ} preserves open and closedness of a set as stated in the following proposition.

Proposition 6.3 *Suppose that Assumption 6.1 or 6.2 holds.*

(i) If $\Omega \subseteq X$ is closed (compact), then $Pre^{\delta}(\Omega)$ is closed (compact).

(ii) If $\Omega \subseteq X$ is open, then $Pre^{\delta}(\Omega)$ is open.

Proof: See Appendix C.5. ■

Similarly, the set of valid control inputs for each state in the controlled predecessor of a given subset in the state space is compact.

Proposition 6.4 *Suppose that Assumption 6.1 or 6.2 holds. The set $K^{\delta}_{\Omega}(x)$ is compact for all $x \in Pre^{\delta}(\Omega)$, where $\Omega \subseteq X$ is compact.*

Proof: See Appendix C.6. ■

In particular, for a decreasing sequence of compact subsets of the state space of system (6.2), the following distributive property of map Pre^{δ} under countable intersections can be shown.

Proposition 6.5 *Suppose that Assumption 6.1 or 6.2 holds. Let $\{A_i\}_{i=0}^{\infty}$ be a decreasing sequence of compact subsets of X. Then*

$$\bigcap_{i=0}^{\infty} Pre^{\delta}(A_i) = Pre^{\delta}\left(\bigcap_{i=0}^{\infty} A_i\right). \tag{6.5}$$

Proof: We prove (6.5) by showing

$$\bigcap_{i=0}^{\infty} \mathrm{Pre}^{\delta}(A_i) \subseteq \mathrm{Pre}^{\delta}\left(\bigcap_{i=0}^{\infty} A_i\right), \tag{6.6}$$

$$\mathrm{Pre}^{\delta}\left(\bigcap_{i=0}^{\infty} A_i\right) \subseteq \bigcap_{i=0}^{\infty} \mathrm{Pre}^{\delta}(A_i). \tag{6.7}$$

We show (6.7) first. For any $x \in \mathrm{Pre}^{\delta}(\bigcap_{i=0}^{\infty} A_i)$, there exists $u \in U$ such that $f(x,u) + w \in A_i$ for all $w \in W$ and $i \in \mathbb{Z}_{>0}$. By definition we also have $x \in \bigcap_{i=0}^{\infty} \mathrm{Pre}^{\delta}(A_i)$, which means that $\mathrm{Pre}^{\delta}(\bigcap_{i=0}^{\infty} A_i) \subseteq \bigcap_{i=0}^{\infty} \mathrm{Pre}^{\delta}(A_i)$.

To see (6.6), we aim to show that $x \in \mathrm{Pre}^{\delta}(\bigcap_{i=0}^{\infty} A_i)$ for all $x \in \bigcap_{i=0}^{\infty} \mathrm{Pre}^{\delta}(A_i)$. Let $x \in \bigcap_{i=0}^{\infty} \mathrm{Pre}^{\delta}(A_i)$ be arbitrary. Then there exists $u_i \in U$ such that $a_i = f(x, u_i) \in (A_i - W)$ for any fixed $i \in \mathbb{Z}_{>0}$. Now consider the sequences $\{u_i\}_{i=1}^{\infty}$ and $\{a_i\}_{i=1}^{\infty}$.

Under Assumption 6.1, there exists a convergent sub-sequence $\{u_{i_j}\}_{j=0}^{\infty}$ with $\lim_{j\to\infty} u_{i_j} = u \in U$, and $\{a_{i_j}\}_{j=1}^{\infty}$ is the corresponding sub-sequence of $\{a_i\}_{i=1}^{\infty}$. We can also find a convergent sub-sequence $\{a_{i_{j_k}}\}_{k=0}^{\infty}$ of $\{a_{i_j}\}_{j=1}^{\infty}$ with the limit point a, i.e.,

$$\lim_{k\to\infty} a_{i_{j_k}} = a, \quad a_{i_{j_k}} \in \left(A_{i_{j_k}} - W\right).$$

Since $A_i - W$ is closed for all $i \in \mathbb{Z}_{>0}$, we have $a \in \bigcap_{k=0}^{\infty} \left(A_{i_{j_k}} - W\right) = \bigcap_{i=0}^{\infty} (A_i - W)$. Then $a \in (\bigcap_{i=0}^{\infty} A_i) - W$ according to Proposition C.2 (iv). By the continuity of $f(x,\cdot)$, we have

$$a = \lim_{k\to\infty} f(x, u_{i_{j_k}}) = f(x, \lim_{k\to\infty} u_{i_{j_k}}) = f(x,u) \in \left(\bigcap_{i=0}^{\infty} A_i\right) - W.$$

Under Assumption 6.2, U is finite, and thus there exists a constant sub-sequence $\{u_{i_j}\}_{j=0}^{\infty}$ with $u_{i_j} = u$ for all $j \in \mathbb{Z}_{>0}$ such that $f(x,u) \in A_i - W$ for infinitely many i. Then $f(x,u) \in (\bigcap_{i=0}^{\infty} A_i) - W$.

Both assumptions all imply that there exists $u \in U$ such that $f(x,u) \in (\bigcap_{i=0}^{\infty} A_i) - W$ for the arbitrary $x \in \bigcap_{i=0}^{\infty} \mathrm{Pre}^{\delta}(A_i)$. Hence $x \in \mathrm{Pre}^{\delta}(\bigcap_{i=0}^{\infty} A_i)$. This completes the proof. ■

Similarly, a distributive property of map Pre^{δ} under countable unions of subsets of X can also be concluded.

Proposition 6.6 *Let $\{A_i\}_{i=0}^{\infty}$ be an increasing sequence of open subsets of X. Then*

$$\bigcup_{i=0}^{\infty} \mathit{Pre}^{\delta}(A_i) = \mathit{Pre}^{\delta}\left(\bigcup_{i=0}^{\infty} A_i\right). \tag{6.8}$$

Proof: It is easy to see that $\bigcup_{i=0}^{\infty} \mathrm{Pre}^{\delta}(A_i) \subseteq \mathrm{Pre}^{\delta}(\bigcup_{i=0}^{\infty} A_i)$ by Proposition 6.2 (ii). Hence, we only need to show that $\mathrm{Pre}^{\delta}(\bigcup_{i=0}^{\infty} A_i) \subseteq \bigcup_{i=0}^{\infty} \mathrm{Pre}^{\delta}(A_i)$.

Let $x \in \mathrm{Pre}^{\delta}(\bigcup_{i=0}^{\infty} A_i)$ be arbitrary. Then, by definition, there exists $u \in U$ such that $f(x,u) + w \in \bigcup_{i=0}^{\infty} A_i$ for all $w \in W$. By the Heine-Borel theorem (see C.2), there exists $i \in \mathbb{N}$ such that $f(x,u) + w \in A_i$ for all $w \in W$ since the set $\{x \in X \mid f(x,u) + w, \forall w \in W\}$ is compact and $A_0 \subseteq A_1 \subseteq \cdots$. It follows that $x \in \mathrm{Pre}^{\delta}(A_i) \subseteq \bigcup_{i=0}^{\infty} \mathrm{Pre}^{\delta}(A_i)$. Therefore, $\mathrm{Pre}^{\delta}(\bigcup_{i=0}^{\infty} A_i) \subseteq \bigcup_{i=0}^{\infty} \mathrm{Pre}^{\delta}(A_i)$, which completes the proof. ■

Note that Assumption 6.1 or 6.2 is not necessary for Proposition 6.6, but the sets A_i's (for all $i \in \mathbb{N}$) have to be open for (6.8) to hold.

It is nontrivial to exactly compute the controlled predecessor $\mathrm{Pre}(Y)$ because of the nonlinear dynamics. Only for some special cases, e.g., controlled predecessors of polyhedral sets with respect to linear dynamics, which can be characterized by linear inequalities, the exact computation is possible. Even for linear systems with polyhedral or ellipsoidal constraints, set operations such as Minkowski difference are likely to introduce irregular shapes, which makes computation of accurate reachable sets impossible. Therefore, one often has to seek approximations of $\mathrm{Pre}^{\delta}(Y)$.

6.3 Invariance Control

In many applications, such as voltage regulation of electrical power converters [78], room temperature stabilization inside a building [179], attitude control in flight control systems [70], the adaptive cruise control [134, 178], and lane-keeping problems for autonomous vehicles [9], the state or output of a control system is often required to be regulated to a *setpoint*. For a dynamical system in the form of (6.2), the regulation condition can be written mathematically as

$$\lim_{t \to \infty} \|x(t) - r\|_2 = 0, \tag{6.9}$$

where $r \in X$ is the setpoint.

According to the definition of the regulation problem as in (6.9), invariance and reachability control is another way to phrase this problem. The objective of *invariance control* is to maintain system state inside a given target area of the state space. In the presence of disturbances, the convergence in (6.9) cannot always be achieved, especially with additive disturbances (see (6.2)). Hence, it is more practical to consider a small region around the given setpoint. Set invariance [32] is also a paramount concept in constrained control where the controlled system trajectories are ideally inside an invariant set that is consistent with the given constraints so that the constraints would not be violated for all time.

Translated into an LTL formula, the invariance control specification can be written as $\Box\Omega$ as we have seen in Chapter 3 in the context of discrete synthesis.

Definition 6.1 *A set $\Omega \subseteq X$ is said to be **δ-robustly controlled invariant** for system (6.2) if, for any initial state $x_0 \in \Omega$, for all δ-bounded sequences of disturbances $\{w(t)\}_{t=0}^{\infty}$, i.e., $w(t) \in W$ for all $t \in \mathbb{N}$, there exists a control signal $\{u(t)\}_{t=0}^{\infty}$ such that the trace $L(\rho) \vDash \Box\Omega$, where $\rho = \{x(t)\}_{t=0}^{\infty}$ is the resulting solution of (6.2). If $\delta = 0$, then Ω is called controlled invariant for the nominal system (6.1).*

To check whether a set is (robustly) controlled invariant or not, we can rely on the following criterion based on controlled predecessors.

Proposition 6.7 ([32, 112]) *A set $\Omega \subseteq X$ is δ-robustly controlled invariant for system (6.2) if and only if $\Omega \subseteq Pre^{\delta}(\Omega)$, where Pre^{δ} is defined in (4.29).*

The given target set Ω in an invariance control problem is not necessarily (robustly) controlled invariant. If Ω is (robustly) controlled invariant itself, then $K_{\Omega}^{\delta}(x) \neq \emptyset$ for all $x \in \Omega$ and the function K_{Ω}^{δ} is a memoryless invariance control strategy. When Ω is not (robustly) controlled invariant, it is still possible to realize the invariance property by identifying subsets of Ω that are (robustly) controlled invariant. Among all such subsets, it is of interest to determine the maximal one, which constitutes the domain of the invariance control strategy.

Definition 6.2 *Let $\Omega \subseteq X$. The set $\mathcal{I}_{\infty}^{\delta}(\Omega)$ is said to be the **maximal δ-robustly controlled invariant set** inside Ω for system (6.2), if it is δ-robustly controlled invariant and contains all δ-robustly controlled invariant sets inside Ω. Specifically for system (6.1), such a set is called the maximal controlled invariant set inside Ω and denoted by $\mathcal{I}_{\infty}(\Omega)$.*

Even for a nominal system (6.1), finding the maximal controlled invariant set $\mathcal{I}_{\infty}(\Omega)$ is not always helpful in practice, because any degree of uncertainties involved in system dynamics will destroy the invariance property of $\mathcal{I}_{\infty}(\Omega)$. And it is still possible that some part of Ω can be controlled invariant under disturbances. Therefore, we often consider the robust version $\mathcal{I}_{\infty}^{\delta}(\Omega)$ as well. Consistent with Proposition 6.7, we have the following proposition to determine whether a set is robustly controlled invariant.

Definition 6.3 *A set $\Omega \subseteq X$ is said to be a **δ-robustly controlled invariant set** ($\delta \geq 0$) for system (6.1) if*

$$\Omega \subseteq Pre(\Omega - \mathbb{B}_{\delta}). \tag{6.10}$$

*We call Ω robustly controlled invariant if $\delta > 0$. The supremum of δ satisfying (6.10) is called the **robust invariance margin** of Ω.*

It is interesting to note that, by definition, the maximal controlled invariant set itself is not robustly controlled invariant. This also indicates that the determination of the maximal invariant set is numerically nontrivial because of approximation errors.

Proposition 6.8 *Let $\Omega \subseteq X$ be compact and $\mathcal{I}_\infty(\Omega)$ be the maximal controlled invariant set in Ω. Suppose that Assumption 6.1 or 6.2 holds and $\mathcal{I}_\infty(\Omega) \neq \Omega$. Then $\mathcal{I}_\infty(\Omega)$ is not robustly controlled invariant.*

Proof: We prove this by showing that some boundary points of $\mathcal{I}_\infty(\Omega)$ will be mapped into the boundary of $\mathcal{I}_\infty(\Omega)$. We only consider the case $\text{int}(\Omega) \neq \emptyset$; otherwise, the conclusion trivially holds by Definition 6.3 because $\text{int}(\mathcal{I}_\infty)(\Omega) = \emptyset$, where $\text{int}(\mathcal{I}_\infty)(\Omega)$ denotes the interior of set $\mathcal{I}_\infty(\Omega)$.

For the purpose of contradiction, we assume that $x \in (\partial\mathcal{I}_\infty(\Omega) \cap \text{int}(\Omega))$, and there exists a $u \in U$ such that $f(x, u) \in \text{int}(\mathcal{I}_\infty)(\Omega)$. That implies there exists a $r > 0$ such that $f(x, u) + \mathbb{B}_r \subseteq \mathcal{I}_\infty(\Omega)$. By continuity of $f(\cdot, u)$ (from Assumption 6.1 or 6.2), we can find a $\delta(r) > 0$ such that any $x' \in x + \mathbb{B}_{\delta(r)}$ satisfies $f(x', u) \in f(x, u) + \mathbb{B}_r$, and thus $f(x + \mathbb{B}_{\delta(r)}, u) \subseteq \mathcal{I}_\infty(\Omega)$, which means x is an interior point of $\mathcal{I}_\infty(\Omega)$. This is a contradiction. ∎

We now consider the determination of (robustly) controlled invariant sets, which is crucial in the construction of invariance control strategies.

Let Inv^δ ($\delta \geq 0$) be a map between subsets of $\mathbb{R}^n$ defined as

$$\text{Inv}^\delta(Y) = \text{Pre}^\delta(Y|Y), \quad Y \subseteq X. \tag{6.11}$$

We show in the next proposition that the maximal δ-robustly controlled invariant set inside a given compact set $\Omega \subseteq X$ can be obtained by using the following fixed-point iteration:

$$\begin{aligned} &\text{Inv}_0^\delta(\Omega) = \Omega, \\ &\text{Inv}_j^\delta(\Omega) = \text{Inv}^\delta(\text{Inv}_{j-1}^\delta(\Omega)), \end{aligned} \tag{6.12}$$

where Inv_j^δ ($j \in \mathbb{Z}_{>0}$) is the jth iterate of the map Inv^δ.

Proposition 6.9 ([27]) *Let $\Omega \subseteq X$ be closed and $\delta \geq 0$. Given Assumption 6.1 or 6.2,*

$$\mathcal{I}_\infty^\delta(\Omega) = \lim_{j \to \infty} Inv_j^\delta(\Omega) = \bigcap_{j=0}^{\infty} Inv_j^\delta(\Omega), \tag{6.13}$$

where $\mathcal{I}_\infty^\delta(\Omega)$ is the maximal δ-robustly controlled invariant set in Ω. Furthermore, $\mathcal{I}_\infty^\delta(\Omega)$ is the maximal fixed point of Inv^δ.

Proof: According to Proposition 6.3, if Ω is closed, then $\text{Pre}(\Omega - \mathbb{B}_\delta)$, and therefore $\text{Inv}_j^\delta(\Omega)$ ($\forall j \in \mathbb{Z}_{>0}$) is closed. By (6.11) and (6.12), $\{\text{Inv}_j^\delta\}_{j=0}^\infty$ is decreasing. Then Proposition C.1 shows that $\lim_{j\to\infty} I^j(\Omega) = \bigcap_{j=1}^{\infty} \mathrm{Inv}_j^{\delta}(\Omega)$ is closed and nonempty if $\mathrm{Inv}_j^{\delta}(\Omega) \neq \emptyset$ for all $j \in \mathbb{N}$.

First, we claim $\bigcap_{j=1}^{\infty} \mathrm{Inv}_j^{\delta}(\Omega) \subseteq \mathcal{I}_{\infty}^{\delta}(\Omega)$ by showing that $\bigcap_{j=1}^{\infty} \mathrm{Inv}_j^{\delta}(\Omega)$ is δ-robustly controlled invariant. For all $j \in \mathbb{Z}_{>0}$, we have

$$\mathrm{Inv}_j^{\delta}(\Omega) = \mathrm{Inv}_{j-1}^{\delta}(\Omega) \cap \mathrm{Pre}^{\delta}(\mathrm{Inv}_{j-1}^{\delta}(\Omega)) \subseteq \mathrm{Pre}^{\delta}(\mathrm{Inv}_{j-1}^{\delta}(\Omega)).$$

Then $\bigcap_{j=0}^{\infty} \mathrm{Inv}_j^{\delta}(\Omega) \subseteq \bigcap_{j=1}^{\infty} \mathrm{Pre}^{\delta}(\mathrm{Inv}_{j-1}^{\delta}(\Omega))$. Since the sequence $\{\mathrm{Inv}_j^{\delta}(\Omega)\}_{j=0}^{\infty}$ is decreasing, by Proposition 6.5,

$$\bigcap_{j=1}^{\infty} \mathrm{Pre}^{\delta}(\mathrm{Inv}_{j-1}^{\delta}(\Omega)) = \mathrm{Pre}^{\delta}(\bigcap_{j=0}^{\infty} \mathrm{Inv}_j^{\delta}(\Omega)) = \mathrm{Pre}(\bigcap_{j=0}^{\infty} \mathrm{Inv}_j^{\delta}(\Omega) - \mathbb{B}_{\delta}).$$

Hence, $\bigcap_{j=0}^{\infty} \mathrm{Inv}_j^{\delta}(\Omega) \subseteq \mathrm{Pre}(\bigcap_{j=0}^{\infty} \mathrm{Inv}_j^{\delta}(\Omega) - \mathbb{B}_{\delta})$, which proves the claim.

Next, we show that $\mathcal{I}_{\infty}^{\delta}(\Omega) \subseteq \bigcap_{j=1}^{\infty} \mathrm{Inv}_j^{\delta}(\Omega)$. We assume that $\mathcal{I}_{\infty}^{\delta}(\Omega) - \mathbb{B}_{\delta} \neq \emptyset$; otherwise, $\mathcal{I}_{\infty}^{\delta}(\Omega) = \emptyset$, which means the conclusion trivially holds. We now use induction. For $j = 0$, we have $\mathcal{I}_{\infty}^{\delta}(\Omega) \subseteq \mathrm{Inv}_0^{\delta}(\Omega) = \Omega$. Suppose that $\mathcal{I}_{\infty}^{\delta}(\Omega) \subseteq \mathrm{Inv}_j^{\delta}(\Omega)$ for some $j \in \mathbb{N}$. By Proposition 6.7, for any $x \in (\mathrm{Inv}_j^{\delta}(\Omega) \setminus \mathrm{Inv}_{j+1}^{\delta}(\Omega))$, $f(x,u) \notin (\mathrm{Inv}_j^{\delta}(\Omega) - \mathbb{B}_{\delta})$ for all $u \in U$, which also means $f(x,u) \notin (\mathcal{I}_{\infty}^{\delta}(\Omega) - \mathbb{B}_{\delta})$. By the definition of $\mathcal{I}_{\infty}^{\delta}(\Omega)$, $x \notin \mathcal{I}_{\infty}^{\delta}(\Omega)$. It follows that $\mathcal{I}_{\infty}^{\delta}(\Omega) \subseteq I_r^{j+1}(\Omega)$. Hence, $\mathcal{I}_{\infty}^{\delta}(\Omega) \subseteq \bigcap_{j=1}^{\infty} \mathrm{Inv}_j^{\delta}(\Omega)$.

Last, to see that $\mathcal{I}_{\infty}^{\delta}(\Omega)$ is a maximal fixed point of Inv^{δ}, it is sufficient to show that a set $Y \subseteq \Omega$ is a fixed point of Inv^{δ} if and only if Y is a δ-robustly controlled invariant set. If $Y \subseteq \Omega$ is a δ-robustly controlled invariant set, i.e., $Y \subseteq \mathrm{Pre}(Y - \mathbb{B}_{\delta})$, then $\mathrm{Inv}^{\delta}(Y) = \mathrm{Pre}(Y - \mathbb{B}_{\delta}|Y) = Y$. On the other side, if $\mathrm{Inv}^{\delta}(Y) = \mathrm{Pre}(Y - \mathbb{B}_{\delta}|Y) = Y$, then we have $Y \subseteq \mathrm{Pre}(Y - \mathbb{B}_{\delta})$, which means that Y is a δ-robustly controlled invariant set. ■

Proposition 6.9 gives a fixed-point algorithm to determine the maximal (robustly) controlled invariant set inside a given target set. The actual computation, however, relies on how sets are represented and set operations are performed.

Consider linear time-invariant (LTI) systems in which (6.1) is of the form

$$f(x(t), u(t)) + w(t) = Ax(t) + Bu(t) + w(t),$$

where $A \in \mathbb{R}^{n\times n}$, $B \in \mathbb{R}^{n\times m}$, and $w(t) \in W = \mathbb{B}_{\delta}$, as given in (6.3).

If both the subtracted target set $\Omega - W$ and the set of control inputs U are polyhedra that are given by

$$\begin{aligned} \Omega - W &= \{x \in \mathbb{R}^n \mid Hx \leq h\}, \quad H \in \mathbb{R}^{l_1\times n}, h \in \mathbb{R}^{l_1} \\ U &= \{u \in \mathbb{R}^m \mid Gu \leq g\}, \quad G \in \mathbb{R}^{l_2\times m}, g \in \mathbb{R}^{l_2}, \end{aligned}$$

where $l_1, l_2 \in \mathbb{Z}_{>0}$ are the numbers of inequalities determining the polyhedra

Ω and U, respectively, then the controlled predecessor of Ω is

$$\mathrm{Pre}^{\delta}(\Omega) = \mathrm{Pre}(\Omega - W) = \left\{ x \in \mathbb{R}^n \;\middle|\; \begin{bmatrix} HA & HB \\ 0 & G \end{bmatrix} \begin{bmatrix} x \\ u \end{bmatrix} \leq \begin{bmatrix} h \\ g \end{bmatrix},\ u \in \mathbb{R}^m \right\}. \tag{6.14}$$

The set $\mathrm{Pre}^{\delta}(\Omega)$ and hence $\mathrm{Inv}^{\delta}(\Omega)$ is also polyhedral, because the intersection of polyhedra is still a polyhedron. Then the iterations of (6.12) can go on without losing the polyhedral properties. However, the maximal (robustly) controlled invariant set is not necessarily polyhedral. The following example illustrates such a case.

Example 6.1 *Consider an LTI system* $x(t+1) = Ax(t)$, *where*

$$A = \begin{bmatrix} 1.0810 & 0.4517 \\ -0.0903 & 0.7197 \end{bmatrix}.$$

With a pair of complex eigenvalues $0.9003 \pm 0.0903i$, *this LTI system is globally stable. Hence, there exists a (controlled) invariant set inside* $\Omega = [-1,1] \times [-1,1]$. *However,* Ω *itself is not (controlled) invariant. This is because the system trajectories are spiral and some of them will leave* Ω *provisionally although they will eventually converge to the origin* $(0,0)$.

Represented by a polyhedron, $\Omega = \{x \in \mathbb{R}^n \mid Hx \leq h\}$, *where*

$$H = \begin{bmatrix} 1 & 0 \\ -1 & 0 \\ 0 & 1 \\ 0 & -1 \end{bmatrix}, \quad h = \begin{bmatrix} 1 \\ 1 \\ 1 \\ 1 \end{bmatrix}.$$

By (6.14), we have

$$Inv_1(\Omega) = Pre(\Omega|\Omega) = \left\{ x \in \mathbb{R}^n \;\middle|\; \begin{bmatrix} HA \\ H \end{bmatrix} x \leq \begin{bmatrix} h \\ h \end{bmatrix} \right\},$$

which is a new polyhedron $\{x \in \mathbb{R}^n \mid H_1 x \leq h_1\}$, *where*

$$H_1 = \begin{bmatrix} HA \\ H \end{bmatrix} = \begin{bmatrix} 1.0810 & 0.4517 \\ -1.0810 & -0.4517 \\ -0.0903 & 0.7197 \\ 0.0903 & -0.7197 \\ 1 & 0 \\ -1 & 0 \\ 0 & 1 \\ 0 & -1 \end{bmatrix}, \quad h_1 = \begin{bmatrix} h \\ h \end{bmatrix} = \begin{bmatrix} 1 \\ 1 \\ 1 \\ 1 \\ 1 \\ 1 \\ 1 \\ 1 \end{bmatrix}.$$

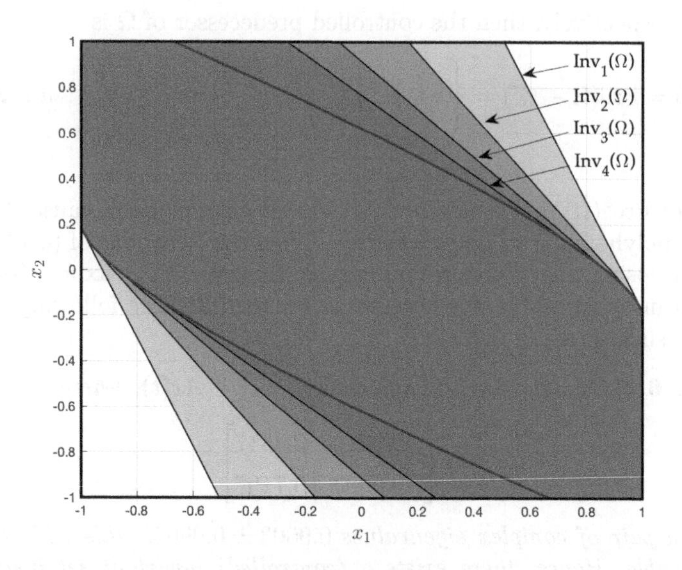

FIGURE 6.1: The set $\text{Inv}_i(\Omega)$ for $i = 0, 1, 2, 3, 4$. The outermost box $[-1, 1] \times [-1, 1]$ represents the initial set $\text{Inv}_0(\Omega) = \Omega$. The two innermost lines are the real boundaries of the maximal (controlled) invariant set inside Ω.

The polyhedral sets obtained within the first 4 iterations are shown in Figure 6.1. It can be observed that the polyhedral set $Inv_i(\Omega)$ ($i \in \mathbb{N}$) keeps shrinking toward the real maximal (controlled) invariant set $\mathcal{I}_\infty(\Omega)$, which is bounded by two innermost boundary lines. It is also clear that $\mathcal{I}_\infty(\Omega)$ is not a polyhedron, and this implies that we will never achieve $\mathcal{I}_\infty(\Omega)$ within a finite number of iterations.

For a general nonlinear form of (6.1), the computation of $\text{Pre}(\Omega)$ is not as easy as for the LTI case, let alone the possibility of terminating in a finite number of iterations.

Example 6.2 *Consider a discrete-time version of a second-order nonlinear system taken from [114, Example 8.6] as follows:*

$$\begin{aligned} x(t+1) &= x(t) + 0.1y(t) \\ y(t+1) &= -0.1x(t) + 0.033x(t)^3 + 0.9y(t). \end{aligned}$$

It has three isolated equilibrium points at $(0, 0)$, $(\sqrt{3}, 0)$, and $(-\sqrt{3}, 0)$. The region between the manifolds that pass through $(\sqrt{3}, 0)$ and $(-\sqrt{3}, 0)$ is the maximal positively invariant set, which is difficult to express analytically.

Finding the (robustly) maximal controlled invariant set is equivalent to determining the winning set for system (6.2) with respect to invariance specification φ_s. If we keep track of the valid control inputs during iterations, the corresponding control strategy can be constructed.

As we have seen, controlled predecessors, which are the main elements of (6.12), are not easy to compute precisely under nonlinear dynamics. Even for linear systems, the fixed-point iteration (6.12) may not terminate in finite time (see Example 6.1). So a workaround is to use under-approximations of controlled predecessors instead, so that the resulting control strategy is well defined for all the states inside the approximated winning set. The approximated invariance control algorithm had a better guarantee of some bound for the approximation error of the winning set. Looking back at Section 4.4, the interval under-approximation (4.32) of the controlled predecessor $\mathrm{Pre}(B|A)$ works for general control systems and provides a lower bound of the approximation.

Algorithm 7 $[Y]^\varepsilon = \textsc{InvControl}([f], \Omega, U, \varepsilon)$

1: $Y \leftarrow \emptyset$, $\widetilde{Y} \leftarrow \Omega$
2: **while** $Y \neq \widetilde{Y}$ **do**
3: $\quad Y \leftarrow \widetilde{Y}$
4: $\quad \underline{Y}, \Delta Y, Y_c = \textsc{Pre}([f], Y, Y, U, \varepsilon)$
5: $\quad \widetilde{Y} \leftarrow \underline{Y}$
6: **end while**
7: **return** Y

Algorithm 7 is an interval-based invariance control algorithm, which calls Algorithm 5 in each iteration. We shall see from the following theorems that it is sound and complete for system (6.1) (in the sense that the invariance specification can be satisfied by system (6.2)).

Theorem 6.1 (Soundness and Completeness via Robustness) *Assume that the control space U is finite for system (6.1). Let $\Omega \subseteq X$ be compact and Assumption 4.1 holds with the Lipschitz constant ρ. Then Algorithm 7 terminates in a finite number of steps. Furthermore, if $\rho\varepsilon \leq \delta$, the output $[Y]^\varepsilon$ for a given precision $\varepsilon > 0$ satisfies:*

(i) If $[Y]^\varepsilon = \emptyset$, then no δ-robustly controlled invariant subset of Ω exists for system (6.1).

(ii) If $[Y]^\varepsilon \neq \emptyset$, then $[Y]^\varepsilon$ is controlled invariant, and

$$\mathcal{I}^\delta_\infty(\Omega) \subseteq [Y]^\varepsilon \subseteq \mathcal{I}_\infty(\Omega).$$

Proof: Let $Y_{c,j}$, ΔY_j, and Y_j be Y_c, ΔY and Y in the jth iteration ($j \in \mathbb{N}$), respectively. Let $Y_0 = \Omega$, and $Y_{j+1} = Y_j \setminus (\Delta Y_j \cup Y_{c,j})$. If $Y_{j+1} \neq Y_j$ for any j, then $(\Delta Y_j \cup Y_{c,j}) \neq \emptyset$, and $\{Y_j\}$ is strictly decreasing. Under a given precision

ε and the compactness of Ω, Y_j contains a finite number of intervals. Then there must exists an $N \in \mathbb{Z}_{>0}$ such that $Y_N = \emptyset$, which results in $Y_{N+1} = \emptyset$. It means that $Y_j = \widetilde{Y_j}$ for all $j \geq N$. Hence, this algorithm will terminate in finite iterations.

To prove (i) and (ii), suppose $\{X_j\}$ and $\{X_j^\delta\}$ is the sequence generated by (6.12) for system (6.1) and (6.2), respectively. Then $X_0 = X_0^\delta = \Omega$, $X_{j+1} = \text{Pre}(X_j|X_j)$, and $X_{j+1}^\delta = \text{Pre}(X_j - \mathbb{B}_\delta|X_j)$. By Lemma 4.2, $Y_0 = X_0 = \Omega$, $\text{Pre}(Y_j - \mathbb{B}_{\rho\varepsilon}|Y_j) \subseteq Y_{j+1} \subseteq \text{Pre}(Y_j|Y_j)$. Hence, $\{X_j\}$,$\{X_j^\delta\}$, and $\{Y_j\}$ are non-increasing, and $X_j^\delta \subseteq Y_j \subseteq X_j$ for all $j \in \mathbb{Z}_{\geq 0}$.

If $\underline{Y}^\varepsilon = \emptyset$, then there exists some integer $N > 0$ such that $Y_N = \emptyset$. It follows that $\mathcal{I}_\infty^\delta(\Omega) = \bigcap_{j=1}^N \text{Inv}_j^\delta(\Omega) = \emptyset$. Hence, (i) is proved.

If $\underline{Y}^\varepsilon \neq \emptyset$, then there exists an integer $J > 0$ such that $\underline{Y}^\varepsilon = Y_J = Y_{J+1} \supseteq X_J \supseteq (\bigcap_{j=1}^\infty \text{Inv}^\delta(\Omega)) = \mathcal{I}_\infty^\delta(\Omega)$. Since $Y_J = Y_{J+1} \subseteq \text{Pre}(Y_J|Y_J)$, we have $Y_J \subseteq \text{Pre}(Y_J)$, i.e., Y_J is controlled invariant by definition. Hence $Y_J \subseteq \mathcal{I}_\infty(\Omega)$. That completes the proof. ∎

Without extra effort, one can show the following theorem for system (6.1) with compact control space based on Lemma 4.3.

Theorem 6.2 (Soundness and Completeness via Robustness) *Assume that the control space U is infinite but compact for system (6.1). Let $\Omega \subseteq X$ be compact and Assumptions 4.1 and 4.2 hold with the Lipschitz constants ρ_1 and ρ_2, respectively. Then Algorithm 7 with $U = [U]_\mu$ ($\mu > 0$) terminates in a finite number of steps. Furthermore, if $(\rho_1\varepsilon + \rho_2\mu) \leq \delta$, the output $[Y]^\varepsilon$ satisfies:*

(i) If $[Y]^\varepsilon = \emptyset$, then no δ-robustly controlled invariant subset of Ω exists for system (6.1).

(ii) If $[Y]^\varepsilon \neq \emptyset$, then $[Y]^\varepsilon$ is controlled invariant, and

$$\mathcal{I}_\infty^\delta(\Omega) \subseteq [Y]^\varepsilon \subseteq \mathcal{I}_\infty(\Omega).$$

Theorem 6.1 or 6.2 additionally suggests that robustly controlled invariance is a sufficient condition for under-approximating the maximal controlled invariant set. It can also be inferred that, for any nonempty controlled invariant set that is not robustly invariant, the approximation of controlled predecessors under any precision fails to give a solid approximation. This is because there does not exist a positive real number as the tolerance for the set approximation error. The following example is such a scenario.

Example 6.3 *Consider a discrete-time system $x(t+1) = A_\theta x(t)$, where*

$$A_\theta = \begin{bmatrix} \cos\theta & -\sin\theta \\ \sin\theta & \cos\theta \end{bmatrix}.$$

Every state moves on a circle centered at the origin. By using any under-approximation of Pre, the result of iteration (6.12) will be an empty set, since Ω is not robustly invariant for this system.

If Algorithm 7 returns a non-empty set, then an invariance controller can be extracted by using the valid control inputs recorded when performing Algorithm 5 each time.

Corollary 6.1 *Let the assumptions in Theorem 6.1 (or 6.2) hold and $M = U$ (or $M = [U]_\mu$) be the finite input set. If Ω has a δ-robustly ($\delta > 0$) controlled invariant set, then there exists a partition $\mathcal{P} = \{P_1, P_2, \cdots, P_N\}$ of Ω and an invariance control strategy $\kappa : \Omega \to 2^M$ with*

$$\kappa(x) = \bigcup_{i \in N} \psi_{P_i}(x), \quad x \in \Omega. \tag{6.15}$$

The map ψ_{P_i} is given by

$$\psi_{P_i}(x) = \begin{cases} \emptyset & x \notin P_i, \\ U_i & x \in P_i, \end{cases} \tag{6.16}$$

where $U_i \subseteq M$ for $i \in \{1, \cdots, N\}$.

Proof: By Theorem 6.1 (or 6.2), for any $\delta > 0$, there exists ε with $0 < \varepsilon \leq \delta/\rho$ (or $\varepsilon\rho_1 + \mu\rho_2 \leq \delta$) such that Algorithm 7 returns a controlled invariant set that are represented by union of intervals, which forms a partition $\mathcal{P}$ of Ω with valid control inputs $[K]_P^\varepsilon(x)$, which is defined in (4.35), for any $P \in \mathcal{P}$. Then $\bigcup_{P \in \mathcal{P}} [K]_P^\varepsilon(x)$ can be used as the set U_i of control inputs for Y_i to realize the controlled invariance according to Algorithm 7. ■

Remark 6.1 *In the following sections, we will focus on the results for systems with compact but infinite control space since the finite control space is a special case of the former with down-sampled control space.*

6.4 Reachability Control

Reachability control deals with the situations where the initial condition x_0 is outside a prescribed target set $\Omega \subseteq X$, and the goal is to steer the system state to Ω in finite time. The corresponding LTL formula for reachability control is $\Diamond\Omega$ (see also Chapter 3). Similar to invariance control problems, we wish to determine the winning set of system (6.2). The following definition interprets the winning set in this special case.

Definition 6.4 *Let Ω be a subset of the state space X of system (6.2). A set $\mathcal{BR}^{\delta}_{\infty}(\Omega) \subseteq X$ is said to be the* ***maximal δ-robustly backward reachable set*** *of system (6.2) from Ω if it contains (and only contains) any initial state $x_0 \in X$ that satisfies: for all δ-bounded sequences of disturbances $\{w(t)\}_{t=0}^{\infty}$, there exists a control signal $\{u(t)\}_{t=0}^{\infty}$ such that $x(t) \in \Omega$ for some $t \in \mathbb{N}$.*

For nominal system (6.1), the winning set with respect to the reachability specification is the backward reachable set $\mathcal{BR}_{\infty}(\Omega)$.

It is worth noting that the integer k and the control signal are dependent on the sequence of disturbances in Definition 6.4. In other words, the minimum time step for any initial state $x_0 \in \mathcal{BR}^{\delta}_{\infty}(\Omega)$ to be controlled into Ω can be different given different sequences of disturbances.

Definition 6.5 *A set $\Omega \subseteq X$ is said to be* ***δ-robustly reachable*** *for system (6.2) (or* ***reachable*** *for system (6.1)) if $\mathcal{BR}^{\delta}_{\infty}(\Omega) \neq \emptyset$.*

We now introduce the following definition for the characterization of $\mathcal{BR}^{\delta}_{\infty}(\Omega)$.

Definition 6.6 *The* ***N-step δ-robustly backward reachable set*** *of system (6.2) from a target set $\Omega \subseteq X$ is a set of initial states from which Ω can be reached within N ($N \in \mathbb{N}$) steps for any possible sequence of disturbances, i.e.,*

$$\begin{aligned}\mathcal{BR}^{\delta}_{N}(\Omega) = \{x \in X \mid \forall \{w_i\}_{i=0}^{N} (w_i \in W), \exists \{u_i\}_{i=0}^{N} \text{ s.t.} \\ x_0 = x,\ x_k \in \Omega, 0 \leq k \leq N \text{ for } \{x_i\}_{i=0}^{N} \text{ by (6.2)}\}.\end{aligned} \tag{6.17}$$

For nominal system (6.1), the N-step and maximal δ-robustly backward reachable set $\mathcal{BR}^{\delta}_{N}(\Omega)$ and $\mathcal{BR}^{\delta}_{\infty}(\Omega)$ are reduced to the N-step backward reachable set $\mathcal{BR}_{N}(\Omega)$ and maximal backward reachable set $\mathcal{BR}_{\infty}(\Omega)$, respectively.

Define Rch^{δ} as a map between subsets of $\mathbb{R}^n$:

$$\mathrm{Rch}^{\delta}(Y) = \mathrm{Pre}^{\delta}(Y) \cup Y, \quad Y \subseteq \mathbb{R}^n. \tag{6.18}$$

Let us now consider the fixed-point iteration ($i \in \mathbb{Z}_{>0}$):

$$\begin{aligned}&\mathrm{Rch}^{\delta}_{0}(\Omega) = \Omega, \\ &\mathrm{Rch}^{\delta}_{j}(\Omega) = \mathrm{Rch}^{\delta}(\mathrm{Rch}^{\delta}_{j-1}(\Omega)).\end{aligned} \tag{6.19}$$

Straightforwardly, the sequence $\left\{\mathrm{Rch}^{\delta}_{j}(\Omega)\right\}_{j=0}^{\infty}$ is increasing and with a slight use of induction, we can conclude the following result.

Proposition 6.10 *Given system (6.2) and a subset $\Omega \subseteq X$, we have*

$$\mathcal{BR}^{\delta}_{N}(\Omega) = \mathit{Rch}^{\delta}_{N}(\Omega), \quad \forall N \in \mathbb{N}, \tag{6.20}$$

where Rch^{δ} and $\mathit{Rch}^{\delta}_{N}(\Omega)$ are defined in (6.18) and (6.19), respectively.

Proof: We show it by induction. The base case holds because by (6.19) $\mathrm{Rch}_0^\delta(\Omega) = \Omega$, which is the set of states that can be controlled into Ω under any allowable disturbance in 0 steps. Suppose that $\mathrm{Rch}_j^\delta(\Omega)$ is the j-step δ-robustly backward reachable set. By (6.19) and Definition 4.15, we have

$$\mathrm{Rch}_{j+1}^\delta(\Omega) = \mathrm{Pre}^\delta(\mathrm{Rch}_j^\delta(\Omega)) \cup \mathrm{Rch}_j^\delta(\Omega),$$

which additionally includes all the states that can be controlled inside $\mathrm{Rch}_j^\delta(\Omega)$ in one step under any allowable disturbance. Hence, $\mathrm{Rch}_{j+1}^\delta(\Omega)$ is the $(j+1)$-step δ-robustly backward reachable set and the claim is proved. ■

As we have seen in Example 6.1, for LTI systems and a given polyhedral target set Ω, the set $\mathrm{Inv}_j^\delta(\Omega)$ in each iteration $j \in \mathbb{N}$ can be exactly computed. Computing robustly backward reachable sets $\mathrm{Rch}_j^\delta(\Omega)$, however, is not as easy. This is because the set union in (6.18) for the computation of $\mathrm{Rch}_j^\delta(\Omega)$ does not keep the shape of polyhedra.

Example 6.4 *Consider a discrete-time double integrator under disturbances [112]:*

$$x(t+1) = \begin{bmatrix} 1 & 1 \\ 0 & 1 \end{bmatrix} x(t) + \begin{bmatrix} 0.5 \\ 1 \end{bmatrix} u(t) + w(t), \tag{6.21}$$

where the state x represents the position and velocity, and

$$u(t) \in U = \left\{ u \in \mathbb{R}^2 \mid \|u\|_\infty \le 1 \right\},$$

$$w(t) \in W = \left\{ w \in \mathbb{R}^2 \mid \|w\|_\infty \le 0.1 \right\}.$$

System (6.21) is a sampled-data version (with sampling time $\tau = 1$) of the following ordinary differential equations that model the acceleration of an object

$$\dot{x} = \begin{bmatrix} 0 & 1 \\ 0 & 0 \end{bmatrix} x + \begin{bmatrix} 0 \\ 1 \end{bmatrix} u, \quad u \in [-1, 1]. \tag{6.22}$$

We consider a target reach set $\Omega = [-0.3, 0.3] \times [-0.3, 0.3]$ on the state space $X = [-8, 8] \times [-4, 4]$. As in Example 6.1, the first 4 iterations are plotted in Figure 6.2. The 1-step δ-robustly backward reachable set $Rch_1^\delta(\Omega) = Pre^\delta(\Omega) \cup \Omega$ is the union of the rectangle and the innermost polytope, which is concave and hence not a polytope. Specifically in this case, the backward reachable sets are not polyhedral until $Rch_4^\delta(\Omega)$, which includes all the previous sets.

Based on (6.20), we can further characterize the maximal backward reachable set $\mathcal{BR}_\infty^\delta(\Omega)$ according to the following proposition.

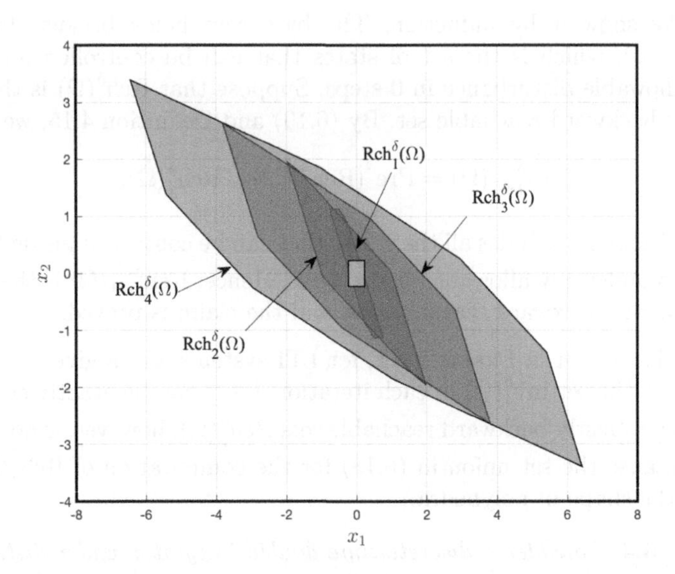

FIGURE 6.2: The 4-step δ-robustly backward reachable set. The small rectangle in the middle is the target set Ω.

Proposition 6.11 *For system (6.2), let $\Omega \subseteq X$ be open and $\delta \geq 0$. Then*

$$\mathcal{BR}_\infty^\delta(\Omega) = \bigcup_{j=0}^{\infty} Rch_j^\delta(\Omega), \tag{6.23}$$

and $\mathcal{BR}_\infty^\delta(\Omega)$ is a fixed point of the map Rch^δ.

Proof: The direction $\mathcal{BR}_\infty^\delta(\Omega) \supseteq \bigcup_{i=0}^{\infty} \mathrm{Rch}_j^\delta(\Omega)$ is clear because $\mathrm{Rch}_j^\delta(\Omega) \subseteq \mathcal{BR}_\infty^\delta(\Omega)$ for all $j \in \mathbb{N}$.

To show that $\mathcal{BR}_\infty^\delta(\Omega) \subseteq \bigcup_{j=0}^{\infty} \mathrm{Rch}_j^\delta(\Omega)$, we first claim that $\bigcup_{j=0}^{\infty} \mathrm{Rch}_j^\delta(\Omega)$ is a fixed point of Rch^δ. Let $A_j = \mathrm{Rch}_j^\delta(\Omega) \subseteq X$. Then $\{A_j\}_{j=0}^{\infty}$ is open and increasing, and by Proposition 6.6, we have

$$\begin{aligned}\mathrm{Rch}^\delta(\bigcup_{j=0}^{\infty} \mathrm{Rch}_j^\delta(\Omega)) &= \mathrm{Pre}^\delta\left(\bigcup_{j=0}^{\infty} A_j\right) \cup \bigcup_{j=0}^{\infty} A_j = \bigcup_{i=0}^{\infty} \mathrm{Pre}^\delta(A_j) \cup \bigcup_{j=0}^{\infty} A_j \\ &= \bigcup_{j=0}^{\infty} \left(\mathrm{Pre}^\delta(A_j) \cup A_j\right) = \bigcup_{j=0}^{\infty} \mathrm{Rch}_{j+1}^\delta(\Omega) = \bigcup_{j=0}^{\infty} \mathrm{Rch}_j^\delta(\Omega).\end{aligned}$$

We now show $x \notin \mathcal{BR}_\infty^\delta(\Omega)$ for all $x \notin \bigcup_{j=0}^{\infty} \mathrm{Rch}_j^\delta(\Omega)$. Let $x_0 \notin \bigcup_{j=0}^{\infty} \mathrm{Rch}_j^\delta(\Omega)$ be arbitrary. Then $x_0 \notin \mathrm{Pre}^\delta(\bigcup_{j=0}^{\infty} \mathrm{Rch}_j^\delta(\Omega))$ because $\bigcup_{j=0}^{\infty} \mathrm{Rch}_j^\delta(\Omega)$ is a fixed point of Rch^δ. This means that for all $u_0 \in U$ there exists $w_0 \in W$ (depending on u_0) such that $x_1 = f(x_0, u_0) + w_0 \notin \bigcup_{j=0}^{\infty} \mathrm{Rch}_j^\delta(\Omega)$

and thus $x_1 \notin \Omega$. Similarly for x_1, for all $u_1 \in U$ we can find $w_1 \in W$ such that $x_2 = f(x_1, u_1) + w_1 \notin \Omega$. Therefore, we can construct a sequence of disturbances $\{w(t)\}_{t=0}^{\infty}$ such that $x(t) \notin \Omega$ for all $\{u(t)\}_{t=0}^{\infty}$ and all $t \in \mathbb{N}$, which implies that $x \notin \mathcal{BR}_{\infty}^{\delta}(\Omega)$.

Hence (6.23) is proved, and the result that $\mathcal{BR}_{\infty}^{\delta}(\Omega)$ is a fixed point of the map Rch^{δ} follows straightforwardly. ■

Note that the target set Ω has to be open in order that algorithm (6.23) yields the maximal robustly backward reachable set, as opposed to Proposition 6.9. This is not surprising because reachability is the dual of invariance, which requires compactness of the target set. To explain why (6.20) does not apply to a closed target set in general, we give the following counter example.

Example 6.5 *Let $a_1 \approx 0.1127$ and $a_2 \approx 0.8873$ be the roots of $a = a^2 + 0.1$. We consider a target set $\Omega = [0, 0.2] \cup \{a_2\}$ for the system*

$$x(t+1) = \begin{cases} x(t)^2 + w(t) & x(t) \in [0, a_2], \\ a_2^2 + w(t) & x(t) \in (a_2, 1], \end{cases}$$

where $x(t) \in X = [0, 1]$, $w(t) \in W = [0, 0.1]$.

Then the trajectories with the initial condition $x_0 \in [0, a_2)$ will enter the region $[0, a_2]$ asymptotically under all possible sequences of disturbances. Hence, we have

$$\bigcup_{j=0}^{\infty} Rch_j^{\delta}(\Omega) = [0, a_2).$$

However, the real maximal robustly backward reachable set is the entire state space X since all $x \in (a_2, 1]$ is mapped within $[0, a_2]$ and a_2 is the point backward reachable from Ω.

Controlling the system state into an open set is usually required in the applications of reachability control. Even if the target set Ω is sometimes given as a closed set, it is always safer to design the reachability control strategy with respect to the interior part of Ω.

Facing the same problem as in invariance control, the exact N-step backward reachable set $\text{Rch}_N^{\delta}(\Omega)$ are often difficult to obtain even for linear systems with polyhedral target set and state, control constraints (see also Example 6.4). However, with interval approximations of controlled predecessors, an approximation of the maximal backward reachable set of a given target set is not hard to find by using the following Algorithm 8, where the input set X is the system state space.

Similar to invariance control synthesis, reachability control can also be sound and robustly complete.

Theorem 6.3 (Soundness and Robust Completeness) *Let $\Omega \subseteq X$ be open and Assumptions 4.1 and 4.2 hold with the Lipschitz constants ρ_1 and*

Algorithm 8 $[Y]^\varepsilon = \text{REACHCONTROL}([f], \Omega, X, U, \varepsilon)$

1: $Y \leftarrow \emptyset$, $\widetilde{Y} \leftarrow \Omega$, $A \leftarrow X$
2: **while** $Y \neq \widetilde{Y}$ **do**
3: $\quad Y \leftarrow \widetilde{Y}$
4: $\quad \underline{Y}, \Delta Y, Y_c = \text{PRE}([f], Y, A, U, \varepsilon)$
5: $\quad \widetilde{Y} \leftarrow \widetilde{Y} \cup \underline{Y}$
6: $\quad A \leftarrow \Delta Y \cup Y_c$
7: **end while**
8: **return** Y

ρ_2 for system (6.1), respectively. Then Algorithm 8 with $U = [U]_\mu$ terminates in a finite number of steps. Furthermore, if ε, μ satisfy $(\rho_1\varepsilon + \rho_2\mu) \leq \delta$, then

$$\mathcal{BR}^\delta_\infty(\Omega) \subseteq [Y]^\varepsilon \subseteq \mathcal{BR}_\infty(\Omega). \tag{6.24}$$

Proof: Let Y_j and A_j be the jth iteration of Y and A in Algorithm 8, respectively. The proof of finite termination is similar to Theorem 6.1: If $Y_{j+1} \neq Y_j$ for any j, then $\{A_j\}$ is strictly decreasing. Since all the intervals in A_j are bisected only if their width are greater than ε ($\varepsilon > 0$). Then there must exist an $N \in \mathbb{Z}_{>0}$ such that $A_N = \emptyset$, which also implies $Y_N = Y_{N+1}$. Hence, Algorithm 8 terminated in a finite number of steps.

By Proposition 6.11, we have $\mathcal{BR}^\delta_\infty(\Omega) = \bigcup_{j=0}^\infty \text{Rch}^\delta_j(\Omega)$ and $\mathcal{BR}_\infty(\Omega) = \bigcup_{j=0}^\infty \text{Rch}_j(\Omega)$. To prove (6.24), we only need to show $\text{Rch}^\delta_j(\Omega) \subseteq Y_j \subseteq \text{Rch}_j(\Omega)$ for all $j \in \mathbb{N}$.

For $j = 0$, we have $\text{Rch}^\delta_0(\Omega) = Y_0 = \text{Rch}_0(\Omega) = \Omega$. For $j = 1$,

$$\begin{aligned}
\text{Rch}^\delta_1(\Omega) &= \text{Pre}^\delta(\text{Rch}^\delta_0(\Omega)) \cup \text{Rch}^\delta_0(\Omega) = \text{Pre}^\delta(\Omega) \cup \Omega, \\
Y_1 &= \underline{[\text{Pre}]}^\varepsilon(Y_0) \cup Y_0 = \underline{[\text{Pre}]}^\varepsilon(\Omega) \cup \Omega, \\
\text{Rch}_1(\Omega) &= \text{Pre}(\text{Rch}_0(\Omega)) \cup \text{Rch}_0(\Omega) = \text{Pre}(\Omega) \cup \Omega.
\end{aligned}$$

By Lemma 4.3, we can conclude $\text{Rch}^\delta_1(\Omega) \subseteq Y_1 \subseteq \text{Rch}_1(\Omega)$ with $(\rho_1\varepsilon + \rho_2\mu) \leq \delta$. Assume that $\text{Rch}^\delta_j(\Omega) \subseteq Y_j \subseteq \text{Rch}_j(\Omega)$ for some $j \in \mathbb{Z}_{>0}$. Then

$$\begin{aligned}
\text{Rch}^\delta_{j+1}(\Omega) &= \text{Pre}^\delta(\text{Rch}^\delta_j(\Omega)) \cup \text{Rch}^\delta_j(\Omega) \\
&\subseteq \underline{[\text{Pre}]}^\varepsilon(Y_j) \cup Y_j = Y_{j+1} \\
&\subseteq \text{Pre}(\text{Rch}_j(\Omega)) \cup \text{Rch}_j(\Omega) = \text{Rch}_{j+1}(\Omega).
\end{aligned}$$

Hence, $\text{Rch}^\delta_{j+1}(\Omega) \subseteq Y_{j+1} \subseteq \text{Rch}_{j+1}(\Omega)$. The proof is now complete by induction. ■

The relation (6.24) indicates that we can obtain an under-approximation of the maximal backward reachable set with a lower bound of δ-robustly backward reachable set if the approximation precision ε is set sufficiently small. Then we can extract a reachability controller that is defined on this approximated maximal backward reachable set.

Corollary 6.2 *Let the assumptions in Theorem 6.3 hold. If there exists a maximal δ-robustly backward reachable set of system (6.1) from Ω, then the system state can be controlled to Ω by using the reachability control strategy $\kappa : X \to 2^U$ in the following form:*

$$\kappa(x) = \begin{cases} [K]^{\varepsilon}_{Y_j}(x) & \forall x \in Y_{j+1} \setminus Y_j, \quad j = 0, \ldots, N, \\ U & \forall x \in \Omega, \\ \emptyset & \text{otherwise}, \end{cases} \tag{6.25}$$

where $[K]^{\varepsilon}_{Y_j}$ is defined in (4.35), $N \in \mathbb{N}$ is the number of iterations of Algorithm 8, and Y_j is the union of intervals in Y after the jth iteration in Algorithm 8.

Proof: For $x \in \Omega$, the formula φ_{r} is always true and hence $\kappa(x) = U$ for all $x \in \Omega$. By Theorem 6.3, Algorithm 8 terminates in N steps, which is a finite number. With $(\rho_1\varepsilon + \rho_2\mu) \leq \delta$, any control value $u \in [K]^{\varepsilon}_{Y_j}(x)$ will steer system state $x \in Y_{j+1} \setminus Y_j$ $(j = 0, \cdots, N)$ into Y_j under any disturbance in set W. Since $Y_0 = \Omega$, $\kappa(x)$ also realizes φ_{r} for any $x \in Y_{j+1}$. ■

6.5 Reach-and-Stay Control

In *reach-and-stay control*, we expect that the closed-loop solution $\{x_i\}_{i=0}^{\infty}$ of system (6.2) satisfies that, for any disturbance $\{w_i\}_{i=0}^{\infty}$, there exists some $j \in \mathbb{N}$ (depending on $\{w_i\}_{i=0}^{\infty}$) such that $x_k \in \Omega$ for all $k \geq j$ $(k \in \mathbb{N})$. The reach-and-stay specification can be written as the LTL formula $\varphi_{\text{rs}} = \Diamond\Box\Omega$.

We denote by $\mathbf{Win}_{(6.2)}(\varphi_{\text{rs}})$ the largest set of initial conditions from which the specification φ_{rs} can be satisfied (or the winning set). It can also be written as $\mathbf{Win}^{\delta}_{(6.1)}(\varphi_{\text{rs}})$ since (6.2) is (6.1) with additional δ-bounded disturbances.

Intuitively, reach-and-stay property is a combination of reachability and invariance. A control strategy that controls the system state to reach a controlled invariant set would serve this purpose [30]. The control synthesis is to first perform the invariance control iteration

$$\begin{aligned} X_0 &= \Omega, \\ X_{i+1} &= \text{Pre}^{\delta}(X_i | X_i), \end{aligned} \tag{6.26}$$

until it reaches the fixed point $X_{\infty} := \bigcap_{i=0}^{\infty} X_i$, and then perform the reachability control synthesis with X_{∞} as the target set:

$$\begin{aligned} Z_0 &= X_{\infty}, \\ Z_{i+1} &= \text{Pre}^{\delta}(Z_i). \end{aligned} \tag{6.27}$$

The fixed point $Z_\infty := \bigcup_{i=0}^{\infty} Z_i$ is then considered as the winning set for the reach-and-control specification.

The resulting control strategy is also a combination of reachability and invariance control strategies:

$$\kappa(x) = \begin{cases} K_{X_\infty}(x) & \forall x \in X_\infty, \\ K_{Z_{i+1}}(x) & \forall x \in Z_{i+1} \setminus Z_i. \end{cases} \tag{6.28}$$

The completeness of the above control synthesis relies on the assumption that *the target set is compact and convex.* For general dynamics and compact target set without this assumption, it fails to yield the real winning set, which can be illustrated by the following example.

Example 6.6 *Consider a target set* $\Omega = [-0.3, 0.3] \cup [0.8, 1.1]$ *and the dynamics*

$$x(t+1) = -x(t)(x(t)^2 - 2.05x(t) + 0.05) + u(t) + w(t), \tag{6.29}$$

where $x(t) \in X = [-0.65, 1.1]$, $u(t) \in U = \{0, 10\}$, *and* $w(t) \in W = [-5, 5] \times 10^{-4}$ $(\delta = 5 \times 10^{-4})$ *for* $t \in \mathbb{N}$.

For all state $x \in X$, *using the control value* 10 *will make the state in the next time step out of domain* X. *Let* $u(t) = 0$ *for all* t. *There are three fixed points* 0, 1, *and* 1.05. *The fixed points* 0 *and* 1.05 *are stable while* 1 *is unstable.*

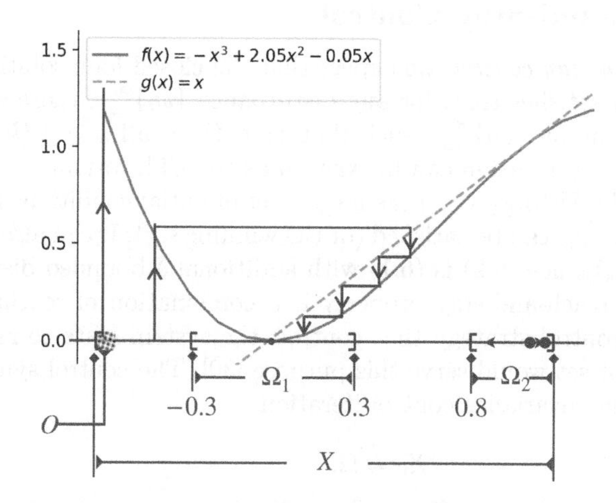

FIGURE 6.3: The evolution of the state of system (6.29) without the disturbance term.

As shown in Figure 6.3, any system state in the interval $O = [-0.65, -0.6311)$ *cannot be controlled inside the state space* X *under arbitrary disturbances, because, for all* $x(t) \in O$, *there exists* $s \in W$ *such that the system state at the next time step* $x(t+1) > 1.1$. *For the nominal system of (6.29), system state* x *evolves to* 0 *for all* $x \in (0,1)$.

The target set is a union of two disconnected intervals $\Omega_1 = [-0.3, 0.3]$ *and* $\Omega_2 = [0.8, 1.1]$. *The set* Ω_1 *is* δ*-robustly controlled invariant since* $\Omega_1 \subseteq Pre^{\delta}(\Omega_1) = [-0.3435, 0.4483]$. *Since* $x = 0.9914$ *satisfies* $-x(x^2 - 2.05x + 0.05) + d = x$ *and the difference* $x(t) - x(t+1)$, *which is negative, between two sequential states is decreasing as* x *increases between* 0.3414 *and* 1.0253, *any state* $x \in [0.3, 0.9914)$ *can be controlled inside* Ω_1. *Because of the overlap between* Ω_1 *and* $[0.3, 0.9914)$, *we can see that* $[-0.3, 1.1] \subseteq \mathbf{Win}_{(6.29)}(\varphi_{rs})$. *In addition,* $x(t+1) \in (0, 1.1]$ *for any state* $x(t) \in [-0.6311, 0)$. *Hence, the real winning set is*

$$\mathbf{Win}_{(6.29)}(\varphi_{rs}) = [-0.6311, 1.1].$$

However, with (6.26) and (6.26), the maximal controlled invariant set inside Ω *is* $X_{\infty} = [-0.3, 0.3] \cup [1.0370, 1.1]$, *and*

$$Z_{\infty} = [-0.6311, -0.6082) \cup (-0.6021, 0.9914) \cup (1.0135, 1.1].$$

Since X_{∞} *is a union of two disconnected intervals* $[-0.3, 0.3]$ *and* $[1.0370, 1.1]$ *with an unstable fixed point* 1 *in between, the interval* $[0.9914, 1.0135]$ *is not included in* Z_{∞}. *The interval* $[-0.6082, -0.6021]$ *is also missing because it is mapped to* $[0.9914, 1.0135]$ *by (6.29).*

To characterize the winning set $\mathbf{Win}_{(6.2)}(\varphi_{\text{rs}})$ with a more general target set, we couple those two fixed-point iterations into

$$\begin{aligned}
&Y_0 = \emptyset, X_0^{\infty} = \emptyset \\
&\left.\begin{aligned} X_{i+1}^{0} &= Y_i \cup \Omega \\ X_{i+1}^{j+1} &= \text{Pre}^{\delta}(X_{i+1}^{j} | X_{i+1}^{j}) \end{aligned}\right\} X_{i+1}^{\infty} \triangleq \bigcap_{j=0}^{\infty} X_{i+1}^{j} \\
&Y_{i+1} = \text{Pre}^{\delta}(X_{i+1}^{\infty})
\end{aligned} \tag{6.30}$$

with the control strategy

$$\kappa(x) = \begin{cases} K_{X_{i+1}^{\infty}}(x) & \forall x \in \Omega \cap \left(X_{i+1}^{\infty} \setminus X_i^{\infty}\right), \\ K_{X_{i+1}^{\infty}}(x) & \forall x \in Y_{i+1} \setminus (Y_i \cup \Omega). \end{cases} \tag{6.31}$$

The invariance control iteration (line 2-3) in (6.30) is now an inner loop inside the reach control iteration. It starts with computing the maximal controlled invariant set X_1^{∞} inside Ω ($Y_0 = \emptyset$) and the controlled predecessor Y_1 from X_1^{∞}. Instead of continuing the backward reachable set computation all the way down, it goes back to look for the maximal controlled invariant set X_2^{∞} of a bigger set $Y_1 \cup \Omega$ than Ω. In this way, Y_i is larger than its equivalent Z_i in (6.27), and the state that jumps in and out of Ω several times before finally stay inside Ω can be captured.

Proposition 6.12 *Suppose that the state space X and $\Omega \subseteq X$ are compact. Let $Y_\infty = \bigcup_{i=0}^{\infty} Y_i$ be a fixed point of (6.30), where $\{Y_i\}_{i=0}^{\infty}$ is a sequence of subsets of X generated from (6.30). Then,*

(i) $Y_\infty = \mathbf{Win}_{(6.2)}(\varphi_{rs})$, and

(ii) the strategy κ (6.31) is a memoryless control strategy that realizes φ_{rs}.

Proof: We only consider $\Omega \neq \emptyset$. Otherwise the results trivially hold.

We first show $Y_\infty \subseteq \mathbf{Win}_{(6.2)}(\varphi_{\rm rs})$ by induction. For the base case, $Y_0 = \emptyset \subseteq \mathbf{Win}_{(6.2)}(\varphi_{\rm rs})$. The induction step aims to show that, for all $i \in \mathbb{Z}_{>0}$, $Y_{i+1} \subseteq \mathbf{Win}_{(6.2)}(\varphi_{\rm rs})$ if $Y_i \subseteq \mathbf{Win}_{(6.2)}(\varphi_{\rm rs})$. Assume that X_i^∞ is compact. Then $Y_i = \mathrm{Pre}^\delta(X_i^\infty)$ and thus $X_{i+1}^0 = \Omega \cup Y_i$ is compact. The sequence $\{X_{i+1}^j\}_{j=0}^{\infty}$ is compact and decreasing by induction, using Proposition 6.1 (i) and Proposition 6.3 since $X_{i+1}^0 = \Omega \cup Y_i$ is compact. It is also easy to show that $\{Y_i\}_{i=0}^{\infty}$ is increasing by induction. Furthermore, by Proposition C.1, we have

$$X_{i+1}^\infty = \lim_{j\to\infty} X_{i+1}^j = \bigcap_{j=0}^{\infty} X_{i+1}^j,$$

which is the maximal controlled invariant set inside $\Omega \cup Y_i$ by Proposition 6.9 and compact. If $Y_i \subseteq \mathbf{Win}_{(6.2)}(\varphi_{\rm rs})$, then $X_{i+1}^\infty \subseteq \mathbf{Win}_{(6.2)}(\varphi_{\rm rs})$ because X_{i+1}^∞ is a controlled invariant set inside $\Omega \cup Y_i$, which gives

$$Y_{i+1} = \mathrm{Pre}^\delta(X_{i+1}^\infty) \subseteq \mathbf{Win}_{(6.2)}(\varphi_{\rm rs})$$

by Definition 4.15. Hence, $Y_\infty = \bigcup_{i=0}^{\infty} Y_i \subseteq \mathbf{Win}_{(6.2)}(\varphi_{\rm rs})$.

To see $\mathbf{Win}_{(6.2)}(\varphi_{\rm rs}) \subseteq Y_\infty$, we aim to show that $x \notin \mathbf{Win}_{(6.2)}(\varphi_{\rm rs})$ for all $x \notin Y_\infty$. Let $x \notin Y_\infty$ be arbitrary. Then $x \notin Y_\infty = \mathrm{Pre}^\delta(\mathcal{I}_\infty(Y_\infty \cup \Omega))$, where $\mathcal{I}_\infty(Y_\infty \cup \Omega)$ is the maximal controlled invariant set inside $Y_\infty \cup \Omega$, since Y_∞ is a fixed point of (6.30). This means that for all $\{u_t\}_{t=0}^{\infty}$ there exists k and $\{w_t\}_{t=0}^{k}$ such that the resulting sequence of (6.1) satisfies $x_k \notin (\Omega \cup Y_\infty)$. Since $x_k \notin Y_\infty$, we can show in the same manner that for all $\{u_t\}_{t=k}^{\infty}$ there exists $k' \geq k$ and $\{w_t\}_{t=k}^{k'}$ such that the k'th state $x_{k'}$ of the resulting solution satisfies $x_{k'} \notin (\Omega \cup Y_\infty)$. In this way, for all $\{u_t\}_{t=0}^{\infty}$, we can find an infinite sequence $\{w_t\}_{t=0}^{\infty}$ for any $x \notin Y_\infty$ so that the resulting solution of (6.1) goes outside of Ω infinitely often. Hence, $x \notin \mathbf{Win}_{(6.2)}(\varphi_{\rm rs})$, which shows $\mathbf{Win}_{(6.2)}(\varphi_{\rm rs}) \subseteq Y_\infty$.

Now we prove (ii). The control strategy κ is constructed by $K^\delta_{X_{i+1}^\infty}$, which is only dependent on the current state x of the system (6.1), and thus κ is memoryless. By the definition of $K^\delta_{X_{i+1}^\infty}$ in (4.30), for all $x \in \Omega \cap (X_{i+1}^\infty \setminus X_i^\infty)$ and $x \in Y_{i+1} \setminus (Y_i \cup \Omega)$ ($i \in \mathbb{N}$), the state x will be controlled inside $\Omega \cup Y_i$ and Y_i in one step, respectively. That means any state $x \in Y_{i+1}$ will be controlled into Y_i until it enters $X_1^\infty = \mathcal{I}_\infty(\Omega) \subseteq \Omega$, which is controlled invariant. Hence, we have also shown that κ realizes $\varphi_{\rm rs}$. ■

The information of the target set Ω is used for every iteration of Y_i in (6.30) while such information is lost after the computation of the maximal (robustly) controlled invariant set X_∞ in (6.26). Hence, some of the system states inside Ω, which would leave Ω but will be controlled back to X_∞ or stay inside Ω under some disturbance, will be missing if we run invariance and reachability iterations sequentially. The real winning set in Example 6.6 can be obtained by using the algorithm (6.30).

Remark 6.2 *The completeness result in Proposition 6.12, i.e., Y_∞ captures all the states that can be controlled to stay in Ω ($\mathbf{Win}_{(6.2)}(\varphi_{rs}) \subseteq Y_\infty$), relies on the assumption that Y_∞ is a fixed point of (6.30). To satisfy such an assumption, the following properties regarding Ω and the predecessor map are required:*

$$Pre^\delta\left(\bigcup_{i=0}^{\infty} Y_i\right) = \bigcup_{i=0}^{\infty} Pre^\delta(Y_i), \tag{6.32}$$

$$\left(\bigcup_{i=0}^{\infty} Pre^\delta(Y_i \cup \Omega)\right) \cap \left(\bigcup_{i=0}^{\infty} (Y_i \cup \Omega)\right) = \bigcup_{i=0}^{\infty} \left(Pre^\delta(Y_i \cup \Omega) \cap (Y_i \cup \Omega)\right). \tag{6.33}$$

However, the condition (6.32) does not generally hold.

There are some computational redundancies in (6.30): the sequence $\{Y_i\}_{i=0}^\infty$ is increasing and so is $\{X_i^j\}_{i=0}^\infty$ for all $j \in \mathbb{N}$. Hence, it is only necessary to compute the incremental parts between two adjacent sets in the sequences. Considering that controlled predecessors cannot be precisely computed, we can also replace all the controlled predecessors by Algorithm 5, which gives us the approximated control synthesis algorithm.

Algorithm 9 $[Y]^\varepsilon = \text{REACHSTAYCONTROL}([f], \Omega, X, U, \varepsilon)$

1: $\widetilde{Y} \leftarrow \emptyset$, $Y \leftarrow \Omega$, $V \leftarrow X \setminus \Omega$
2: **while** $Y \neq \widetilde{Y}$ **do**
3: $\quad Y \leftarrow \widetilde{Y}$
4: $\quad \widetilde{Z} \leftarrow \Omega \setminus Y$, $Z \leftarrow \emptyset$
5: $\quad$ **while** $Z \neq \widetilde{Z}$ **do**
6: $\quad\quad Z \leftarrow \widetilde{Z}$
7: $\quad\quad \underline{Z}, \Delta Z, Z_c = \text{PRE}([f], Y \cup Z, Z, U, \varepsilon)$ ⊳ $\kappa(x) = [K]^\varepsilon_{Y\cup Z}(x), \forall x \in Z$.
8: $\quad\quad \widetilde{Z} \leftarrow \underline{Z}$
9: $\quad$ **end while**
10: $\quad \underline{A}, \Delta A, A_c = \text{PRE}([f], Y \cup Z, V, U, \varepsilon)$ ⊳ $\kappa(x) = [K]^\varepsilon_{Y\cup Z}(x), \forall x \in \underline{A}$.
11: $\quad V \leftarrow V \setminus \underline{A}$
12: $\quad \widetilde{Y} \leftarrow Y \cup Z \cup \underline{A}$
13: **end while**
14: **return** Y

In Algorithm 9, the set Y keeps track of the states (found by the interval-based Algorithm 5) that can be controlled into the target set Ω and stays there forever after each iteration. The set Z represents the part of Ω that does not belong to Y, i.e., the states in Ω that cannot be guaranteed to belong to $\mathbf{Win}_{(6.2)}$ in the current iteration. The main difference between Algorithm 9 and (6.30) is that only the states in Z are checked against the reach-and-stay property instead of the entire set $\Omega \cup Y$.

Theorem 6.4 (Soundness and Robust Completeness) *Let the assumptions in Proposition 6.12 hold for system (6.1). Suppose that Assumptions 4.1 and (4.2) hold with Lipschitz constants ρ_1 and ρ_2, respectively. Then Algorithm 9 terminates in finite time and the output $[Y]^\varepsilon$ satisfies*

$$\mathbf{Win}_{(6.2)}(\varphi_{rs}) \subseteq [Y]^\varepsilon \subseteq \mathbf{Win}_{(6.1)}(\varphi_{rs}). \tag{6.34}$$

if $\rho_1\varepsilon + \rho_2\mu \leq \delta$.

Proof: For Algorithm 9, let $\widetilde{Y}_i$ be the set Y after the ith iteration and Z_i^j be the set Z after the jth inner iteration and ith outer iteration for system (6.1). To simplify the notation, we let $\widetilde{X}_{i+1}^j = Z_i^j \cup \widetilde{Y}_i$ and $\widetilde{X}_{i+1}^\infty = Y_i \cup Z_i^\infty$, where Z_i^∞ denotes the output of the inner loop in the ith outer iteration. We also denote by Y_∞ and Y_∞^δ ($\delta > 0$) the outputs of (6.30) for system (6.1) and (6.2), respectively. We prove the theorem in the following steps:

(i) Show that Algorithm 9 is equivalent to (6.30) when the controlled predecessor computation is accurate, i.e., $\widetilde{X}_i^\infty = X_i^\infty$ (or $X_i^{\delta\infty}$) and $\widetilde{Y}_i = Y_i$ (or Y_i^δ) for all $i \in \mathbb{N}$.

(ii) Algorithm 9 terminates in a finite number of steps.

(iii) Show $\widetilde{Y}_\infty^\delta \subseteq [Y]^\varepsilon \subseteq \widetilde{Y}_\infty$ under the given condition.

It is straightforward that $Y_i \subseteq \mathrm{Pre}(Y_i)$ for all i since X_i^∞ is a controlled invariant set, $X_i^\infty \subseteq \mathrm{Pre}(X_i^\infty)$, and $Y_i = \mathrm{Pre}(X_i^\infty) \subseteq \mathrm{Pre}(\mathrm{Pre}(X_i^\infty)) = \mathrm{Pre}(Y_i)$ by the definition of Y_i in (6.30) and monotonicity of Pre.

First of all, we prove (i) by induction. The base case clearly holds since $\widetilde{Y}_0 = Y_0 = \emptyset$ and $\widetilde{X}_0^\infty = X_0^\infty = \emptyset$. Suppose that $Z_i^\infty \cup \widetilde{Y}_i = X_i^\infty$ and $\widetilde{Y}_i = Y_i$ for some $i \in \mathbb{N}$. Then

$$\begin{aligned}
\widetilde{X}_{i+1}^0 &= \widetilde{Y}_i \cup (\Omega \setminus \widetilde{Y}_i) = \widetilde{Y}_i \cup \Omega = X_{i+1}^0, \\
\widetilde{X}_{i+1}^{j+1} &= \widetilde{Y}_i \cup Z_i^{j+1} = \widetilde{Y}_i \cup (\mathrm{Pre}(\widetilde{X}_{i+1}^j) \cap Z_i^j) \\
&= (\widetilde{Y}_i \cup \mathrm{Pre}(\widetilde{X}_{i+1}^j)) \cap (\widetilde{Y}_i \cup Z_i^j) \\
&= (\widetilde{Y}_i \cup \mathrm{Pre}(\widetilde{X}_{i+1}^j)) \cap \widetilde{X}_{i+1}^j.
\end{aligned}$$

Also, $\mathrm{Pre}(\widetilde{X}_{i+1}^j) = \mathrm{Pre}(\widetilde{Y}_i \cup Z_i^j) \supseteq \mathrm{Pre}(\widetilde{Y}_i) \supseteq \widetilde{Y}_i$, which implies that $\widetilde{X}_{i+1}^{j+1} = \mathrm{Pre}(\widetilde{X}_{i+1}^j) \cap \widetilde{X}_{i+1}^j$. This is the same as the iteration step in (6.30), and thus $\widetilde{X}_{i+1}^\infty = X_{i+1}^\infty$. Now consider the sequence $\{V_i\}_{i=0}^\infty$. We have $V_0 = X \setminus \Omega$ and

$$V_{i+1} = V_i \setminus (\mathrm{Pre}(\widetilde{X}_i^\infty) \cap V_i) = V_i \setminus \mathrm{Pre}(\widetilde{X}_i^\infty).$$

Unfolding V_i until V_0 and using that $\text{Pre}(\widetilde{X}_i^\infty) \subseteq \text{Pre}(\widetilde{X}_{i+1}^\infty)$, we can derive

$$V_i = X \setminus (\Omega \cup \text{Pre}(\widetilde{X}_i^\infty)) = X \setminus (\Omega \cup Y_i) = X \setminus \widetilde{X}_{i+1}^0.$$

Then

$$\begin{aligned}\text{Pre}(\widetilde{X}_{i+1}^\infty) &= \text{Pre}(\widetilde{X}_{i+1}^\infty) \cap (\widetilde{X}_{i+1}^0 \cup V_i) \\ &= \left[\text{Pre}(\widetilde{X}_{i+1}^\infty) \cap \widetilde{X}_{i+1}^0\right] \cup \left[\text{Pre}(\widetilde{X}_{i+1}^\infty) \cap V_i\right] \\ &= \widetilde{X}_{i+1}^\infty \cup \text{Pre}(\widetilde{X}_{i+1}^\infty | V_i) = \widetilde{Y}_{i+1}.\end{aligned} \tag{6.35}$$

The equality $\text{Pre}(\widetilde{X}_{i+1}^\infty) \cap \widetilde{X}_{i+1}^0 = \widetilde{X}_{i+1}^\infty$ can be seen by contradiction. If there exists $A \subseteq \widetilde{X}_{i+1}^0 \setminus \widetilde{X}_{i+1}^\infty$ such that $A \subseteq \text{Pre}(\widetilde{X}_{i+1}^\infty)$ then $\widetilde{X}_{i+1}^\infty \cup A \subseteq \text{Pre}(\widetilde{X}_{i+1}^\infty \cup A)$, which indicates $A \cup \widetilde{X}_{i+1}^\infty$ is a larger controlled invariant set inside $\widetilde{X}_{i+1}^0$, but $\widetilde{X}_{i+1}^\infty$ is the maximal one. Therefore, $Y_{i+1} = \widetilde{Y}_{i+1}$. The above argument also applies for system (6.2).

Secondly, for (ii), under a given precision $\varepsilon > 0$, Z_i can only be partitioned to finite number of intervals. Then, for the inner loop, there must exist a positive integer N such that $Z_i^N = \emptyset$ if $Z_i^j \neq Z_i^{j+1}$ for all $j \in \mathbb{Z}_{\geq 0}$ because $\{Z_i^j\}_{j=0}^\infty$ is decreasing. Thus, the inner loop terminates within each outer loop. Likewise, the outer loop is also terminating since $\{V_i\}_{i=0}^\infty$ is decreasing and X only consists of a finite number of intervals.

Lastly, to prove (iii), we aim to show $X_i^{\delta\infty} \subseteq \widetilde{X}_i^\infty \subseteq X_i^\infty$ and $Y_i^\delta \subseteq \widetilde{Y}_i \subseteq Y_i$ for all i. With $\rho_1\varepsilon + \rho_2\mu \leq \delta$ and Proposition 6.1 (ii), we have

$$\text{Pre}^\delta(X_1^{\delta 0} | X_1^{\delta 0}) \subseteq X_1^1 = Y_0 \cup \underline{[\text{Pre}_\mu]}^\varepsilon(\widetilde{X}_1^0 | Z_1^0) \subseteq \text{Pre}(X_1^0 | X_1^0)$$

since $X_1^{\delta 0} = \widetilde{X}_1^0 = X_1^0 = Z_1^0 = \Omega$ and $Y_0 = \emptyset$. This means $X_1^{\delta 1} \subseteq \widetilde{X}_1^1 \subseteq X_1^1$. By induction and (i), it is easy to see that $X_1^{\delta j} \subseteq \widetilde{X}_1^j \subseteq X_1^j$ for any $j \in \mathbb{N}$. Thus $X_1^{\delta\infty} \subseteq \widetilde{X}_1^\infty \subseteq X_1^\infty$. As shown in (6.35), $\widetilde{Y}_i = \underline{[\text{Pre}_\mu]}^\varepsilon(\widetilde{X}_i^\infty)$. Then

$$Y_1^\delta = \text{Pre}^\delta(X_1^{\delta\infty}) \subseteq \text{Pre}^\delta(\widetilde{X}_1^\infty) \subseteq \widetilde{Y}_1 \subseteq \text{Pre}(\widetilde{X}_i^\infty) \subseteq \text{Pre}(X_i^\infty) = Y_1.$$

Therefore, (iii) can also be shown using induction. ■

Corollary 6.3 *Let the assumptions in Theorem 6.4 hold. If the reach-and-stay specification φ_{rs} can be satisfied for system (6.1), then the control strategy $\kappa : X \to 2^U$ given in Algorithm 9 realizes φ_{rs} for all states in the returned winning set.*

6.6 Temporal Logic Specifications

In many situations, control specifications beyond simply invariance and reachability need to be considered. For example, specifications such as "visiting different areas in some order infinitely often and avoid obstacles" are frequently considered in motion planning. Control of an elevator or a network of distributed resources involves a request-response pattern. This motivates the study of control synthesis for dynamical systems to realize the properties that require ordering, liveness, or reactivity. Such properties can be well captured by general LTL formulas [23].

Similar to the previous sections, the goal of this section is to look for a sound and robustly complete control synthesis algorithm with respect to a general class of LTL formulas. As opposed to using the product system of a finite abstraction of the original system and the automaton translated from the given LTL formula, which is discussed in Chapter 5, we look at this control synthesis problem more directly: characterize the winning set by a fixed-point algorithm based on the automata structure of the given formula.

Recall that, in Chapter 3, every LTL formula φ built on a set of atomic propositions Π has an equivalent Büchi automaton, which accepts the words specified by φ. In this section, we focus on the ones that can be translated to DBA and denote by $\mathcal{A}_\varphi = (Q, \Sigma, r, q_0, F)$ an equivalent DBA of an LTL formula φ, where $\Sigma = 2^\Pi$ is the input alphabet. Such LTL formulas are said to be *DBA-recognizable*.

An input symbol $\sigma \in \Sigma$ is usually represented by a propositional formula over the set Π of atomic propositions for φ. Without loss of generality, we assume that $\mathcal{A}_\varphi$ is *total*, i.e., $\mathrm{Out}(q) \neq \emptyset$, $\forall q \in Q$, since we can always construct a total automaton for any BA [23]. An automaton $\mathcal{A} = (Q, \Sigma, r, q_0, F)$ can be presented as a *directed graph* $\mathcal{G} = (V, E)$, where $V = Q$ is a set of nodes and E is a set of directed edges. We further define the set of *outgoing edges* $\mathrm{Out}(v) = \{\sigma \in \Sigma \mid r(v, \sigma) \neq \emptyset\}$.

6.6.1 Winning Set Characterization on the Continuous State Space

To develop the direct control synthesis algorithm, we first of all characterize the winning set of an LTL specification on the continuous state space for system (6.2).

Let $\mathcal{A}_\varphi$ be an equivalent DBA of an LTL formula φ and $q \in Q$ be an arbitrary state of $\mathcal{A}_\varphi$. Since $\mathcal{A}_\varphi$ is deterministic and total, every state has at least one outgoing edge and

$$\bigvee_{\sigma \in \mathrm{Out}(q)} \sigma = \top, \; \sigma \wedge \sigma' = \bot, \; \forall \sigma, \sigma' \in \mathrm{Out}(q) \text{ s.t. } \sigma \neq \sigma'. \tag{6.36}$$

We consider traces of system (6.2) as input words to $\mathcal{A}_\varphi$. Hence, given a control signal $\{u(t)\}_{t=0}^{\infty}$ and a sequence of disturbances $\{w(t)\}_{t=0}^{\infty}$, the resulting run $\varrho = \{v_t\}_{t=0}^{\infty}$ of $\mathcal{A}_\varphi$ is obtained explicitly by (for all $t \in \mathbb{Z}_{>0}$)

$$\begin{cases} v_0 = q_0, \; v_t = r(v_{t-1}, L(x_{t-1})), \; v_t \in Q, \\ x_t = f(x_{t-1}, u_{t-1}) + w_{t-1}, \; x_t \in X, \end{cases} \tag{6.37}$$

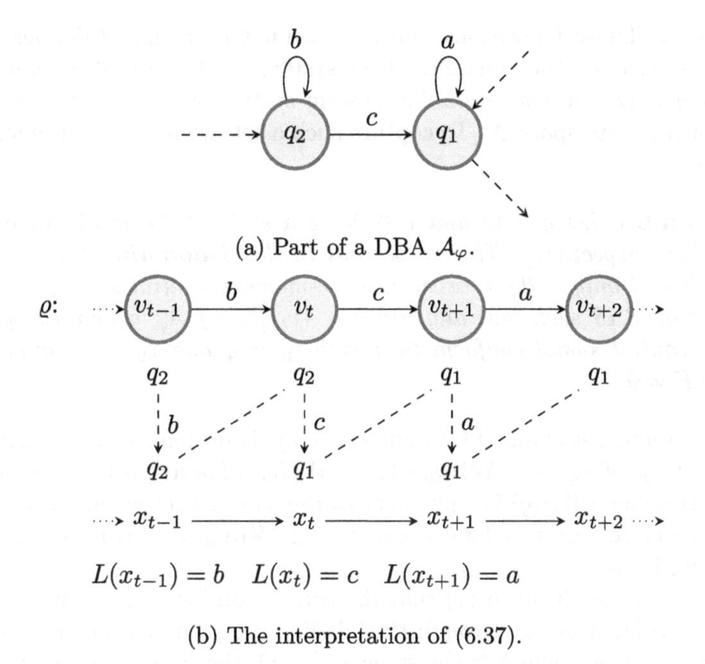

(a) Part of a DBA $\mathcal{A}_\varphi$.

(b) The interpretation of (6.37).

FIGURE 6.4: The connection between system (6.2) and an equivalent DBA $\mathcal{A}_\varphi$ of a given LTL formula φ. Assume $v_{t-1} = q_2$ and $L(x_{t-1}) = b$. (b) shows how the partial sequence $v_{t-1}v_t v_{t+1}v_{t+2}$ is driven by $x_{t-1}x_t x_{t+1}x_{t+2}$ according to the relevant part of $\mathcal{A}_\varphi$ shown in part (a).

where the system and DBA states are updated only at discrete time instances. An intuitive illustration of (6.37) is given in Figure 6.4.

It is easy to see from (6.37) that the connection between the dynamical system (6.2) and the targeted DBA $\mathcal{A}_\varphi$ is through the labeling function L, which divides the system state space into finitely many areas of interest.

Let $\{\alpha_1, \cdots, \alpha_N\} \in 2^\Sigma$ be such that

$$\bigvee_{i=1}^{N} \alpha_i = \top, \quad \alpha_i \wedge \alpha_j = \bot, \, i \neq j. \tag{6.38}$$

Then we can obtain a partition $\mathcal{P}_0 = \{P_1, P_2, \cdots, P_N\}$ (see Definition 4.16) of the state space X, where

$$P_i := L^{-1}(\alpha_i) = \{x \in X \mid L(x) = \alpha_i\}. \tag{6.39}$$

This is to say that all the states inside a cell are assigned the same atomic proposition. Additionally, $L^{-1}(\top) = X$.

In order to control system (4.44) so that the resulting traces are accepted by $\mathcal{A}_\varphi$, each transition along the successful runs of $\mathcal{A}_\varphi$ needs to be executed sequentially. Those transitions, however, cannot be assigned deliberately as they have to satisfy the dynamics of the system (4.44). This also implies that, for each $q \in Q$, the corresponding system state x is restricted to a certain subset of the state space X. To capture such a set, we introduce the following definition.

Definition 6.7 *Let $q \in Q$ and $x \in X$ be a state of the DBA $\mathcal{A}_\varphi$ and system (4.44), respectively. Then x belongs to the* **$\mathcal{S}$-domain** *of q, written as $W_\mathcal{S}(q)$, if and only if there exists a finite-memory control strategy κ defined in Defintion 3.18 such that any run $\varrho = \{v_t\}_{t=0}^{\infty}$ of $\mathcal{A}_\varphi$ generated by (6.37) under a control signal conform to κ with $v_0 = q$ and $x_0 = x$ satisfies that $\mathrm{Inf}(\varrho) \cap F \neq \emptyset$.*

The winning set of an LTL formula φ is, by definition, the $\mathcal{S}$-domain of the initial state q_0 of $\mathcal{A}_\varphi$, i.e., $\mathbf{Win}_{(6.2)}(\varphi) = W_\mathcal{S}(q_0)$. For any other state $q \in \mathcal{A}_\varphi$, the (4.44) domain $W_\mathcal{S}(q)$ is equivalent to the winning set when q is the initial state. Therefore, the problem of computing $\mathbf{Win}_{(6.2)}(\varphi)$ can be reduced to computing $W_\mathcal{S}(q_0)$.

It is easy to see from (6.37) that the connection between system (6.2) and the targeted DBA $\mathcal{A}_\varphi$ is through the labeling function L. Since the outgoing edges satisfy (6.36), which is the same as (6.38), the set $\mathrm{Out}(q)$ of every state $q \in Q$ forms a partition $\mathcal{P}(q) = \{L^{-1}(\sigma)\}_{\sigma \in \mathrm{Out}(q)}$ of the state space X through the labeling function L as defined in (6.39).

Because of the graph structure of a DBA $\mathcal{A}_\varphi$, $\mathcal{S}$-domains of different automaton states are related with one another by the transitions among them (see Figure 6.4). Any state $x \in W_\mathcal{S}(q)$ can be controlled to the $\mathcal{S}$-domain of one of the succeeding states of q in $\mathcal{A}_\varphi$. Suppose that the current state of the DBA in Figure 6.4a is q_2 and the system state is x. Let $W_\mathcal{S}(q_2)$ and $W_\mathcal{S}(q_1)$ be the $\mathcal{S}$-domains of q_2 and q_1, respectively. Then $x \in W_\mathcal{S}(q_0)$ if and only if $\exists u \in U$ such that $\forall w \in W$,

$$\begin{cases} f(x,u) + w \in W_\mathcal{S}(q_0), & \text{if } L(x) = b, \\ f(x,u) + w \in W_\mathcal{S}(q_1), & \text{if } L(x) = c. \end{cases}$$

Let $\mathbf{M}$ be an n_1 by n_2 ($n_1, n_2 > 0$) matrix of symbols from Σ, and V and W be two vectors of subsets of X of length n_2. Denote by m_{ij} the element at the ith row and jth column of $\mathbf{M}$. Define

$$W + V \triangleq \begin{bmatrix} W[1] \cup V[1] & \cdots & W[n_2] \cup V[n_2] \end{bmatrix}^T, \tag{6.40}$$

$$W - V \triangleq \begin{bmatrix} W[1] \setminus V[1] & \cdots & W[n_2] \setminus V[n_2] \end{bmatrix}^T, \tag{6.41}$$

$$V \preceq W \triangleq V[i] \subseteq W[i],\ i = 1, \ldots, n_2, \tag{6.42}$$

$$W = V \triangleq W[i] = V[i],\ i = 1, \ldots, n_2, \tag{6.43}$$

$$\mathbf{T}^{\delta}(\mathbf{M}, W) = W', \tag{6.44}$$

where

$$W'[i] = \bigcup_{j=1}^{n_2} \mathrm{Pre}^{\delta}\left(W[j] \,|\, L^{-1}(m_{ij})\right), \; i = 1, \ldots, n_1.$$

For the nominal system (6.1), we use $\mathbf{T}$ for the sake of simplicity. Based on the properties of predecessor maps, the operator $\mathbf{T}^{\delta}$ satisfies the following properties.

Proposition 6.13 *Given a matrix* $\mathbf{M}$ *of symbols and vectors* V, W *of subsets of* X *that match in dimension for operator* $\mathbf{T}^{\delta}$ *defined in (6.44) with* $\delta \geq 0$ *and* $0 \leq \delta_1 \leq \delta_2$*,*

(i) $\mathbf{T}^{\delta}(\mathbf{M}, V) \preceq \mathbf{T}^{\delta}(\mathbf{M}, W)$ *if* $V \preceq W$*,*

(ii) $\mathbf{T}^{\delta_2}(\mathbf{M}, W) \preceq \mathbf{T}^{\delta_1}(\mathbf{M}, W) \preceq \mathbf{T}(\mathbf{M}, W)$.

Proof: To show (i), assume that $\mathbf{M}$ is of size $n_1 \times n_2$ and W, V of size $n_2 \times 1$. As defined in (6.42), $V \preceq W$ means $V[j] \subseteq W[j]$. By the monotonicity of Pre^{δ}, we have $\bigcup_{i=1}^{n_1} \mathrm{Pre}^{\delta}(V[j] \,|\, L^{-1}(m_{ij})) \subseteq \bigcup_{i=1}^{n_1} \mathrm{Pre}^{\delta}(W[j] \,|\, L^{-1}(m_{ij}))$, which gives (i).

Property (ii) is straightforward by the fact that $\mathrm{Pre}^{\delta}(A - \mathbb{B}_{\delta_2}) \subseteq \mathrm{Pre}^{\delta}(A - \mathbb{B}_{\delta_1}) \subseteq \mathrm{Pre}^{\delta}(A)$ for all $A \subseteq X$. ■

The graph representation of a DBA can be coded into a matrix of symbols, which is given in the following definition.

Definition 6.8 *Given a DBA* $\mathcal{A}_{\varphi}$ *with the set of indexed states* Q *with* $Q[i]$ *being the* i*th element* ($i \in \{1, \cdots, |Q|\}$)*, the* ***transition matrix*** $\mathbf{M}_{\varphi}$ *of* $\mathcal{A}_{\varphi}$ *is a* $|Q|$ *by* $|Q|$ *matrix of symbols from* Σ*. The element* m_{ij} *in the* i*th row and* j*th column* ($i, j = 1, \ldots, |Q|$) *of* $\mathbf{M}_{\varphi}$ *is given by*

$$m_{ij} = \begin{cases} \sigma, & Q[j] = r(Q[i], \sigma), \sigma \in \Sigma, \\ e, & \textit{otherwise}, \end{cases} \tag{6.45}$$

where $e \in \Sigma$ *denotes an empty symbol with* $L^{-1}(e) = \emptyset$*.*

As defined in (6.44), the operator $\mathbf{T}^{\delta}$ computes controlled predecessors according to the transition relation coded in $\mathbf{M}_{\varphi}$.

To track the control values that can activate the transitions, we also define a vector $\mathcal{K} = \begin{bmatrix} \kappa_1 & \ldots & \kappa_{n_1} \end{bmatrix}$ of memoryless control strategies, where ($i = 1, \ldots, n_1$)

$$\kappa_i(x) = \bigcup_{j=1}^{n_2} K^{\delta}_{W[j]}(x), \; \forall x \in W'[i]. \tag{6.46}$$

For a DBA $\mathcal{A}_\varphi$, the dependencies among the $\mathcal{S}$-domains can be captured by using the operator $\mathbf{T}^\delta$ and the transition matrix $\mathbf{M}_\varphi$. Suppose V is a vector of subsets of states in the state space X, where each element $V[i]$ of V is a goal region for $Q[i]$. Then $\mathbf{T}^\delta(\mathbf{M}_\varphi, V)$ by definition is the vector of sets that can be controlled into V for system $\mathcal{S}^\delta$ in one step.

Proposition 6.14 *Let $\mathbf{M}_\varphi$ be the transition matrix of a DBA $\mathcal{A}_\varphi$ and $\mathbf{W}_\mathcal{S}$ be the vector of $\mathcal{S}$-domains of $\mathcal{A}_\varphi$ for $\mathcal{S}^\delta$. Then $\mathbf{W}_\mathcal{S} = \mathbf{T}^\delta(\mathbf{M}_\varphi, \mathbf{W}_\mathcal{S})$.*

Proof: Let $V = \mathbf{T}^\delta(\mathbf{M}_\varphi, \mathbf{W}_\mathcal{S})$ and $i \in \{1, \dots, |Q|\}$. We first show that $V \preceq \mathbf{W}_\mathcal{S}$. By the definition of $\mathbf{T}^\delta$, any state $x \in V[i]$ can be controlled into $\bigcup_{j=1}^{|Q|} W_\mathcal{S}(Q[j])$ after one step by some $u \in U$ for all $d \in W$. This implies that for any state $x \in W_\mathcal{S}(q)$ (any $q \in Q$), there exists a run ϱ with $\varrho[t] = q$ at some $t \in \mathbb{Z}_{>0}$, which is generated according to (6.37), such that ϱ visits F infinitely often. Hence, $x \in W_\mathcal{S}(Q[i])$ by Definition 6.7, and $V[i] \subseteq W_\mathcal{S}(Q[i])$. Since i is arbitrary, $V \preceq \mathbf{W}_\mathcal{S}$.

Next we show that $\mathbf{W}_\mathcal{S} \preceq V$. Suppose that there is an $x \in W_\mathcal{S}(Q[i])$ but $x \notin V[i] = \bigcup_{j=1}^{|Q|} \mathrm{Pre}^\delta\left(W_\mathcal{S}(Q[j]) \mid L^{-1}(m_{ij})\right)$. Then by Definition 4.15, for all $j \in \{1, \dots, |Q|\}$ and $u \in U$, there exists $d_{j,u} \in W$ such that $x' = f(x, u) + d_{j,u} \notin W_\mathcal{S}(Q[j])$ or $x \notin L^{-1}(m_{ij})$. If $x \notin L^{-1}(m_{ij})$, then the transition from $Q[i]$ to $Q[j]$ will not happen. If $x' = f(x, u) + d_{j,u} \notin W_\mathcal{S}(Q[j])$, then by Definition 6.7 there exists no control strategy that the rest of the run with system state starting from x' will visit F infinitely often. Hence, $x \notin W_\mathcal{S}(Q[i])$, which contradicts the assumption. Therefore, by (6.42) and (6.43), $V \preceq \mathbf{W}_\mathcal{S}$ and $\mathbf{W}_\mathcal{S} \preceq V$ gives $\mathbf{W}_\mathcal{S} = V$. ■

Proposition 6.14 is a necessary condition for a vector W to be $\mathbf{W}_\mathcal{S}$, and $\mathbf{W}_\mathcal{S}$ may not be the unique fixed point of $\mathbf{T}^\delta(\mathbf{M}_\varphi, \cdot)$ with respect to a transition matrix $\mathbf{M}_\varphi$. By Definition 6.7, $\mathbf{W}_\mathcal{S}$ is the maximal fixed point of $\mathbf{T}^\delta(\mathbf{M}_\varphi, \cdot)$ that satisfies the Büchi condition.

Let us look at the following example for a more specific illustration of the above concepts and results.

Example 6.7 *Consider the following system*

$$x(t+1) = u(t)(x(t) - 1) + 1 + w(t), \tag{6.47}$$

where $x(t), x(t+1) \in X = [0, 2]$, $u(t) \in U = [-0.9, -0.8]$, $w(t) \in W = [-0.01, 0.01]$. The set of atomic propositions is $\Pi = \{a_1, a_2\}$, and the labeling function is given by $L(a_1) = [0.1, 0.2]$, $L(a_2) = [0.5, 0.6]$.

Let the specification be $\varphi = \Diamond(a_1 \wedge \Diamond a_2)$, which can be translated into a DBA shown in Figure 6.5.

By Definition 6.7, $W_\mathcal{S}(q_0) = X$, $W_\mathcal{S}(q_1)$ and $W_\mathcal{S}(q_2)$ are the sets of states that can reach $L(a_2)$ and $L(a_1) \cap W_\mathcal{S}(q_1)$, respectively. They can be computed exactly: $W_\mathcal{S}^0(q_1) = [0, 0.6] \cup [1.444, 2]$, $W_\mathcal{S}^0(q_2) = [0, 0.012] \cup [0.1, 0.2] \cup [1.889, 2]$, $W_\mathcal{S}^\delta(q_1) = [0, 0.483] \cup [0.5, 0.6] \cup [1.456, 2]$, and $W_\mathcal{S}^\delta(q_2) = [0.1, 0.2] \cup [1.9, 2]$.

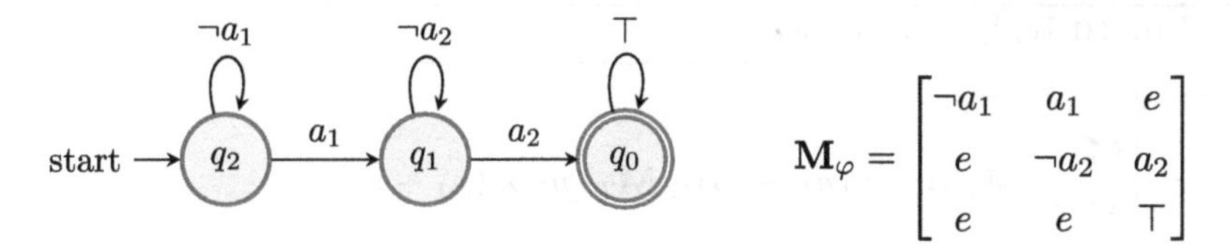

FIGURE 6.5: The DBA of φ and $\mathbf{M}_\varphi$ with the order q_2, q_1, q_0.

One can verify Proposition 6.14 for both system (6.47) and its nominal system ($\delta = 0$):

$$\begin{aligned} W_S^\delta(q_1) &= Pre^\delta(W_S(q_0)|L(a_2)) \cup Pre^\delta(W_S(q_1)|L(\neg a_2)), \\ W_S(q_2)^\delta &= Pre^\delta(W_S(q_1)|L(a_1)) \cup Pre^\delta(W_S(q_2)|L(\neg a_1)). \end{aligned}$$

6.6.2 Completeness via Robustness

Same as previous sections, the determination of $\mathcal{S}$-domains for system (6.1) involves the computation of the operator $\mathbf{T}$, which is nontrivial under general nonlinear dynamics. Even if $\mathbf{T}(\mathbf{M}_\varphi, V)$ can be computed precisely for a vector V, the computation of finding a fixed point is usually iterative and not guaranteed to terminate in a finite number of iterations, because system (6.1) contains infinite number of states. Since the definition of $\mathbf{T}$, however, has a strong connection to the controlled predecessor operator Pre, the sound and robustly complete solution to the DBA control synthesis problem is not difficult to construct based on Algoritm 5, which we have used many times.

The input arguments of Algorithm 10 are the transition matrix $\mathbf{M}_\varphi$ of $\mathcal{A}_\varphi$ that reflects the transition relation of system $\mathcal{S}$ and an operator $[\mathbf{T}_\mu]^\varepsilon$, which is defined by replacing Pre by $[\underline{\mathrm{Pre}}_\mu]^\varepsilon$. We assume that the nodes of $\mathcal{A}_\varphi$ are sorted so that accepting nodes rank after non-accepting ones, and the transition matrix $\mathbf{M}_\varphi$ is divided into 2 matrix blocks $\mathbf{M}_1$, $\mathbf{M}_2$ which represent the transitions from non-accepting and accepting nodes, respectively. Let $\bar{F} = Q \setminus F$.

Let l and ν denote the indices of the inner loop and outer loop of Algorithm 10, respectively. $\{Y_\nu^l\}_{l=0}^\infty$ denotes the sequence of vectors generated by the inner loop of the νth outer loop, and $\{Z_\nu\}_{\nu=0}^\infty$ denotes the sequence of vectors generated by the outer loop. It is clear that $Y_\nu^0 = \emptyset$ and $Y_\nu^l \preceq Y_\nu^{l+1}$ by (6.40) for all $l \in \mathbb{N}$. It follows that $\{Y_\nu^l\}_{l=0}^\infty$ is increasing for all $\nu \in \mathbb{N}$. Let Y_ν be the fixed point of the νth outer loop, i.e., $Y_\nu = \bigcup_{l=0}^\infty Y_\nu^l$.

Intuitively, by setting $Y_\nu^0 = \emptyset$ for all ν, the inner loop generates the set of states for each non-accepting node that can be controlled to reach one of the elements in Z_ν. The outer loop keeps the states in each element of Z_ν that can still be controlled to reach any of the elements in Z_ν. The sequences $\{Y_\nu\}_{\nu=0}^\infty$ and $\{Z_\nu\}_{\nu=0}^\infty$ approach the $\mathcal{S}$-domains of the non-accepting and accepting

Algorithm 10 $W, \mathcal{K} = \text{SDOM}(\mathbf{M}_\varphi, [\mathbf{T}_\mu]^\varepsilon)$

1: $n_1 = |Q| - |F|$, $n_2 = |F|$

2: $\mathbf{M}_\varphi = \begin{bmatrix} \mathbf{M}_1 \\ \mathbf{M}_2 \end{bmatrix}$, $\mathbf{M}_1$ $(n_1 \times (n_1 + n_2))$, $\mathbf{M}_2$ $(n_2 \times (n_1 + n_2))$

3: $\widetilde{Y}(n_1 \times 1)$, $\mathcal{K}(1 \times |Q|)$

4: $\widetilde{Z}(n_2 \times 1)$, $\widetilde{Z}[j] \leftarrow X$, $j \in \{1, \cdots, n_2\}$

5: **repeat**

6: $\quad Z \leftarrow \widetilde{Z}$

7: $\quad \widetilde{Y}[i] \leftarrow \emptyset$, $i \in \{1, \ldots, n_1\}$

8: $\quad$ **repeat**

9: $\quad\quad Y \leftarrow \widetilde{Y}$

10: $\quad\quad \widetilde{Y} \leftarrow Y + [\mathbf{T}_\mu]^\varepsilon(\mathbf{M}_1, \begin{bmatrix} Y \\ Z \end{bmatrix})$ $\quad \triangleright$ (6.40) (6.44)

11: $\quad\quad$ assign $\mathcal{K}[i](x)$ by (6.46) for all $x \in \widetilde{Y}[i] \setminus Y[i]$ and $i \in \{1, \cdots, n_1\}$

12: $\quad$ **until** $Y = \widetilde{Y}$

13: $\quad \widetilde{Z} \leftarrow [\mathbf{T}_\mu]^\varepsilon(\mathbf{M}_2, \begin{bmatrix} Y \\ Z \end{bmatrix})$ $\quad \triangleright$ (6.44)

14: $\quad$ assign $\mathcal{K}[n_1 + j](x)$ by (6.46) for all $x \in \widetilde{Z}[j]$ and $j \in \{1, \cdots, n_2\}$

15: **until** $Z = \widetilde{Z}$ $\quad \triangleright$ (6.43)

16: **Return** $W \leftarrow \begin{bmatrix} Y \\ Z \end{bmatrix}$, $\mathcal{K}$

nodes, respectively. Define

$$W = \begin{bmatrix} Y \\ Z \end{bmatrix} = \bigcap_{\nu=0}^{\infty} \begin{bmatrix} Y_\nu \\ Z_\nu \end{bmatrix} = \bigcap_{\nu=0}^{\infty} W_\nu, \quad W_\nu = \begin{bmatrix} Y_\nu \\ Z_\nu \end{bmatrix}. \tag{6.48}$$

The proof of the main completeness result is based on the following lemma.

Lemma 6.1 *Consider system (6.1) with an under-sampled control set defined in (4.38) with the parameter* μ. *Let* $\mathbf{M}$ *be a matrix of symbols from* Σ *and* W *be a vector of subsets of* X, *with* $\mathbf{M}$ *and* W *having the same dimension. If Assumptions 4.1 and 4.2 hold with Lipschitz constants* ρ_1 *and* ρ_2, *respectively, then*

$$\mathbf{T}^r(\mathbf{M}, W) \preceq [\mathbf{T}_\mu]^\varepsilon(\mathbf{M}, W) \preceq \mathbf{T}(\mathbf{M}, W),$$

where $r = \rho_1\varepsilon + \rho_2\mu$.

Proof: Let $V = \mathbf{T}(\mathbf{M}, W)$, $V' = [\mathbf{T}_\mu]^\varepsilon(\mathbf{M}, W)$ and $V'' = \mathbf{T}^r(\mathbf{M}, W)$, where $r = \rho_1\varepsilon + \rho_2\mu$. Assume that $\mathbf{M}$ and W are $n_1 \times n_2$ and $n_2 \times 1$, respectively.

Then for all $i \in \{1, \cdots, n_1\}$,

$$V[i] = \bigcup_{j=1}^{n_2} \text{Pre}\left(W[j] \| L^{-1}(m_{ij})\right),$$

$$V'[i] = \bigcup_{j=1}^{n_2} [\underline{\text{Pre}_\mu}]^\varepsilon \left(W[j] \| L^{-1}(m_{ij})\right),$$

$$V''[i] = \bigcup_{j=1}^{n_2} \text{Pre}^r \left(W[j] \| L^{-1}(m_{ij})\right),$$

where $V[i]$ is the ith element of V. Lemma 4.3 gives

$$\text{Pre}^r \left(W[j]\right) = \text{Pre}\left(W[j] - \mathbb{B}_r\right) \subseteq [\underline{\text{Pre}_\mu}]^\varepsilon \left(W[j]\right) \subseteq \text{Pre}\left(W[j]\right)$$

for all $j \in \{1, \cdots, n_2\}$. Then we have $V''[i] \subseteq V'[i] \subseteq V[i]$, which shows that $V'' \preceq V' \preceq V$ by (6.42). ■

Based on Lemma 6.1, we show in Theorem 6.5 that, by using a sufficiently small precision parameter ε and a sampling grid size μ for the control set in Algorithm 10, control synthesis for system (6.1) with respect to DBA-recognizable LTL formulas can be made sound and robustly complete.

Theorem 6.5 *Consider the system (6.1) and a DBA $\mathcal{A}_\varphi$. Denote by $\mathbf{M}_\varphi$ the transition matrix of $\mathcal{A}_\varphi$. Let $[\mathbf{T}_\mu]^\varepsilon$ be an interval implementation of $\mathbf{T}$ for system (6.1), where ε is the precision and μ is the sampling parameter in (4.38). Suppose that Assumptions 4.1 and 4.2 hold with Lipschitz constants ρ_1 and ρ_2, respectively. Then Algorithm 10 terminates in a finite number of iterations. Furthermore, if $\rho_1\varepsilon + \rho_2\mu \leq \delta$, then*

$$\mathbf{Win}_{(6.2)}(\varphi) \subseteq W(q_0) \subseteq \mathbf{Win}_{(6.1)}(\varphi), \tag{6.49}$$

where $W(q_0)$ is the element of W corresponding to the initial state q_0 of $\mathcal{A}_\varphi$.

Proof: Let $\mathbf{W}_{(6.1)}$ and $\mathbf{W}_{(6.2)}$ be the $\mathcal{S}$-domains of systems (6.1) and (6.2), respectively. Let $i, i' \in \{1, \cdots, n_1\}$ and $j, j' \in \{1, \cdots, n_2\}$ denote the indices of the elements in $\bar{F}$ and F, respectively. We assume i, i', j, j' to be arbitrary throughout the proof.

Claim: For all $\nu \in \mathbb{Z}_{>0}$, $Y_\nu \preceq Y_{\nu-1}$ if $Z_\nu \preceq Z_{\nu-1}$. By Proposition 6.13 (i), $Y_\nu^0 = Y_{\nu-1}^0 = \emptyset$, and $Z_\nu \preceq Z_{\nu-1}$, we have

$$\begin{aligned} Y_\nu^1 &= Y_\nu^0 + [\mathbf{T}_\mu]^\varepsilon \left(\mathbf{M}_1, \begin{bmatrix} Y_\nu^0 \\ Z_\nu \end{bmatrix}\right) \\ &\preceq Y_{\nu-1}^0 + [\mathbf{T}_\mu]^\varepsilon \left(\mathbf{M}_1, \begin{bmatrix} Y_{\nu-1}^0 \\ Z_{\nu-1} \end{bmatrix}\right) = Y_{\nu-1}^1. \end{aligned}$$

This gives $Y_\nu^l \preceq Y_{\nu-1}^l$ for all $l \in \mathbb{N}$. It follows that $Y_\nu = \bigcup_{l=0}^\infty Y_\nu^l \preceq \bigcup_{l=0}^\infty Y_{\nu-1}^l = Y_{\nu-1}$. Hence, the claim is proved.

To see the finite termination of Algorithm 10, we first show that $\{Z_\nu\}_{\nu=0}^\infty$ and $\{Y_\nu\}_{\nu=0}^\infty$ are decreasing by induction. We have $Z_1 \preceq Z_0$ since $Z_0[j] = X$. Then $Y_1 \preceq Y_0$ by the claim. Assume $Z_\nu \preceq Z_{\nu-1}$ and $Y_\nu \preceq Y_{\nu-1}$ for some $\nu \in \mathbb{Z}_{>0}$. Then $Z_{\nu+1} = [\mathbf{T}_\mu]^\varepsilon(\mathbf{M}_2, W_\nu) \preceq [\mathbf{T}_\mu]^\varepsilon(\mathbf{M}_2, W_{\nu-1}) = Z_\nu$ by Proposition 6.13 (i), and $Y_{\nu+1} \preceq Y_\nu$ by the claim, which shows that $\{Z_\nu\}_{\nu=0}^\infty$ and $\{Y_\nu\}_{\nu=0}^\infty$ are decreasing.

Since the widths of the intervals that partition the state space are lower bounded by ε, each element in Y and Z contains finitely many intervals. Hence, Algorithm 10 will terminate in a finite number of steps because $\{Y_\nu^l\}_{l=0}^\infty$ and $\{Z_\nu\}_{\nu=0}^\infty$ are decreasing. This implies that $\exists N_Z \in \mathbb{N}$ such that $Z_{N_Z} = Z_{N_Z+1}$, and for all ν, $\exists N_Y(\nu) \in \mathbb{N}$ such that $Y_\nu^{N_Y(\nu)} = Y_\nu^{N_Y(\nu)+1} = Y_\nu$ and $Y_{N_Z} = Y_{N_Z+1}$. It follows that

$$Y_\nu = Y_\nu + [\mathbf{T}_\mu]^\varepsilon(\mathbf{M}_1, W_\nu) \ \forall \nu \in \mathbb{N}, \tag{6.50}$$

$$Z = [\mathbf{T}_\mu]^\varepsilon(\mathbf{M}_2, W), \tag{6.51}$$

where $Z = Z_{N_Z}$, $Y = Y_{N_Z}^{N_Y}$, and $W = \begin{bmatrix} Y & Z \end{bmatrix}^T$.

By Proposition 6.13 (ii), if $\rho_1\varepsilon + \rho_2\mu \le \delta$, then $\mathbf{T}^\delta(\mathbf{M}, V) \preceq \mathbf{T}^{\rho_1\varepsilon+\rho_2\mu}(\mathbf{M}, V)$ for any consistent vector V of sets. Together with Lemma 6.1, we have

$$\mathbf{T}^\delta(\mathbf{M}, V) \preceq [\mathbf{T}_\mu]^\varepsilon(\mathbf{M}, V) \preceq \mathbf{T}(\mathbf{M}, V). \tag{6.52}$$

Let ϱ be a run of $\mathcal{A}_\varphi$ resulting from a solution $\{x_t\}_{t=0}^\infty$ and $\varrho[t]$ denote the DBA state at time t.

We now show the soundness, i.e., $W \preceq \mathbf{W}_{(6.1)}$. Let $A_\nu^0 = Y_\nu^0 = \emptyset$ and $A_\nu^l = A_\nu^{l-1} + \mathbf{T}(\mathbf{M}_1, \begin{bmatrix} A_\nu^l & Z_\nu \end{bmatrix}^T)$ for $l \in \mathbb{Z}_{>0}$. By the definition of $\mathbf{T}$, if $\varrho[t] = \bar{F}[i]$, any $x_t \in A_\nu^l[i]$ for $\mathcal{S}^0$ can trigger the transition in $\mathcal{A}_\varphi$ to $F[j]$ within l steps. By (6.52), $Y_\nu^l \preceq A_\nu^l$ for all $l \in \mathbb{N}$, which leads to $Y \preceq A_{N_Z}^{N_Y(N_Z)}$. That is to say, for $\mathcal{S}^0$, $\forall x_t \in Y[i]$, $\exists \{u_k\}_{k=t}^{t+N_Y(N_Z)-1}$ such that $\varrho[t'] \in F$ for some $t' \in [t, t+N_Y(N_Z)]$. Based on (6.51) and (6.52), we have $Z \preceq \mathbf{T}(\mathbf{M}_2, W)$. This means if $\varrho[t] = F[j]$ and $x_t \in Z[j]$ for $\mathcal{S}^0$, then either (i) $x_{t+1} \in Y[i]$ if $\varrho[t+1] = \bar{F}[i]$, and F can be visited again from $Y[i]$, or (ii) $x_{t+1} \in Z[i']$ if $\varrho[t+1] = F[j']$. Hence, we can conclude that $W \preceq \mathbf{W}_{(6.1)}$.

To see $\mathbf{W}_{(6.2)} \preceq W$, we aim to show $x \notin Y[i] \Rightarrow x \notin \mathbf{W}_{(6.2)}(\bar{F}[i])$ and $x \notin Z[j] \Rightarrow x \notin \mathbf{W}_{(6.2)}(F[j])$. Consider arbitrary $t \in \mathbb{N}$ and arbitrary $\nu \in \{1, \cdots, N_Z\}$. We first discuss two situations based on (6.52) and the definition of $\mathbf{T}^\delta$ in (6.44):

(i) $\varrho[t]$ is accepting, i.e., $\varrho[t] = \bar{F}[i]$: if $x_t \notin Y_\nu[i]$, then $\forall u_t \in U$, $\exists w_t$ with $\|w_t\|_\infty \le \delta$ such that

$$\begin{cases} x_{t+1} \notin Y_\nu[i'], & \text{if } \varrho[t+1] = r(\bar{F}[i], L(x_t)) \notin F, \\ x_{t+1} \notin Z_\nu[j], & \text{if } \varrho[t+1] = r(\bar{F}[i], L(x_t)) = F[j]. \end{cases}$$

In addition, by (6.50), if $x_t \notin Y_\nu[i]$ and $\varrho[t+1] \notin F$, then $x_t \notin [\mathbf{T}_\mu]^\varepsilon(\mathbf{M}_1, W_\nu)$. Otherwise $Y_\nu \neq Y_\nu + [\mathbf{T}_\mu]^\varepsilon(\mathbf{M}_1, W_\nu)$. This implies $\forall u_t \in U$, $\exists w_t$ with $\|w_t\|_\infty \leq \delta$ such that $x_{t+1} \notin Y_\nu$ if $\varrho[t+1] \notin F$.

(ii) $\varrho[t]$ is non-accepting, i.e., $\varrho[t] = F[j]$: if $x_t \notin Z_\nu[j]$, then $\forall u_t \in U$, $\exists w_t$ with $\|w_t\|_\infty \leq \delta$ such that

$$\begin{cases} x_{t+1} \notin Y_{\nu-1}[i], & \text{if } \varrho[t+1] = r(F[j], L(x_t)) \notin F, \\ x_{t+1} \notin Z_{\nu-1}[j'], & \text{if } \varrho[t+1] = r(F[j], L(x_t)) = F[j']. \end{cases}$$

If $x_0 = x \notin Y[i]$ (i.e., $x_0 \notin Y_{N_Z}[i]$), then $x_1 \notin Y_{N_Z}[i']$ if $\rho[1] \notin F$ and $x_1 \notin Z_{N_Z}[j]$ if $\rho[1] \in F$ by (i). If $\rho[k] = F[j]$ for some $k_0 \in \mathbb{Z}_{>0}$, then $x_{k_0} \notin Z_{N_Z}[j]$; otherwise $\rho[t] \notin F$ for all t, which implies $x \notin \mathbf{W}_{(6.2)}(\bar{F}[i])$. Considering (ii) for $x_{k_0} \notin Z_{N_Z}[j]$, we have $x_{k_0+1} \notin Y_{N_Z-1}[i']$ if $\varrho[k_0+1] = \bar{F}[i']$ and $x_{k_0+1} \notin Z_{N_Z-1}[j']$ if $\varrho[k_0+1] = F[j']$. Combining (i) and (ii) in this manner, ν decreases from N_Z by 1 every time ϱ visits F until $\nu = 0$ at some $k_{N_Z} \in \mathbb{Z}_{>0}$, and $x_{k_{N_Z}} \notin Y_0[i']$. If $\varrho[k_{N_Z}+1] = F[j]$ for some $j \in \{1, \cdots, n_2\}$, then $x_{k+1} \notin Z_0[j]$, which means is impossible since Z_0 is a vector of the full state space of the system. Hence, we have $\varrho[t'] \notin F$ for all $t' \geq k_{N_Z}$, which gives $x \notin \mathbf{W}_{(6.2)}(\bar{F}[i])$. The same argument applies to the case in which $x_0 = x \notin Z_\nu[j]$ and proves $x \notin \mathbf{W}_{(6.2)}(F[j])$. Therefore, $\mathbf{W}_{(6.2)} \preceq W$. ■

Example 6.8 *Consider again Example 6.7. We can choose $\varepsilon = \mu = 0.005$ by Theorem 6.5 since the Lipschitz constant $\rho_1 = \rho_2 = 1$. Let $Y = [W[2], W[1]]^T$ and $Z = W[0]$ with $W[i]$ approximating $W_\mathcal{S}(q_i)$ ($i = 0, 1, 2$) in Algorithm 10. Initially, $Y^0 = [\emptyset, \emptyset]^T$ and $Z^0 = X$. For the lth inner iteration,*

$$W^l[1] = W^{l-1}[1] \cup [\underline{Pre}_\mu]^\varepsilon(W^{l-1}[0]|L(a_2)) \cup [\underline{Pre}_\mu]^\varepsilon(W^{l-1}[1]|L(\neg a_2)),$$
$$W^l[2] = W^{l-1}[2] \cup [\underline{Pre}_\mu]^\varepsilon(W^{l-1}[1]|L(a_1)) \cup [\underline{Pre}_\mu]^\varepsilon(W^{l-1}[2]|L(\neg a_1)).$$

Algorithm 10 terminates after 7 inner iterations and 1 outer iteration, and it returns $W[0] = X$, $W[1] = [0, 0.6] \cup [1.448, 2]$ and $W[2] = [0, 0.003] \cup [0.1, 0.2] \cup [1.893, 2]$, which shows (6.49).

Similar to the classic Büchi game algorithm designed for finite-state systems, Algorithm 10 is composed of two nested fixed-point iterations of $\mathbf{T}^\delta(\mathbf{M}_\varphi, \cdot)$. The outer loop computes $\mathbf{Win}_{(6.1)}(F)$, and the inner loop computes $\mathbf{Win}_{(6.1)}(\bar{F})$, where $\mathbf{Win}_{(6.1)}(F)$ and $\mathbf{Win}_{(6.1)}(\bar{F})$ denote the vectors of $\mathcal{S}$-domains of states in F and $\bar{F}$ ($\bar{F} = Q \backslash F$), respectively. But simply applying the classic algorithm directly to continuous-state systems (using the exact operator $\mathbf{T}$) will not give the actual winning set. This is because the sequence of sets $\{Y_\nu\}$ are required to be compact to conclude $\lim_{\nu \to \infty} Z_\nu = \mathbf{Win}_{(6.1)}(F)$, which is similar to computing maximal controlled invariant sets, while $\{Y_\nu\}$ need to be open so that $\lim_{\nu \to \infty} Y_\nu = \mathbf{Win}_{(6.1)}(\bar{F})$, which is equivalent to the backward reachable set computation in Section 6.4. By using an approximation $[\mathbf{T}_\mu]^\varepsilon$ of $\mathbf{T}$, however, Algorithm 10 can terminate in finite time and yield an approximation of the real winning set that is lower-bounded by the winning set with δ-perturbation. An arbitrarily accurate approximation of the winning set $\mathbf{Win}_{(6.1)}(\varphi)$ can be achieved if $\lim_{\delta\to 0}\mathbf{Win}_{(6.2)}(\varphi) = \mathbf{Win}_{(6.1)}(\varphi)$.

Algorithm 10 essentially induces a finite-memory control strategy that is embedded with the given DBA.

Definition 6.9 *Let $\mathcal{A}_\varphi = (Q, \Sigma, r, q_0, F)$ be an equivalent DBA of an LTL formula φ. An* automaton-embedded control strategy *for system (6.1) is*

$$\mathcal{C}_\varphi = \langle X_c, U_c, Q_c, \Sigma_c, r_c, q_0, H\rangle : \tag{6.53}$$

- $X_c \subseteq X$ *is a set of inputs;*
- $Q_c = Q$ *is a finite set of states;*
- $\Sigma_c = \Sigma = 2^{AP}$ *is an alphabet;*
- $r_c = r \subseteq Q_c \times \Sigma_c \times Q_c$ *is a transition relation that updates the controller state;*
- q_0 *is the initial state;*
- $U_c \subseteq 2^U$ *is a set of outputs;*
- $H : Q_c \times X_c \to U_c$ *is an output function defined by*

  $$H(q, x) = \kappa_{Id(q)+1}(x),\ x \in X_c, q \in Q_c,$$

  where $\kappa_{Id(q)}(x)$ *belongs to the set of memoryless control strategies* $\{\kappa_i\}_{i=1}^{|Q_c|}$ *returned by Algorithm 10 and* $Id(q)$ *is the index of the state* q.

The components Q_c, Σ_c, r_c, q_0 originally given in $\mathcal{A}_\varphi$ are embedded into $\mathcal{C}_\varphi$. One can use a single variable that takes values in a subset of $\mathbb{N}$ to represent Q. Such a variable is called a *memory variable.* A memoryless control strategy κ from $\mathcal{K}$ is activated by the function H, which outputs the index of current state q of $\mathcal{A}_\varphi$ by the transition relation r of $\mathcal{A}_\varphi$ according to the previous automaton state and the labels $L(x)$ of the current system state x. Therefore, the embedded $\mathcal{A}_\varphi$ manages the control memory, and the structure in Definition 6.9 is visualized in Figure 6.6.

Corollary 6.4 *Consider system (6.1) and an LTL formula φ. Let $\mathcal{A}_\varphi$ be an equivalent DBA for an LTL specification φ. If φ is robustly realizable for system (6.1), then there exists a finite-memory control strategy (6.53) to realize φ for system (6.1).*

It is worth noting that the output W of Algorithm 10 is a vector of unions of non-uniform intervals that under-approximate the $\mathcal{S}$-domains of $\mathcal{A}_\varphi$. This is a result of using an interval implementation $[\mathbf{T}_\mu]^\varepsilon$ of $\mathbf{T}$, in which the approximation of controlled predecessors are obtained by adaptively partition the state space X with respect to both the dynamics and the target sets. Such a computation scheme can lead to an efficiency gain by avoiding discretizing X uniformly with a high precision except in constraint-critical areas.

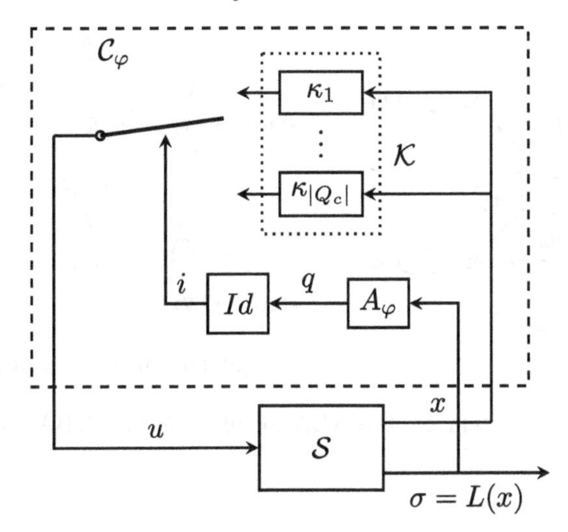

FIGURE 6.6: The finite-memory controller structure.

6.6.3 Sound Control Synthesis for Full LTL

As we have seen, a DBA not only guides control synthesis but also serves as the memory update mechanism in the resulting finite-memory controller. Unlike DBA, NBA cannot be used to update controller memory variables because the updating rule is nondeterministic. For this reason, determinization of NBA (into deterministic automata such as deterministic Rabin automata (DRA)) is usually applied in general LTL control synthesis. Similar to Algorithm 1, an interval implementation of a Rabin game algorithm [182] would, in principle, give robust completeness guarantees for full LTL specifications. Nonetheless, the inherent complexity of a complete algorithm for solving a Rabin game would render such an approach impractical.

In most of the control applications, it is usually unnecessary to seek a complete set of control solutions. Sound solutions that can be found in a reasonable time are more appealing. A simple and direct treatment is to trim off redundant transitions under the same input propositions in an NBA, which generates a DBA that specifies a subset of behaviors of the original NBA. Based on the DBA trimmed from the NBA translated from a general LTL formula, Algorithm 10 can provide sound solutions to the corresponding control synthesis problem. The following example demonstrates such a process.

Example 6.9 *Consider a discrete system S' (Figure 6.7a) and the correspond NBA (Figure 6.7b) of the formula $\Diamond\Box b$. The NBA can be trimmed into the DBA in Figure 6.7c. $L(s_1) = L(s_2) = \{b\}$ and $L(s_0) = L(s_3) = L(s_4) = \{\neg b\}$. The real winning set of system S' with respect to $\Diamond\Box b$ is $\{s_0, s_1, s_2, s_3, s_4\}$. By using Algorithm 10, the vector Y, which will be returned as a vector $[W_{\mathcal{S}}(q_0), W_{\mathcal{S}}(q_2)]^T$ of $\mathcal{S}$-domains, is initialized to $Y_0^0 = [\emptyset, \emptyset]^T$, and Z, which*

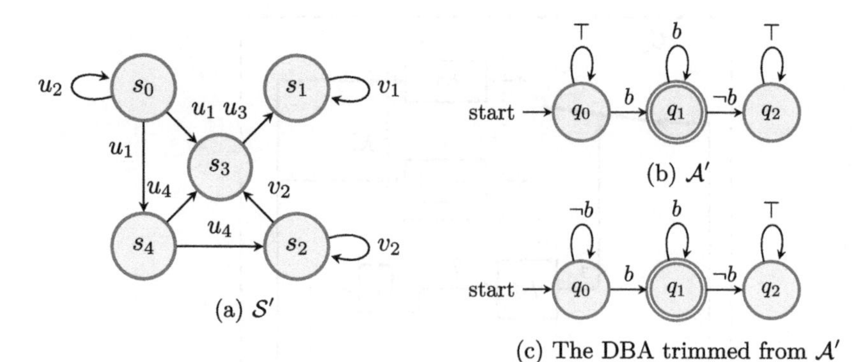

FIGURE 6.7: The system $\mathcal{S}'$, the NBA $\mathcal{A}'$ of $\Diamond\Box b$, and DBA trimmed from $\mathcal{A}'$.

represents $W_{\mathcal{S}}(q_1)$, is initialized to $Z_0 = \{s_0, s_1, s_2, s_3, s_4\}$. By the transition matrix of $\mathcal{A}'$ and (6.44), for the lth iteration of the 1st inner loop (line 8-12),

$$Y_0^l[1] = Y_0^{l-1} \cup Pre^{\delta}(Z_0|L^{-1}(b)) \cup Pre^{\delta}(Y_0^{l-1}[1]|L^{-1}(\neg b)) = \{s_1, s_2\},$$

and the first inner loop finishes in 4 iterations with

$$Y_0[1] = Y_0^4[1] = \{s_0, s_1, s_2, s_3, s_4\}$$
$$Y_0[2] = \emptyset.$$

Then

$$Z_1 = Pre^{\delta}(Z_0|L^{-1}(b)) \cup Pre^{\delta}(Y_0[2]|L^{-1}(\neg b)) = \{s_1, s_2\}.$$

Using Z_1 in the 2nd inner loop gives $Y_1[1] = \{s_1, s_3\}$ and $Y_1[2] = \emptyset$. As such, the returned vectors are $Z = \{s_1\}$ and $Y = [\{s_1, s_3\}, \emptyset]^T$. Since q_0 is the initial node, the winning set obtained by Algorithm 10 is $Y[1] = \{s_1, s_3\}$, which is a subset of the real one.

6.6.4 Specification Pre-Processing

We are also concerned with the computational complexity of control synthesis. In Algorithm 10, the vector Z updates only after the vector Y remains unchanged in the inner loop. In addition, at the beginning of each computation in the outer loop, the value of Y needs to be reinitialized since the value of Z has changed from the last iteration. In this sense, the interdependency between Z and Y increases the complexity.

Suppose that the transition matrix $\mathbf{M}_{\varphi}$ of a DBA $\mathcal{A}_{\varphi}$ is in the form:

$$\mathbf{M}_{\varphi} = \begin{bmatrix} \mathbf{M}_{11} & \mathbf{M}_{12} \\ \mathbf{M}_{e} & \mathbf{M}_{22} \end{bmatrix}_{|Q|\times|Q|}, \tag{6.54}$$

where $\mathbf{M}_{11}$ and $\mathbf{M}_{22}$ are n_L by n_L and n_R by n_R matrices, respectively, and $n_L + n_R = |Q|$, $n_L, n_R \in \mathbb{Z}_{>0}$. $\mathbf{M}_e$ is a matrix of empty symbols (i.e., e's). Let Q_L be the set of states of $\mathcal{A}_\varphi$ with the first n_L indices and Q_R be the set of the rest of the states. Define

$$F_L = \{q \in Q \mid q \in F \wedge q \in Q_L\},$$
$$F_R = \{q \in Q \mid q \in F \wedge q \in Q_R\}.$$

Denote $n_{L2} = |F_L|$, $n_{R2} = |F_R|$, $n_{L1} = n_L - n_{L2}$, and $n_{R1} = n_R - n_{R2}$. We also assume that the states in Q_L and Q_R are sorted so that the accepting states always rank after non-accepting ones.

If Q_R contains accepting nodes, then $\mathbf{M}_{22}$ can be treated as a sub-transition matrix based on which $\{\mathbf{W}_\mathcal{S}(q)\}_{q \in Q_R}$ can be approximated firstly by Algorithm 10, independent of other parts of $\mathcal{A}_\varphi$. If Q_R has no accepting nodes, then computing $\{\mathbf{W}_\mathcal{S}(q)\}_{q \in Q_R}$ is pointless because there is no transition from any $q \in Q_R$ to $q' \in Q_L$ and any run that contains q does not satisfy the Büchi accepting condition. The approximation of $\{\mathbf{W}_\mathcal{S}(q)\}_{q \in Q_L}$, starts after the computation with respect to $\mathbf{M}_{22}$ completes. In this way, the repetitive initialization and computation of W_L caused by the updates in W_R can be avoided. Therefore, if we can arrange the transition matrix $\mathbf{M}_\varphi$ into a triangular matrix or triangular block matrix without changing the original transition relations in $\mathcal{A}_\varphi$, then Algorithm 10 can reduce to a single loop or several smaller nested loops. We now compare the complexities of Algorithm 10 applying directly to $\mathbf{M}_\varphi$ and sequentially to the blocks in (6.54).

Suppose that the number of accepting and non-accepting nodes in $\mathcal{A}_\varphi$ is n_2 and $n_1 = |Q| - n_2$, respectively, and that the resulting numbers of outer-loop and inner-loop iterations by using Algorithm 10 directly are K_2 and K_1. Then the complexity is $\mathcal{O}(n_2 K_2 n_1 K_1)$ for the control synthesis without using its triangular form. Let the numbers of outer and inner-loop iterations for block $\mathbf{M}_{22}$ be K_{R2} and K_{R1}, respectively, and the ones for block $\begin{bmatrix}\mathbf{M}_{11} & \mathbf{M}_{12}\end{bmatrix}$ be K_{L2} and K_{L1}, respectively. If we perform control synthesis sequentially to blocks $\mathbf{M}_{22}$ and $\begin{bmatrix}\mathbf{M}_{11} & \mathbf{M}_{12}\end{bmatrix}$, the complexity is $\mathcal{O}(n_{R2} K_{R2} n_{R1} K_{R1} + n_{L2} K_{L2} n_{L1} K_{L1})$. The number of outer and inner-loop iterations is determined by the row that converges the slowest. As defined before, $n_2 = n_{L2} + n_{R2}$ and $n_1 = n_{L1} + n_{R1}$. Then

$$\begin{aligned}
&n_{R2} K_{R2} n_{R1} K_{R1} + n_{L2} K_{L2} n_{L1} K_{L1} \\
&\leq (n_{R2} n_{R1} + n_{L2} n_{L1}) K_2 K_1 \\
&< (n_{R2} + n_{L2})(n_{R1} + n_{L1}) K_2 K_1 = n_2 n_1 K_2 K_1,
\end{aligned}$$

which shows that we can gain computational efficiency by using an upper triangular block matrix.

To reform $\mathbf{M}_\varphi$ so that $\mathbf{M}_\varphi$ is a triangular block matrix, we can follow the PREPROCESS procedure.

Algorithm 11 $\mathbf{M}_\varphi = \text{PREPROCESS}(\mathcal{A}_\varphi)$

1: Detect all SCCs[1] in the graph representation of $\mathcal{A}_\varphi$. Then $\mathcal{A}_\varphi$ is simplified to a DAG[2] $\mathcal{G}_{dag} = (V, E)$ in which each node $v \in V$ is either a single state or an SCC.
2: Perform a *topological sort* on the DAG $\mathcal{G}_{dag}$, which determines a linear ordering of the nodes in $\mathcal{G}_{dag}$ so that v precedes v' for any $(v, \sigma, v') \in E$. Let $v_1 \cdots v_k \cdots v_{|V|}$ $(1 \leq k \leq |V|)$ be the resulting order and v_k is the last node that is or contains an accepting state.
3: List the states in Q in the order of $v_1 \cdots v_k$. No specific order of the states in the same $v \in V$ is required but the accepting states rank after the non-accepting ones.
4: Write $\mathbf{M}_\varphi$ with respect to the current order.

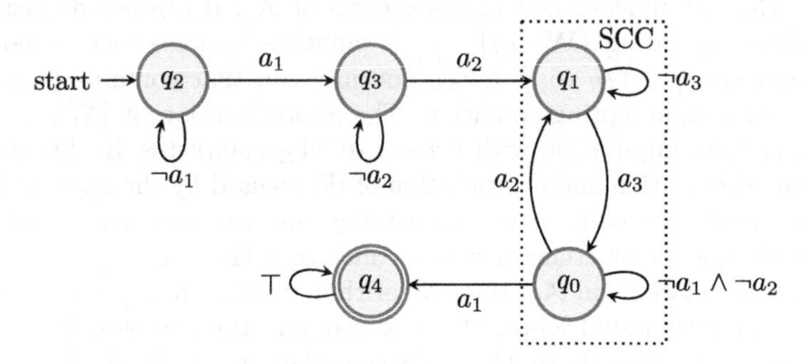

FIGURE 6.8: The translated DBA using Spot [62].

The transition matrix $\mathbf{M}_\varphi$ based on the order of the automaton states obtained by PREPROCESS can be formulated as an upper triangular block matrix. Algorithm 10, as a result, can be performed independently for the sub-matrices in the reversed order of the topological sort.

Example 6.10 *Consider the LTL formula*

$$\varphi = \Diamond(a_1 \wedge \Diamond(a_2 \wedge \Diamond(a_3 \wedge (\neg a_2)\mathbf{U}a_1))).$$

Its translated DBA is shown in Figure 6.8.

The states q_1 and q_0 constitute an SCC $q_{1,0}$, and the rest of the states are trivial SCCs. A topological sort of $\mathcal{A}_\varphi$ is $q_2 q_3 q_{1,0} q_4$, and q_4 is the unique accepting state. Then the PREPROCESS yields an order of the states in Q:

[1] A strongly connected component (SCC) is a (sub)graph where there exists a path between any two nodes. An SCC with one node is called a trivial SCC.

[2] A directed acyclic graph (DAG) is a directed graph without cycles.

$q_2q_3q_1q_0q_4$. *Based on this order, the transition matrix is*

$$\mathbf{M}_\varphi = \left[\begin{array}{c|c|ccc|c} \neg a_1 & a_1 & e & & e & e \\ \hline e & \neg a_2 & a_2 & & e & e \\ \hline e & e & \neg a_3 & & a_3 & e \\ e & e & a_2 & & \neg a_1 \wedge \neg a_2 & a_1 \\ \hline e & e & e & & e & \top \end{array}\right]$$

Since q_4 is the only accepting state and its corresponding block matrix $\mathbf{M}_4 = \top$, $\mathbf{W}_S(q_4) = X$ and $\kappa_4(x) = U$ for all $x \in X$. Algorithm 10 is then applied to $\mathbf{M}_\varphi$ backwardly and blockwisely.

6.7 Extension to Continuous-Time Dynamical Systems

We have shown in Chapter 5 that sampled-data control of nonlinear systems with temporal logic specifications is robustly decidable in the sense that, given a continuous-time nonlinear control system and a temporal logic formula, one can algorithmically decide whether there exists a robust sampled-data control strategy to realize this specification when the right-hand side of the system is slightly perturbed by a small disturbance. The same result can also be achieved by using direct interval computation. As opposed to using uniform discretization parameters that are sufficiently small, in the interval-based method presented in this section, the system state space is adaptively partitioned in sync with control synthesis. This can reduce the number of intervals involved in computation and therefore is more suitable for practical implementation.

Recall that τ-backward reachable sets for system (4.45) can be under-approximated by Algorithm 6. If we can also control the lower bound of such an under-approximation by the precision parameter ε just as for discrete-time systems, then there is a great chance that we can extend all the previous results for discrete-time systems to continuous-time systems. The following Proposition 6.15 gives us an affirmative answer by showing that the τ-backward reachable set $\mathcal{R}^\delta_\tau(X_0, u)$ from X_0 for system (4.44) contains the one of the nominal system (4.45) (i.e., $\mathcal{R}_\tau(X_0, u)$) with at least a margin of $\delta\tau$. If we can over-approximate $\mathcal{R}_\tau(X_0, u)$ within $\delta\tau$, then the approximated τ-controlled predecessor $[\mathrm{Pre}_\tau]^\varepsilon(X_0)$ returned by Algorithm 6 is guaranteed to contain $\mathrm{Pre}^\delta_\tau(X_0)$.

Proposition 6.15 *Let $D \subseteq X$. Suppose that Assumption 4.1 holds with the Lipschitz constant ρ_L. The τ-forward reachable set of (4.45) from an initial* *set of states $X_0 \subseteq D$ under a constant control input $u \in U$ satisfies*

$$\mathcal{R}_\tau(X_0, u) + \mathbb{B}_{r_1} \subseteq \mathcal{R}_\tau^\delta(X_0, u) \subseteq \mathcal{R}_\tau(X_0, u) + \mathbb{B}_{r_2}, \tag{6.55}$$

where $r_1 = \delta\tau$ and $r_2 = \delta\rho_L^{-1}(e^{\rho_L \tau} - 1)$.

Proof: Consider solutions $x(t)$ and $y(t)$ of $\dot{x}(t) = f(x(t), u(t)) + w(t)$ and $\dot{y}(t) = f(y(t), u(t))$ with $x(0) = y(0)$, respectively. Then

$$\begin{aligned} \|\dot{x}(t) - \dot{y}(t)\|_\infty &= \|f(x(t), u(t)) - f(y(t), u(t)) + w(t)\|_\infty \\ &\le \rho_L \|x(t) - y(t)\|_\infty + \|w(t)\|_\infty. \end{aligned}$$

Letting $z(t) = \|x(t) - y(t)\|_\infty \ge 0$ gives $\dot{z}(t) \le \rho_L z(t) + \delta$. By Gronwall's Lemma, we obtain that $\|z(t)\|_\infty \le \delta\rho_L^{-1}(e^{\rho_L \tau} - 1)$, which proves the right part of (6.55).

To prove the left part, let

$$w(t) = \delta \frac{f(x(t), u(t)) - f(y(t), u(t))}{\|f(x(t), u(t)) - f(y(t), u(t))\|_\infty}.$$

It follows that

$$\dot{z}(t) = \delta + \|f(x(t), u(t)) - f(y(t), u(t))\|_\infty \ge \delta.$$

Hence $z(\tau) \ge \delta\tau$ and the left part is proved. ■

So far, we have only considered reachable sets at sampling time instances. The inter-sample correctness, however, needs to be guaranteed to satisfy an LTL specification φ for a continuous-time system. Same as the abstraction-based methods presented in Chapter 5, we use the ϵ-strengthening L_ϵ of the labeling function L for control synthesis and choose the sampling time τ that satisfies

$$\epsilon \ge (M + \delta)\tau/2. \tag{6.56}$$

In this way, the satisfaction of φ by checking only the reachable sets obtained by using L_ϵ at sampling time instances can guarantee the satisfaction of φ for continuous-time systems.

Theorem 6.6 *Consider an LTL formula φ that can be translated into a DBA $\mathcal{A}_\varphi$ and an ϵ-strengthening of the corresponding labeling function for system (4.45). Suppose that Assumption 4.3 holds. Let W be the output of Algorithm 10 based on Algorithm 6 and q_0 is the initial state of $\mathcal{A}_\varphi$. Then*

$$\mathbf{Win}_{(4.44)}(\varphi) \subseteq W(q_0) \subseteq \mathbf{Win}_{(4.45)}(\varphi), \tag{6.57}$$

if τ satisfies (6.56) and the condition in Lemma 4.4, and the a priori *enclosure $\widehat{[x_0]}$ and the corresponding order $\bar{k}$ are constructed by Lemma 4.4 for any interval $[x_0] \subseteq X$ with $\mathrm{w}([x_0]) < \varepsilon$, and additionally,*

$$k \geq \max\left\{\bar{k}-1, \left\lceil \frac{\log \frac{(1-\alpha)\delta}{K\bar{\mathrm{w}}} + \log(\bar{k}+1)!}{\log \tau} \right\rceil \right\}, \tag{6.58}$$

$$\varepsilon \leq \frac{\alpha \tau}{2Ke^{\tau}}\delta, \tag{6.59}$$

where $\lceil \cdot \rceil$ is the ceiling function, $\alpha \in (0,1)$, $\bar{\mathrm{w}} = \mathrm{w}(\widehat{[x_0]})$.

Proof: For any interval $[x_0] \subseteq X$, by Lemma 4.4, there exists an order $\bar{k}$ and an *a priori* enclosure $\widehat{[x_0]}$ such that $\widehat{\mathcal{R}}^k_\tau([x_0], u)$ obtained by (4.23) is an over-approximation of the reachable set $\mathcal{R}_\tau([x_0], u)$.

For $B \subseteq A \subseteq X$, we first derive a sufficient condition such that

$$\mathrm{Pre}^\delta_\tau(B|A) \subseteq [\mathrm{Pre}_\tau]^\varepsilon(B|A) \subseteq \mathrm{Pre}_\tau(B|A). \tag{6.60}$$

It is trivial that $[\mathrm{Pre}_\tau]^\varepsilon(B|A) \subseteq \mathrm{Pre}_\tau(B|A)$ for all k and τ, so we only consider the conditions such that $\mathrm{Pre}^\delta_\tau(B|A) \subseteq [\mathrm{Pre}_\tau]^\varepsilon(B|A)$ here. Let x_0 be the center point of an arbitrary interval $[x_0]$ with $\mathrm{w}([x_0]) \leq 2\varepsilon$. Under Assumption 4.3, we rewrite (4.23) in the following centered form

$$\widehat{\mathcal{R}}^k_\tau([x_0], u) = \sum_{i=0}^{k} f^{[i]}(x_0, u)\frac{\tau^i}{i!} + \underbrace{[f]^{[k+1]}(\widehat{[x_0]}, u)\frac{\tau^{k+1}}{(k+1)!}}_{\text{truncation error}}$$
$$+ \underbrace{\sum_{i=0}^{k} K([x_0] - x_0)\frac{\tau^i}{i!}}_{\text{propagated enclosure}}.$$

For the propagated enclosure,

$$\mathrm{w}\left(\sum_{i=0}^{k} K([x_0] - x_0)\frac{\tau^i}{i!}\right) \leq 2K\varepsilon \sum_{i=0}^{k} \frac{\tau^i}{i!} \leq 2K\varepsilon \sum_{i=0}^{\infty} \frac{\tau^i}{i!} = 2K\varepsilon e^{\tau}.$$

For the truncation error, we have

$$\mathrm{w}([f]^{[k+1]}(\widehat{[x_0]}, u)\frac{\tau^{k+1}}{(k+1)!}) \leq K\bar{\mathrm{w}}\frac{\tau^{k+1}}{(k+1)!}.$$

Let $\alpha \in (0,1)$, $k \geq \bar{k}$ and

$$K\bar{\mathrm{w}}\frac{\tau^{k+1}}{(\bar{k}+1)!} \leq (1-\alpha)\delta\tau, \tag{6.61}$$

$$2K\varepsilon e^{\tau} \leq \alpha\delta\tau. \tag{6.62}$$

Then $\mathrm{w}\left(\widehat{\mathcal{R}}_\tau^k([x_0],u)\right) \leq (1-\alpha)\delta\tau + \alpha\delta\tau = \delta\tau$, which leads to $\widehat{\mathcal{R}}_\tau^k([x_0],u) \subseteq \mathcal{R}_\tau(x_0,u) + \mathbb{B}_{\delta\tau}$. Solving for k and ε in (6.61) and (6.62) gives

$$k \geq \lceil \log(K\bar{\mathrm{w}})^{-1}(1-\alpha)\delta + \log(\bar{k}+1)!/\log\tau \rceil$$

and (6.59). We take the maximum of k and $\bar{k}-1$ to guarantee that $\widehat{[x_0]}$ is an *a priori* enclosure. Hence, we arrive at (6.58).

Suppose that $x_0 \in \mathrm{Pre}_\tau^\delta(B|A)$, i.e., $\mathcal{R}_\tau^\delta(x_0,u) \subseteq B$ for some $u \in U$. Assumption 4.3 implies that $f(\cdot,u)$ is Lipschitz over X for all $u \in U$. Then we have $\widehat{\mathcal{R}}_\tau^k([x_0],u) \subseteq \mathcal{R}_\tau(x_0,u) + \mathbb{B}_{\delta\tau} \subseteq \mathcal{R}_\tau^\delta(x_0,u) \subseteq B$ by Proposition 6.15. It implies that $x \notin [x] \in A_c$, because $\widehat{\mathcal{R}}_\tau^k([x],u) \cap B \neq \emptyset$. Any interval $[x] \in \Delta A$ that contains x_0 satisfies $[x] \subseteq [x_0]$. It then follows that $\widehat{\mathcal{R}}_\tau^k([x],u) \subseteq B$, but $\widehat{\mathcal{R}}_\tau^k([x],u) \nsubseteq B$ by Algorithm 6. Hence, $x_0 \notin [x] \in \Delta A$ and thus it is only possible that $x_0 \in [x] \in \underline{A}$, which means $x_0 \in [\mathrm{Pre}_\tau]^\varepsilon(B|A)$.

Since the operator $[\mathbf{T}_\mu]^\varepsilon$ is based off of Algorithm 6, which satisfies (6.60) with proper choice of parameters as shown above, the arguments in Theorem 6.5 can be used to prove (6.49) for system behaviors at sampling time instances. The inter-sample correctness is guaranteed by using the ϵ-strengthening of the labeling function. ■

The fraction α is used to distribute the error allowed in interval approximation for the first k terms and the remainder. As indicated in Theorem 6.6, a larger α allows us to use a lower partition precision (i.e., larger ε), but it leads to a higher-order Taylor expansion. With lower partition precision, fewer number of iterations in control synthesis algorithms will be needed, but computing the reachable sets in each iteration will take longer because of the induced higher order in Taylor expansion. Hence, optimizing α depends on the system dynamics and the control specification. Without much effort, one can also see that Theorem 6.6 holds for reach-and-stay specifications.

According to (6.57), we can always under-approximate the winning set for system (4.45) such that the approximation covers the winning set of a δ-perturbed system (4.44) with any given perturbation bound δ. To demonstrate, let us look at the following example with a reach-and-stay specification.

Example 6.11 *Consider the dynamics*

$$x' = -x(x^2 - 1.5x + 0.5) + w, \tag{6.63}$$

over the state space $X = [-1,2]$ *and a target set* $\Omega = [-0.3, 0.3]$ *for the reach-and-stay specification. For the unperturbed system (6.63), the winning set is* $\mathbf{Win}(\Omega) = [-1, 0.5)$ *as* $x = 0, 1$ *are two stable equilibrium points and there is an unstable equilibrium point at* $x = 0.5$. *For the* δ*-perturbed system with the bound* $\delta = 0.01$ *(*$w \in [-\delta, \delta]$*), the winning set is* $\mathbf{Win}^\delta(\Omega) = [-1, 0.4597)$. *With* ε *chosen by condition (6.59) for Algorithm 6, the approximation* W *will be a subset of* $\mathbf{Win}(\Omega)$ *but a superset of* $\mathbf{Win}^\delta(\Omega)$.

6.8 Complexity Analysis

By using a branch-and-bound scheme (Algorithm 5) in the direct control synthesis algorithms presented in this chapter, the discretization of the state space X is only refined in local areas regarding the satisfaction of the DBA. As a result, the size of the discretized system can be reduced. In addition, no space is required to store the abstraction and the product system with a possibly huge number of transitions. These all lead to lower space complexity for control synthesis. On the other hand, managing nonlinear data structures induces overhead cost in time, which makes the time complexity of the specification-guided method higher than abstraction-based methods in the worst case.

Assume that $\varepsilon > 0$ is the uniform grid size of the state space X for abstraction-based methods and the minimum width of an interval in Algorithm 5. For both methods, $\mu > 0$ is the under-sample parameter of the control space U of system (6.1). Then the number of states and controls in the abstraction are of $N_X = \mathcal{O}((1/\varepsilon)^n)$ and $N_U = \mathcal{O}((1/\mu)^m)$, respectively.

Algorithms 5 and 6 can be implemented by using a binary tree[3]. The branch-and-bound scheme usually results in an unbalanced tree, but in the worst case, each binary tree is balanced and has N_X number of leaves, which means that the tree height is $h = \log_2(N_X)$ and there are at most $\sum_{i=0}^{h} 2^i = 2N_X - 1$ nodes. For a single loop in Algorithms 5, the cost of a membership test of an interval along the tree is of $\mathcal{O}(N_U \log_2 N_X)$ while the cost of computing $[y] = [f]([x], u)$ is of $\mathcal{O}(N_U)$. Hence, the complexity of Algorithm 5, assuming there are N_X number of iterations, is

$$N_{\text{Pre}}^1 = \mathcal{O}(N_X(\log_2 N_X + 1)N_U). \tag{6.64}$$

For Algorithm 6, the Taylor expansion order k can be determined directly by using the conditions in Lemma 4.4 and Theorem 6.6, and therefore the complexity of reachable set over-approximation for an interval is linear to k, i.e., $\mathcal{O}(k)$, which gives

$$N_{\text{Pre}}^2 = \mathcal{O}(N_X(\log_2 N_X + k)N_U) \tag{6.65}$$

as the worst case complexity of Algorithm 6.

In the abstraction for abstraction-based methods, there is a transition from an interval (or a grid cell) $[x]$ to an interval $[x']$ as long as there exists a control $u \in [U]_\mu$ such that the one-step reachable set of $[x]$ under u intersects with $[x']$. Assume that, for the sake of simplicity, Assumption 4.1 and 4.2 hold with the same Lipschitz constant ρ, such a reachable set can be ρ^n times the volume of a single interval, and therefore the reachable set can intersect with as much

[3]A tree is an abstract data structure that has a root node which is linked by children nodes. A binary tree is a tree data structure in which each node has at most two children.

as $(\lceil \rho \rceil + 1)^n$ intervals. By enumerating all intervals and control inputs, there are

$$N_R = \mathcal{O}((\lceil \rho \rceil + 1)^n N_X N_U) \tag{6.66}$$

number of transitions.

6.8.1 Control Synthesis with Simple Specifications

Let N_G be the number of the set of intervals that represents the target set Ω. Then the number of intervals outside of Ω is $N_X - N_G$. Normally for a regulation problem, the target area Ω is rather small compared to the state space X, i.e., $N_G \ll N_X$.

For abstraction-based methods, the space complexity is of $\mathcal{O}(N_R)$ for either invariance, reachability or reach-and-stay specifications. The time complexity of constructing an abstraction is of $\mathcal{O}(N_X N_U)$. Based on a transition system, the time complexity of synthesizing the reachability objective $\varphi_\text{r} = \Diamond\omega$ is of $\mathcal{O}(N_R)$. Since invariance objective $\varphi_\text{s} = \Box\omega$ is a dual to φ_r, the time complexity is the same as for reachability control. Therefore, the overall time complexity of abstraction-based control synthesis, including abstraction and synthesis, with respect to invariance or reachability is

$$\mathcal{O}(N_X N_U + (\lceil \rho \rceil + 1)^n N_X N_U). \tag{6.67}$$

For reach-and-stay control objective $\varphi_\text{rs} = \Diamond\Box\omega$, the overall time complexity is of $\mathcal{O}(N_X N_U + N_X N_R)$, i.e.,

$$\mathcal{O}(N_X N_U + (\lceil \rho \rceil + 1)^n N_X^2 N_U). \tag{6.68}$$

We can see that, for abstraction-based methods, both space and time complexities are independent of the size of the target set.

To analyze the complexities of Algorithms 7, 8 and 9, we consider the worst case where the list Y (and Z in Algorithm 9) of intervals changes by one interval after each (inner/outer) iteration.

In Algorithm 7, the procedure PRE (Algorithms 5 or 6) and the partition of system state is only performed to the target set Ω. The complexity of a single PRE is therefore of $\mathcal{O}(N_G(\log_2 N_G + 1)N_U)$ (or $\mathcal{O}(N_G(\log_2 N_G + k)N_U)$ if Algorithm 6 is used). Assuming that $N_G \gg 1$, the overall computational complexity is

$$\mathcal{O}\left(N_G^2(\log_2 N_G + 1)N_U\right). \tag{6.69}$$

If the target set is small enough so that $N_G^2 \log N_X < N_X$, the worst case complexity of the invariance control algorithm with interval implementation is lower than the one for abstraction-based methods (6.67).

For Algorithm 8, we also consider that $N_G, 1 \ll N_S$. Then we have the computational complexity

$$\mathcal{O}\left(N_U N_X^2 + N_U N_X^2 \log_2 N_X\right), \tag{6.70}$$

TABLE 6.1: Complexities for control synthesis with simple specifications.

		Via Abstractions	Direct Interval Computation
Space		$\mathcal{O}((\lceil\rho\rceil+1)^n N_X N_U)$	$\mathcal{O}(N_X)$
	φ_s	$\mathcal{O}(N_X N_U + (\lceil\rho\rceil+1)^n N_X N_U)$	$\mathcal{O}\left(N_G^2(\log_2 N_G + 1)N_U\right)$
Time	φ_r	$\mathcal{O}(N_X N_U + (\lceil\rho\rceil+1)^n N_X N_U)$	$\mathcal{O}\left(N_U N_X^2 + N_U N_X^2 \log_2 N_X\right)$
	φ_{rs}	$\mathcal{O}(N_X N_U + (\lceil\rho\rceil+1)^n N_X^2 N_U)$	$\mathcal{O}\left(\log_2 N_X(N_X N_U + N_X^2 N_U)\right)$

which is higher than (6.67).

Based on the complexity of Algorithms 5 or 6, the rest of the complexity analysis for Algorithm 9 is to evaluate how many times the computation for a single interval, which we have discussed above, can be called. If $N_G << N_X$, then the number of iterations is

$$\sum_{i=0}^{N_\Omega}(i^2+i) + \sum_{i=0}^{N_X-N_\Omega} i = \frac{N_\Omega^3 + 3N_\Omega^2 + 8N_\Omega}{6} + \frac{(N_X-N_\Omega)^2 + (N_X-N_\Omega)}{2}$$
$$\approx \frac{N_X^2 + N_X}{2} \approx N_X^2$$

Hence, the time complexity of the overall algorithm is

$$\mathcal{O}\left(\log_2 N_X(N_X N_U + N_X^2 N_U)\right). \tag{6.71}$$

The following Table 6.1 is a summary of the above complexity analysis.

6.8.2 Control Synthesis with DBA Specifications

For abstraction-based methods, the space complexity is proportional to the number of transitions in the product of the abstraction and the DBA, and the computational time includes the time for abstraction and control synthesis. The product of an abstraction of system (6.1) and the DBA $\mathcal{A}_\varphi$ has $N_X^{\text{prod}} = \mathcal{O}(|Q|\,N_X)$ number of states and $N_R^{\text{prod}} = \mathcal{O}(|Q|\,N_R)$ number of transitions. Therefore, the space and time complexities are $\mathcal{O}(N_R^{\text{prod}})$ and $\mathcal{O}(N_X^{\text{prod}}N_R^{\text{prod}})$, respectively.

For the interval-based method, the space complexity is $\mathcal{O}(2\,|Q|\,N_X)$ since $|Q|$ binary trees are constructed. To analyze the time complexity of Algorithm 10, we assume the worst case that Y and $\widetilde{Y}$ only differ in one of the elements (i.e., $Y[i]$ for some $i \in \{1,\cdots,n_1\}$) by one interval in the inner loop and the same for the outer loop. The operator $[\mathcal{T}_\mu]^\varepsilon$ in Algorithm 10 requires $|Q|$ such binary trees as defined in (6.44). Hence, the maximal numbers of inner and outer iterations are $(|Q|-|F|)N_X$ and $|F|\,N_X$, respectively. The complexity for the operator $[\mathcal{T}_\mu]^\varepsilon$ is then $\mathcal{O}(|Q|\,N_{\text{Pre}})$ by definition. Therefore, the overall time complexity of Algorithm 10 is $\mathcal{O}(((|Q|-|F|)N_X+1)\,|F|\,N_X\,|Q|\,N_{\text{Pre}}) = \mathcal{O}(|Q|^2\,|F|\,N_X^3(\log_2 N_X + c)N_U)$, considering $(|Q|-|F|)N_X+1 \approx |Q|\,N_X$.

The above analysis is summarized in Table 6.2. The interval-based method is more efficient in memory usage because usually $(\lceil\rho\rceil+1)^n N_U \gg 2$. The main reason is that it does not store all the transitions of the product system, which can be exceptionally huge when ε needs to be very small or the system dimension is high. The abstraction-based approach has a lower time complexity than the worst case of the specification-guided approach. This is because the specification-guided approach trades time for saving the space to store transitions.

TABLE 6.2: Complexities for DBA control synthesis.

	Via Abstractions	Direct Interval Computation
Space	$\mathcal{O}(\|Q\|\,N_X + (\lceil\rho\rceil+1)^n\,\|Q\|^2\,N_X^2 N_U)$	$\mathcal{O}(\|Q\|^2\,\|F\|\,N_X^3(\log_2 N_X+1)N_U)$
Time	$\mathcal{O}((\lceil\rho\rceil+1)^n N_U\,\|Q\|\,N_X)$	$\mathcal{O}(2\,\|Q\|\,N_X)$

Theoretically, the worst-case complexity is exponential to the system state and control input dimensions. Although the worst case rarely happens in practice, for high-dimensional system (e.g., system with dimension $n \geq 3$), the computational cost is still expensive. This indicates that our method suffers from the curse of dimensionality, which is indeed the case for all discretization-based approaches for controller synthesis.

The complexity of applying the controller synthesized by the specification-guided approach on the fly is $\mathcal{O}(\log_2 N_X)$ because the control value for the current system state is searched along the binary tree. One way to reduce this complexity is to map the binary tree to a uniform partition of the state space (with the granularity as the size of the smallest interval of the binary tree) through which online querying of valid control inputs is a constant time.

6.9 Summary

This chapter presents a specification-guided approach as another way, in addition to the abstraction-based approach in Chapter 5, to solve formal control synthesis problems for dynamical systems. Instead of discretizing the dynamics in order to reuse the control synthesis algorithms for finite transition systems (in Chapter 3), the specification-guided approach adapts the control synthesis algorithms to the continuous dynamics, and the "adapte" is the interval approximations of controlled predecessors of sets defined in the continuous state-space of a dynamical system. Thanks to the ability to bound the approximation error with arbitrary precision by using the interval analysis tool (discussed in Chapter 6), the control synthesis turns out to be sound and robustly complete, i.e., a control strategy can be found as long as the given specification is realizable for the dynamical system with bounded perturbation, at least for a general class of LTL formulas.

For invariance, reachability, and reach-and-stay control synthesis problems, control algorithms that are similar to the ones for finite transition systems have been established in [27, 28, 30, 31, 96, 211] with additional attention to the finite-termination property. Unlike finite-state systems, the fixed point (or more precisely, the fixed set) usually cannot be reached within a finite number of iterations for continuous-state systems unless certain properties are met. For example, being λ-*contractive* around a compact and convex set for a linear system is a sufficient condition of a finitely terminated invariance control algorithm. The results from Section 6.3 to Section 6.5 are based on [139, 142], which are inspired by those pioneering works. The robust invariance margin (Definition 6.3) is an extension of the λ-contractivity [31] to general nonlinear systems. With robustly invariant property and the use of interval analysis for precision-controlled set approximation, the finitely determined invariance control algorithm in Section 6.3 via direct interval computation can be sound and complete. This is the prototype of the specification-guided framework that applies to a general class of LTL formulas as we have seen in Section 6.6 (based on [147]). Section 6.7, which is based on [143], further extends the method to sampled-data systems.

Complexity greatly increases as we aim to solve more general LTL formulas. A promising finite abstraction of nonlinear dynamics is usually huge in size, and control synthesis on a product system would be intractable because the number of states is the multiplication of the sizes of both the abstraction and the Büchi automata. The complexity comparison in Section 6.8 reveals that the specification-guided method is more memory-efficient than abstraction-based methods. In some use cases, we can have 300-fold reduction in memory consumption. Nevertheless, the curse of dimensionality is a weak point of both methods. This limitation motivates the research on system decomposition [115, 167], hierarchical abstractions [107], and parallel computation [59, 113]. On the other hand, the reduction in the size of a specification can also reduce the complexity, which is demonstrated by the specification pre-process procedure in Section 6.6.

Chapter 7

Applications and Case Studies

Moving from theory to applications, LTL control synthesis has gained great popularity, especially in robotics motion planning where continuous robot dynamics meets discrete planning specifications. For example, while being subject to mechanical constraints and dynamics, robots are designed to fulfill tasks such as pickup-delivery, parts assembly, surveillance, and persistent monitoring. Usually, these tasks have to be completed in specific orders, and robots are required to be reactive to the change of environment.

In this chapter, we present different applications and discuss the performance of formal control synthesis algorithms. Experimental results are obtained by using ROCS[1] [141, 146], a toolbox that provides both abstraction-based and specification-guided control synthesis algorithms.

7.1 DC-DC Boost Converter

A DC-DC boost converter is a power electronic circuit that converts the power from the source V_s to the load r_o, providing a higher voltage V_o than the source V_s. This is accomplished by controlling two metal-oxide-semiconductor field effect transistors (MOSFETs) T_1 and T_2 open and closing. A schematic circuit diagram of a DC-DC boost converter is given in Figure 7.1. Formal methods have been shown effective in controlling such power electronics [87, 213]. By different configurations of T_1 and T_2, the system can be simplified to a switched linear system with two modes.

Let the state x of the converter be a vector of the inductor current i_l and the capacitor voltage V_c. Then the state space model of the DC-DC boost

[1] `https://git.uwaterloo.ca/hybrid-systems-lab/rocs`

DOI: 10.1201/9780429270253-7

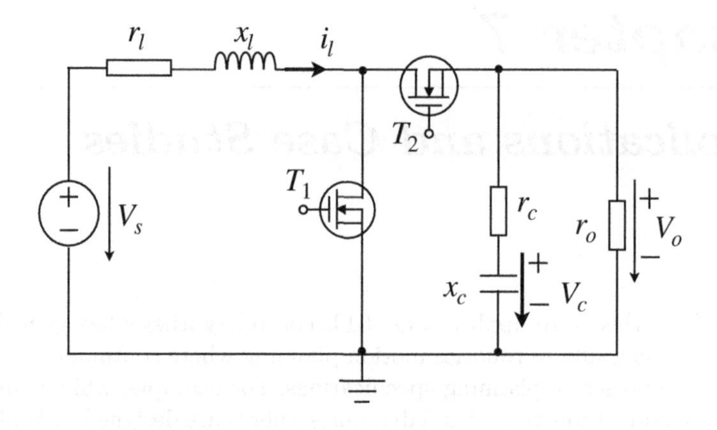

FIGURE 7.1: The circuit of a DC-DC boost converter.

converter is linear affine:

$$\begin{aligned} &\dot{x} = A_p x + b, \quad p = 1, 2, \\ &x = \begin{bmatrix} i_l \\ V_c \end{bmatrix}, \quad b = \begin{bmatrix} \frac{V_s}{V_l} \\ 0 \end{bmatrix}, \\ &A_1 = \begin{bmatrix} -\frac{r_l}{x_l} & 0 \\ 0 & -\frac{1}{x_c(r_c+r_o)} \end{bmatrix}, \quad A_2 = \begin{bmatrix} -\frac{1}{x_l}(r_l + \frac{r_o r_c}{r_o+r_c}) & -\frac{r_o}{x_l(r_o+r_c)} \\ \frac{r_o}{x_c(r_o+r_c)} & -\frac{1}{x_c(r_o+r_c)} \end{bmatrix}. \end{aligned} \tag{7.1}$$

The parameters in (7.1) is provided in Table 7.1.

Given the ordinary differential equations (ODEs) (7.1), we can derive the exact discrete-time model via integrating 7.1 by a sampling time $\tau_s > 0$:

$$x(t+1) = e^{A_p \tau_s} x(t) + \int_0^{\tau_s} e^{\tau_s - s} b \, ds.$$

A typical function of a DC-DC boost converter is to regulate the output voltage V_o within a certain range. Depending on whether the initial

TABLE 7.1: The parameters in (7.1), and "p.u."= *per unit.*

Parameters	Value (p.u.)	Physical meaning
x_c	70	The capacity of the capacitor
r_c	0.005	The resistance of the capacitor
x_l	3	The inductance of the inductor
r_l	0.05	The resistance of the inductor
r_o	1	The load resistance
V_s	1	The source voltage

state x_0 of the system falls inside this range or not, such a control objective can be described as an invariance specification $\varphi_s = \Box\Omega$ or a reach-and-stay specification $\varphi_{rs} = \Diamond\Box\Omega$ for system (4.26) with transition relation determined by (7.1). Hence, we consider two scenarios with the state space $X = [0.6490, 1.6500] \times [0.9898, 1.1900]$. In the first scenario, we aim to maintain system state inside a target region $\Omega_1 = [1.15, 1.55] \times [1.09, 1.17]$, and the second case is to control a state in the state space X to reach a target set $\Omega_2 = [1.10, 1.6] \times [1.08, 1.18]$ and stay there for all future time.

A sampling time $\tau_s = 0.5$ s is used for constructing the discrete-time model (7.2), which gives the discrete-time model

$$\begin{aligned} &\text{Mode 1:} \quad x(t+1) = \begin{bmatrix} 0.9917 & 0 \\ 0 & 0.9929 \end{bmatrix} x(t) + \begin{bmatrix} 0.1660 \\ 0 \end{bmatrix}, \\ &\text{Mode 2:} \quad x(t+1) = \begin{bmatrix} 0.9903 & -0.1645 \\ 0.0070 & 0.9923 \end{bmatrix} x(t) + \begin{bmatrix} 0.1659 \\ 0.0006 \end{bmatrix}. \end{aligned} \tag{7.2}$$

The precision parameter $\varepsilon = 0.001$ is used in both Algorithm 7 and 9.

In the first case, we approximate the maximal controlled invariant set inside Ω_1, which is the winning set $\mathbf{Win}_{(7.2)}(\varphi_s)$ for system (7.2) with respect to the invariance specification φ_s, by the union of intervals marked as the shaded area in Figure 7.2. The Lipschitz constant $\rho = \max\{0.9929, 1.0737\} = 1.0737$. Implied by Theorem 6.1, the target set Ω_1 is not δ-robustly controlled invariant for any $\delta > \rho\varepsilon = 0.0010737$.

Applying the constructed partition-based memoryless control strategy in the form of (6.15), a closed-loop system trajectory from the initial state

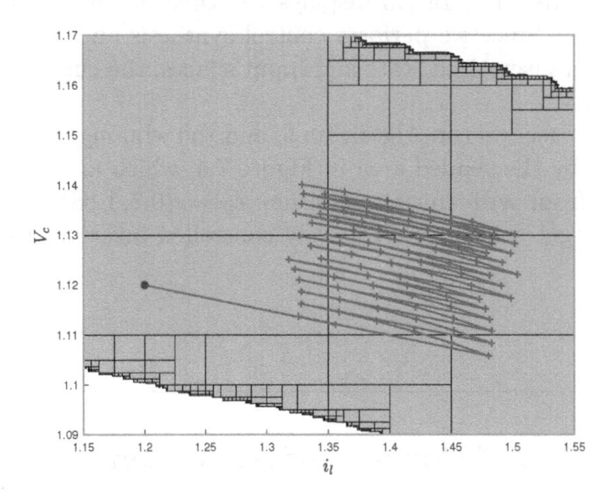

FIGURE 7.2: The phase portrait of a closed-loop trajectory that satisfies the invariance specification φ_s. The dot is the initial condition, and the shaded area is the approximated maximal controlled invariant set inside Ω_1.

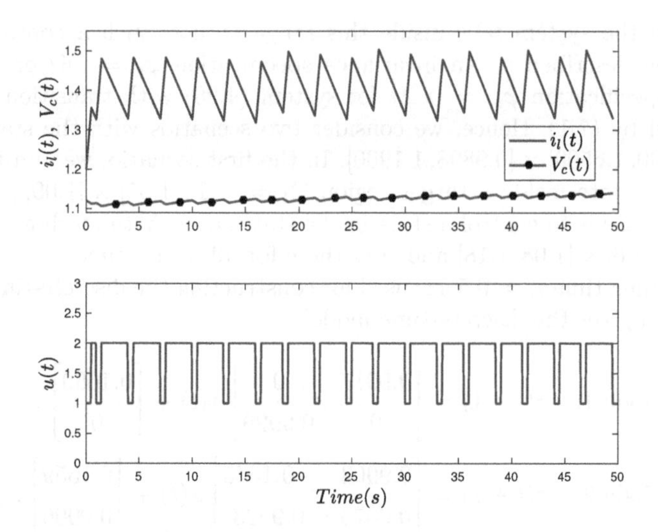

FIGURE 7.3: The corresponding time history of closed-loop states and control variables in Figure 7.2.

$x_0 = (1.2, 1.12)$ is shown in Figure 7.3. Such a control strategy returns (possibly) multiple valid control inputs for a state in the winning set, and any one of the control values realizes the invariance specification. In this example, the valid control inputs can be both modes 1 and 2. The only one of the control values that we select in our closed-loop control simulation is the one closest to the last used control value, which results in less mode switching. It can be seen that the whole trajectory is confined to the controlled invariant set inside Ω_1 as required. Figure 7.3 displays the time history of system states and control inputs. Since we perform control synthesis automatically by formal algorithms on discretized state and input spaces, the curves in Figure 7.3 show discontinuity.

In the second case, we run Algorithm 9, and the winning set $\mathbf{Win}_{(7.2)}(\varphi_{\text{rs}})$ is approximated by the shaded area in Figure 7.4, which also shows a closed-loop trajectory from with initial condition $x_0 = (0.7, 1.08)$. Similar to the first case, the target set is not δ-robustly controlled invariant itself as $\Omega_2 \not\subseteq \mathbf{Win}_{(7.2)}(\varphi_{\text{rs}})$.

7.2 Estimation of Domains-of-Attraction

A problem of interest in the study of dynamical systems is to determine the domain of attraction (DOA) of an equilibrium point (see Definition 1.9).

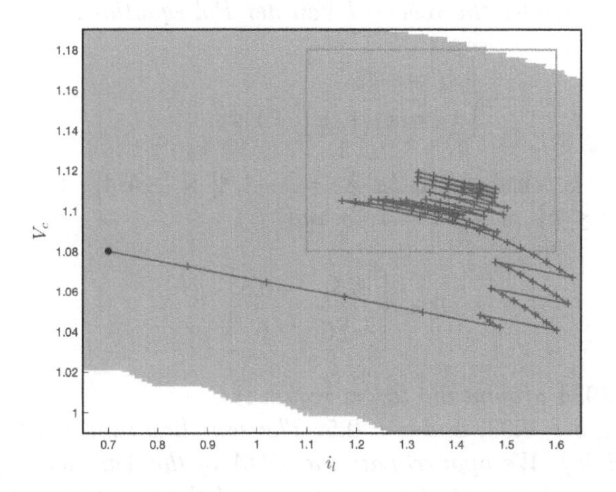

FIGURE 7.4: A closed-loop trajectory with the control strategy that realizes the reach-and-stay specification φ_{rs}. The target set Ω_2 is marked as the rectangle in the upper right corner.

This problem has important applications in safety-critical industries such as aviation and power systems, where determining the operating envelope of an aircraft or a power network is vital. In the literature, computational methods for determining the DOA for nonlinear systems have been developed by way of Lyapunov functions. The key aspect is to search Lyapunov functions that maximize the estimated DOA. For this purpose, linear matrix inequalities [46] and sum-of-square programming techniques [233] are used for the construction of such Lyapunov functions for polynomial systems. Using Lyapunov functions with fixed forms, subsets of the DOAs can also be obtained by solving a constraint satisfaction problem [229]. How to choose the form of Lyapunov functions, however, remains a challenging problem. We now take DOA estimation as an example to illustrate the application of specification-guided control synthesis approach.

Consider the continuous-time system

$$\dot{x}(t) = f(x(t)), \tag{7.3}$$

where $x \in \mathbb{R}^n$, f is continuously differentiable and the origin is a hyperbolic stable equilibrium point.

The DOA approximation problem for system (7.3) can be interpreted as a reach-and-stay control problem with the specification $\varphi(\Omega)$, where $\Omega \subseteq \mathbb{R}^n$ is a subset of the real DOA of system (7.3) that contains the origin. The set Ω can be conveniently captured by using a quadratic Lyapunov function $V(x) = x^T Px$ for the linearization of the system at the origin [114, Theorem 4.7].

Example 7.1 *Consider the reversed Van der Pol equations*

$$\begin{cases} \dot{x}_1 = -x_2, \\ \dot{x}_2 = x_1 + (x_1^2 - 1)x_2. \end{cases} \tag{7.4}$$

The state space is assumed to be $X = [-4, 4] \times [-4, 4]$. *We choose* $\Omega = \{x \in \mathbb{R}^n \mid x^T P x \leq c\}$, *where* $c = 1.43$ *and*

$$P = \begin{bmatrix} 1.5 & -0.5 \\ -0.5 & 1 \end{bmatrix}.$$

Hence, $M = 1.8934$ *around the target region* Ω_c.

Let $\epsilon = 0.04$, $\delta = 0.03$, *and* $\alpha = 0.5$. *The sampling time* $\tau = 0.04s$ *is used according to (6.56). We approximate the DOA of the Van der Pol equations with different precision control parameters, and the results together with the real limit cycle is shown in Figure 7.5. As observed, a higher precision yields a closer under-approximation to the real DOA. By setting* ε *sufficiently small, the estimated boundary of the DOA can be of arbitrarily close to the real limit cycle.*

Formulating the problem of DOA approximation as a reach-and-stay control synthesis problem releases the burden of choosing proper Lyapunov functions. The required smoothness condition is less strict than being polynomial in many of the methods for DOA estimation.

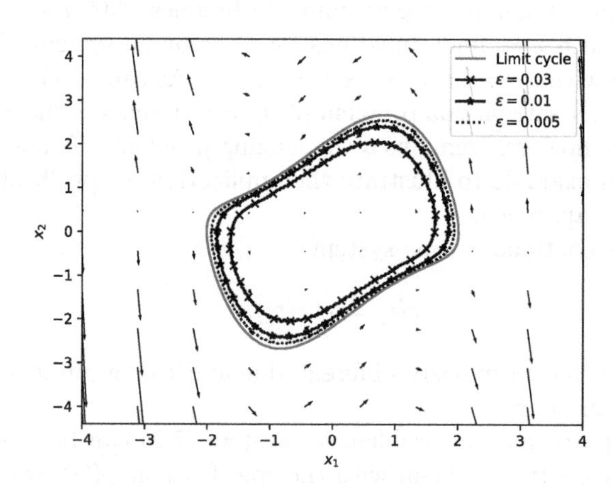

FIGURE 7.5: Comparison of under-approximations of the DOA for reversed Van der Pol sampled-data system with three different precisions.

7.3 Control of the Moore-Greitzer Engine

The reduced Moore-Greitzer model is a commonly used nonlinear model for capturing the average flow Φ and pressure Ψ of an axial-flow jet engine compressor. As the throttle coefficient μ decreases, surge instability occurs and generates a pumping oscillation (Hopf-bifurcation) that can cause flame-out and engine damage [22, 91]. In practice, to deal with the requests from downstream, the operation point needs to be switched during the process. Without any controls, however, operation points are determined by μ and a smaller μ may result in unstable operation points [244]. To alleviate the oscillation and prevent substantial pressure loss during the switch, it is motivated to design controllers to lead the state (Φ, Ψ) reach and stay in a small region around an unstable operation point, and meanwhile avoid touching the region with low average pressure.

The reduced Moore-Greitzer ODE model is given as

$$\begin{cases} \dot{\Phi} = \frac{1}{l_c}(\psi_c - \Psi) + u, \\ \dot{\Psi} = \frac{1}{4 l_c B^2}(\Phi - \mu\sqrt{\Psi}), \end{cases} \tag{7.5}$$

where $\psi_c = a + \iota * [1 + \frac{3}{2}(\frac{\Phi}{\Theta} - 1) - \frac{1}{2}(\frac{\Phi}{\Theta} - 1)^3]$, and for a global parameter set-up, we let

$$l_c = 8,\ \iota = 0.18,\ \Theta = 0.25,\ a = \frac{1}{3.5}.$$

The states Φ and Ψ are the average flow rate and pressure of an axial-flow jet engine compressor, respectively. The control inputs are the throttle coefficient μ and an additional control u.

The reach-avoid-stay specification can be written as the LTL formula $\varphi_{\text{rs}} = \Box\neg\mathcal{O} \wedge \Diamond\Box\Omega$, which can only be translated into an NBA (see Figure 6.7b). The target set Ω and the unsafe set $\mathcal{O}$ are 2-dimensional balls with the radius $r = 0.003$ around an unstable equilibrium point $\omega = (0.4519, 0.6513)$ and the point $a = (0.500, 0.653)$, respectively. The sets are also explicitly given in Table 7.2. The initial condition is $(0.5343, 0.6553)$, the state space is $X = [0.44, 0.6] \times [0.54, 0.7]$, and the control space is $U = \{(u, \mu) \mid u \in [-0.05, 0.05], \mu \in [0.5, 0.8]\}$.

Algorithm 6 is used with a sampling time $\tau_s = 0.1$s to approximate the predecessors in Algorithm 9. The approximated winning sets with precision 1.8×10^{-4} for both scenarios obtained by DBA control synthesis cover more

TABLE 7.2: Reach-avoid-stay specification for Moore-Greitzer Engine control.

$\Omega =$	$\{(\Phi, \Psi) \mid \Phi \in [0.4489, 0.4549], \Psi \in [0.6483, 0.6543]\}$
$\mathcal{O} =$	$\{(\Phi, \Psi) \mid \Phi \in [0.497, 0.503], \Psi \in [0.650, 0.656]\}$

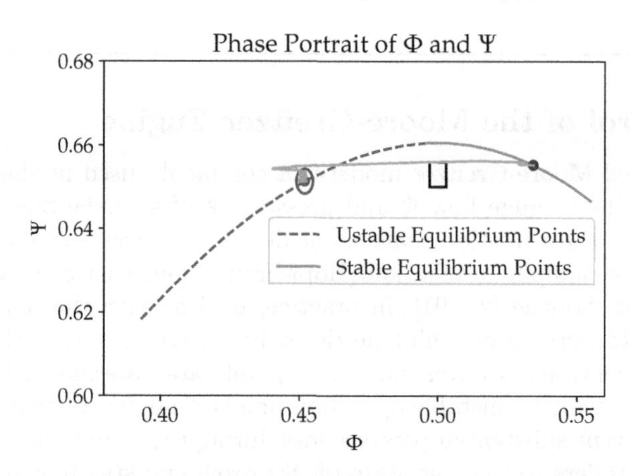

(a) The phase portraits of Φ and Ψ. The circle on the left is the target Ω, and the rectangle is the unsafe area $\mathcal{O}$.

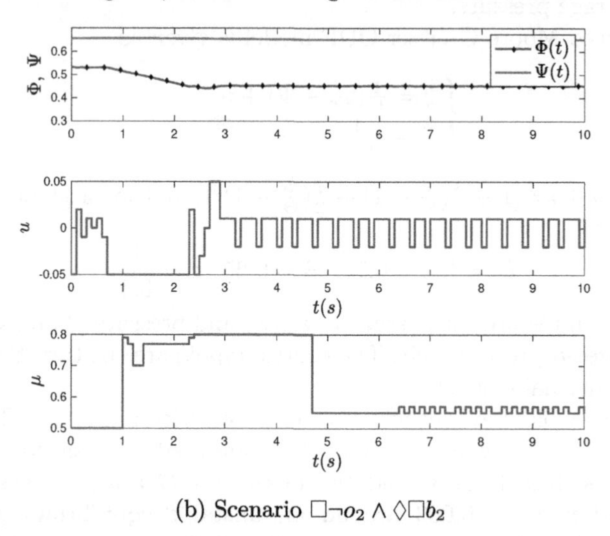

(b) Scenario $\Box\neg o_2 \wedge \Diamond\Box b_2$

FIGURE 7.6: Reach-avoid-stay control synthesis results for the Moore-Greitzer engine.

than 99.6% of the state space. As shown in Figure 7.6, the system state for both scenarios can be controlled from the given initial condition to the target sets.

Another way to solve such a reach-avoid-stay problem is to use control Lyapunov-barrier functions (see Section 1.4). Let $h_1(x) = -\|x - \omega\|_\infty + r$, $h_2(x) = \|x - a\|_\infty - r$, $\Omega = \{x : h_1(x) \geq 0\}$, and $\Lambda = \{x : h_2(x) < 0\}$. To satisfy Theorem 1.4, we choose the open set $D = \{x : h_2(x) > 0\}$, $V(x) = \|x - \omega\|^2$ for all $x \in D \setminus \omega$ and $B(x) = 1/h_2(x)$. The control strategy for the

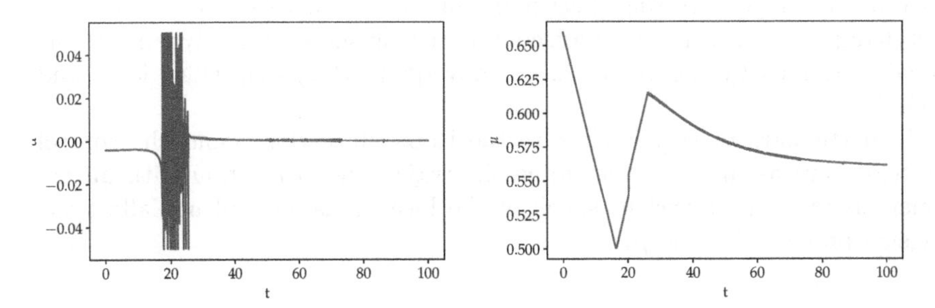

FIGURE 7.7: Control signal u and μ solved by QP with (7.6) as the cost function.

reach-avoid-stay problem based on (V, B) is then obtained as $\kappa(x)$ as shown in Section 1.5. The control strategy $\kappa(x)$ can be solved via a quadratic programming (QP) framework [9] with the cost function

$$\|u\|^2 + \frac{2u}{l_c}(\psi_c - \Psi) + \left(\frac{1}{4l_c}(\Phi - \mu(t))\sqrt{\Psi}\right)^2 \tag{7.6}$$

to minimize the control effort for every $t > 0$.

As a result, the control signals u and μ are shown in Figure 7.7. The sufficient conditions on the signals generated by control Lyapunov-barrier functions are shown to be effectively embedded within the QP with the minimum input energy 7.6. In particular, the extra conditions on the changing rate of signals are reactively included.

The phase portraits of the resulting trajectory are shown in Figure 7.8. The local Lipschitz continuity of u can also be guaranteed in this framework ([9,

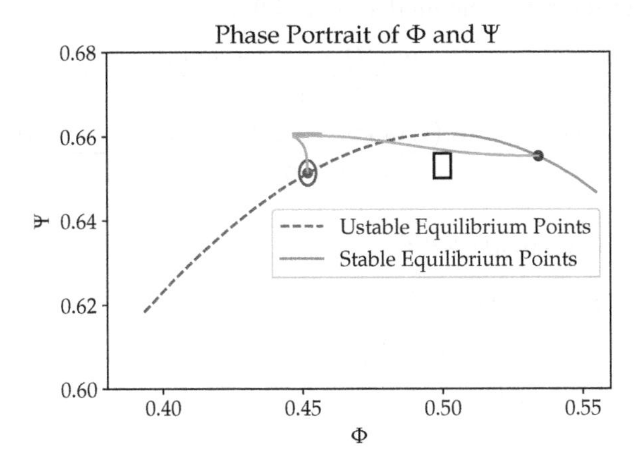

FIGURE 7.8: Phase portrait of (Φ, Ψ) generated based on signal u and μ.

Theorem 3]). The chattering effect of u around time 20 is due to the relatively fast change of μ, which in turn affects the varying speed of the dynamics. The signal μ decided by the QP tends to converge to the $\mu(\omega)$, which is around 0.56.

In particular, when μ is close to the Hopf-bifurcation point, the control strategy $\kappa(x)$ as in (1.28) can force the trajectory to reach ω with an exponential rate, the transient speed of the local dynamic will not affect the decision process of u and μ.

7.4 Mobile Robot Motion Planning

The level of mobile robot modeling involved in motion planning is usually restricted to kinematics, and in this section, the following model [19] is used:

$$\begin{bmatrix} \dot{x} \\ \dot{y} \\ \dot{\theta} \end{bmatrix} = \begin{bmatrix} v\cos(\gamma+\theta)\cos(\gamma)^{-1} \\ v\sin(\gamma+\theta)\cos(\gamma)^{-1} \\ v\tan(\phi) \end{bmatrix}, \tag{7.7}$$

where (x, y) is the planar position of center of the vehicle, θ is its orientation, the control variable v represents the velocity, and ϕ is the steering angle command.

The vehicle structure is shown in Figure 7.9, and the variable $\gamma = \arctan(a\tan(\phi)/b)$, where a is the distance from the gravity center to the rear wheels of the vehicle, and b is the distance between its front and rear wheels. We use $a/b = 1/2$ in the simulation.

Considering constant control inputs during each sampling period, we can derive the exact discrete-time model for $\phi \neq 0$:

$$\begin{aligned} x(t+1) &= \frac{\sin(\gamma(t)+\tau_s v(t)\tan\phi(t)+\theta(t)) - \sin(\gamma(t)+\theta(t))}{\cos\gamma(t)\tan\phi(t)} + x(t), \\ y(t+1) &= \frac{-\cos(\gamma(t)+\tau_s v(t)\tan\phi(t)+\theta(t)) + \cos(\gamma(t)+\theta(t))}{\cos\gamma(t)\tan\phi(t)} + y(t), \\ \theta(t+1) &= \tau_s v(t)\tan\phi(t) + \theta(t). \end{aligned} \tag{7.8}$$

For $\phi = 0$, the discrete-time model becomes

$$\begin{aligned} x(t+1) &= v(t)\cos\theta(t)\tau_s + x(t), \\ y(t+1) &= v(t)\sin\theta(t)\tau_s + y(t), \\ \theta(t+1) &= \theta(t). \end{aligned} \tag{7.9}$$

The discrete-time model can be readily verified Lipschitz continuous over the state space X and the input space U: For (7.9) when the steering angle $\phi = 0$, $\rho_1 = 1.3$ and $\rho_2 = 0.3$ are the Lipschitz constants with respect to state and input, respectively. Letting $\phi \to 0$, (7.8) and (7.9) will be almost equivalent.

7.4.1 Parallel Parking

We now consider an automatic parallel parking problem in which the goal is to control a vehicle to park along the curb between two other vehicles. Such an objective can be expressed by a reach-and-stay specification.

In our simulation, the sampling time is $\tau_s = 0.3$s, the state space is $X = [0, 8] \times [0, 4] \times [-72°, 72°]$, and the set of control values is $U = \{\pm 0.9, \pm 0.6, \pm 0.3, 0\}$, which is sampled by uniform discretization of the space $[-1, 1] \times [-1, 1]$ with grid width $\eta = 0.3$.

Suppose that the length and width of the vehicle are $L = 2$ and $H = 1$, respectively. For the purpose of analysis, we consider two problem settings: parking with a wide marginal space $\Delta = L = 2$ and a narrow marginal space $\Delta = 0.5$. The marginal space is the distance between the front and rear vehicles in addition to L. For both cases, the rear vehicle center is at $(1, 0.5)$, and the front vehicle center is at $(1 + 3L/2 + \Delta, 0.5)$. The target area is

$$\Omega = [1 + L, 1 + L + \Delta] \times [0.5, 0.6] \times [-3°, 3°].$$

The collision area (the center position and orientation of the vehicle that causes collision with the parked vehicles and the curb) needs to be determined before control synthesis. We assume that vehicles and the curb are rectangles. Then the collision area can be interpreted by inequalities, which is derived by checking if two polyhedra intersect. It is clear that the center of the vehicle

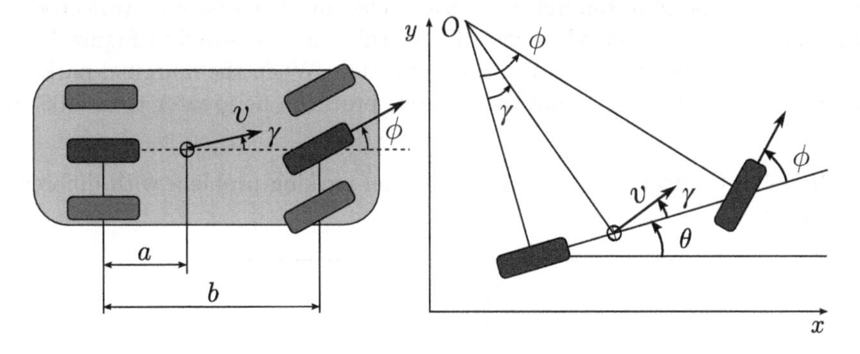

FIGURE 7.9: The vehicle structure [19].

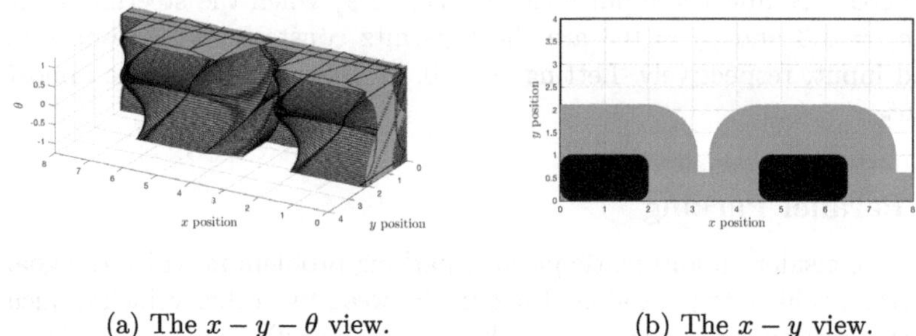

(a) The $x - y - \theta$ view. (b) The $x - y$ view.

FIGURE 7.10: Collision area when $\Delta = 0.5$. In (b), the gray area is the $x - y$ plane projection of the 3D collision area, and the two black rectangles represent the bodies of rear and front vehicle.

has different admissible regions with different orientations. Hence, the collision area is not simply a hyper-rectangle in $\mathbb{R}^3$, as shown in Figure 7.10 (a). The free workspace (the admissible position of the vehicle center in $\mathbb{R}^3$) determined by such a constraint can be handled by Algorithm 5 for approximating controlled predecessors which are used in Algorithm 9.

By Theorem 6.4, if parallel parking is robustly realizable with the given marginal space, we can always synthesize a control strategy using a sufficiently small precision without calculating the Lipschitz constant. To see if the specifications in these two parking scenarios are realizable, we use different precision control parameters. The corresponding control synthesis results regarding the number of partitions ($\#\mathcal{P}_{1,2}$) and the run time ($t_{1,2}$) are summarized in Table 7.3.

For both scenarios, the vehicle can be successfully parked into the target spot from any point of the free workspace. The controlled parking trajectories with the resulting memoryless control strategies are presented in Figure 7.11, which all meet the parallel parking specification. When the marginal parking space Δ is 0.5, we need a control synthesis precision no greater than 0.06 so

TABLE 7.3: Control synthesis of the parallel parking problem with different precisions.

ε	$\#\mathcal{P}_1$	t_1 (s)	$\#\mathcal{P}_2$	t_2 (s)
0.07	176786	102.93	–	–
0.06	176666	103.19	1,797,027	295.68
0.02	203166	127.44	1,832,589	327.50
0.01	274694	176.20	1,920,929	427.48

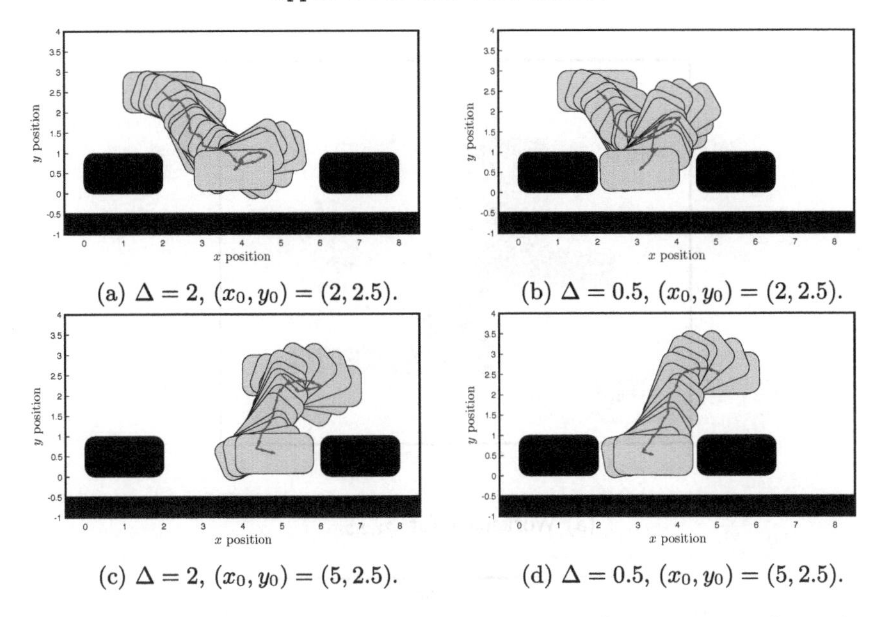

(a) $\Delta = 2$, $(x_0, y_0) = (2, 2.5)$.

(b) $\Delta = 0.5$, $(x_0, y_0) = (2, 2.5)$.

(c) $\Delta = 2$, $(x_0, y_0) = (5, 2.5)$.

(d) $\Delta = 0.5$, $(x_0, y_0) = (5, 2.5)$.

FIGURE 7.11: Controlled parking trajectories from an initial condition (x_0, y_0) with wide and narrow marginal parking spaces.

that a memoryless control strategy can be generated. Additionally for this specific example, using a smaller ε only increases the winning set by adding intervals close to the boundary of the free workspace.

Such a parallel parking task can also be solved by using a piecewise-affine controller defined on a pre-designed triangular partition of the configuration space [180], which contains the initial states of the car. The main advantage of using the specification-guided approach is that the partition of the state space is performed automatically. As a result, we do not need to re-design the partition for a different parking scenario. In addition, the control design is based on the nonlinear model as opposed to different linearizations of the nonlinear model on different polytopes in the state space.

7.4.2 Motion Planning

We now switch to the state space $X = [0, 10] \times [0, 10] \times [-\pi, \pi]$ and the control space $U = [-1, 1] \times [1, 1]$ with the sample grid $\mu = 0.3$ for motion planning. A state space discretization precision ≤ 0.2 in each dimension can yield winning sets that cover more than 84% of the state space.

Two layouts of the workspace are considered, as shown in Figure 7.12a and Figure 7.12b, respectively. With Figure 7.12a layout, we aim to design controllers with respect to three specifications given in Table 7.4

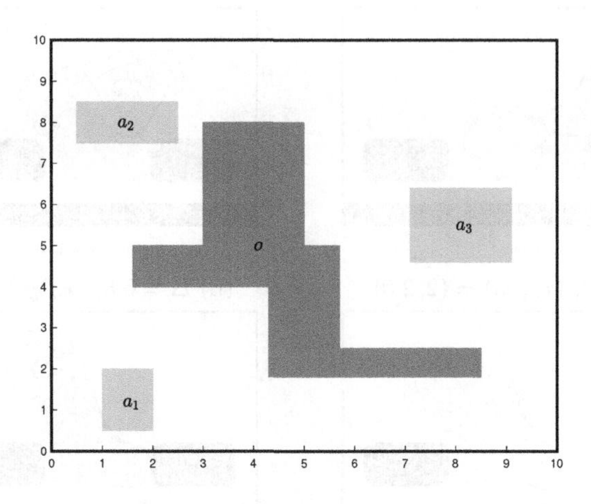

(a) Workspace for $\varphi_{1,2,3}$.

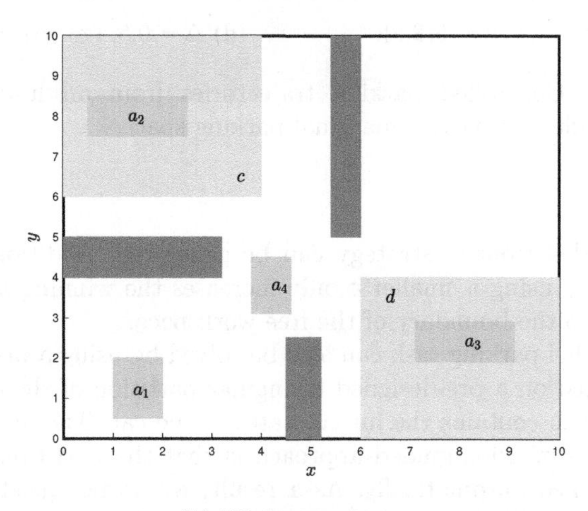

(b) Workspace for φ_4.

FIGURE 7.12: Two motion planning workspaces. (a) $L^{-1}(a_1) = [1,2] \times [0.5,2] \times [-\pi,\pi]$, $L^{-1}(a_2) = [0.5,2.5] \times [7.5,8.5] \times [-\pi,\pi]$, and $L^{-1}(a_3) = [7.1,9.1] \times [4.6,6.4] \times [-\pi,\pi]$. (b) a_1 and a_2 are the same as in (a), $L^{-1}(a_3) = [7.1,9.1] \times [1.9,2.9] \times [-\pi,\pi]$, $L^{-1}(a_4) = [3.8,4.6] \times [3.1,4.5] \times [-\pi,\pi]$, $L^{-1}(c) = [0,4] \times [6,10] \times [-\pi,\pi]$, and $L^{-1}(d) = [6,10] \times [0,4] \times [-\pi,\pi]$

TABLE 7.4: Specifications for scenario 1.

$\varphi_1 = \Diamond(a_1 \wedge \Diamond(a_2 \wedge \Diamond(a_3 \wedge (\neg a_2)\mathbf{U}a_1)))$
$\varphi_2 = \bigwedge_{i=1}^{3} \Box\Diamond a_i$
$\varphi_3 = \bigwedge_{i=1}^{3} \Diamond a_i$

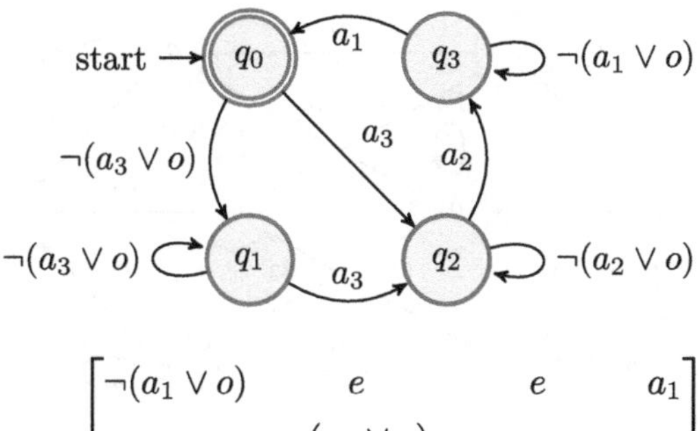

$$\varphi_2 = \begin{bmatrix} \neg(a_1 \vee o) & e & e & a_1 \\ a_2 & \neg(a_2 \vee o) & e & e \\ e & a_3 & \neg(a_3 \vee o) & e \\ e & a_3 & \neg(a_3 \vee o) & e \end{bmatrix}$$

FIGURE 7.13: The DBA of φ_2 and $\mathbf{M}_{\varphi_2}$.

The formula φ_1 (see Example 6.10) specifies the order of areas that the vehicle has to visit:

$$a_1 \to a_2 \to a_3 \to \neg a_2 \to a_1.$$

To satisfy φ_2, the vehicle is expected to visit three isolated areas labeled by a_1, a_2 and a_3 infinitely often. The DBA of φ_2 (Figure 7.13) itself is an SCC, and thus pre-processing is not needed. The transition matrix $\mathbf{M}_{\varphi_2}$ is obtained by arranging the states in the order $q_3q_2q_1q_0$. An equivalent DBA of φ_3 is shown in Figure 7.14, which contains no non-trivial SCC. The paths that lead to the accepting node q_0 enumerate the orders of visiting $a_{1,2,3}$. A topological sort gives the order of DBA nodes for $\mathbf{M}_{\varphi_3}$: $q_3q_2q_6q_7q_4q_5q_1q_0$.

Figure 7.15 shows the controlled trajectory of the mobile robot for φ_1 from an initial state $x_0 = (1.3, 5, 135°)$. The DBA in Figure 6.8 starts from node q_2, and thus the robot turns back to the area $L^{-1}(a_1)$ at the beginning to trigger the transition in the DBA that leads to node q_3. From the trajectory and the change of DBA states, we can see that the specification φ_1 is fulfilled without collision. Simulations of closed-loop trajectories with different initial conditions for $\varphi_{2,3}$ are shown in Figure 7.16.

With the second layout, the task for the robot is

$$\varphi_4 = \Diamond(a_1 \wedge \Diamond a_3) \vee \Diamond(a_2 \wedge (\neg a_4 \mathbf{U} a_3)) \wedge \Box\Diamond c \wedge \Box\Diamond d.$$

To follow φ_4, the robot not only needs to visit areas c and d repeatedly but also has to go to a_1 then a_3 or a_2 then a_3 while avoiding a_4. The corresponding DBA shown in Figure 7.17 contains two SCCs, which indicates a matrix decomposition in the form of (6.54) and the pre-processing can be used to reduce the time for control synthesis. The closed-loop simulation results with initial conditions $x_0 = (1.3, 6.5, 90°)$ and $x_0 = (8, 7, 90°)$ are shown in Figure 7.18.

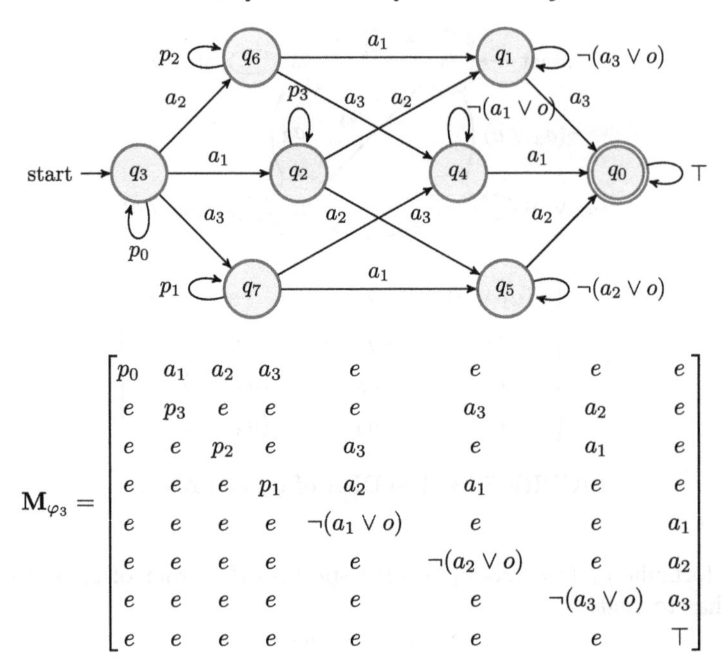

$$
\mathbf{M}_{\varphi_3} = \begin{bmatrix}
p_0 & a_1 & a_2 & a_3 & e & e & e & e \\
e & p_3 & e & e & e & a_3 & a_2 & e \\
e & e & p_2 & e & a_3 & e & a_1 & e \\
e & e & e & p_1 & a_2 & a_1 & e & e \\
e & e & e & e & \neg(a_1 \vee o) & e & e & a_1 \\
e & e & e & e & e & \neg(a_2 \vee o) & e & a_2 \\
e & e & e & e & e & e & \neg(a_3 \vee o) & a_3 \\
e & e & e & e & e & e & e & \top
\end{bmatrix}
$$

FIGURE 7.14: The DBA φ_3 with $p_0 = \neg(a_1 \vee a_2 \vee a_3 \vee o)$, $p_1 = \neg(a_1 \vee a_2 \vee o)$, $p_2 = \neg(a_1 \vee a_3 \vee o)$, $p_3 = \neg(a_2 \vee a_3 \vee o)$.

7.5 Online Obstacle Avoidance

In the motion planning problems in Section 7.4, we assume static environments where the world map is known and obstacles are static. The capability of formal control synthesis, however, is not limited to such situations. In this section, we are going to see a solution to online collision avoidance with moving obstacles without failing the LTL mission.

The central ingredient of this solution is an online replanning scheme, which dynamically modifies a feedback control strategy generated offline with respect to the given LTL task by using abstraction-based methods. Online modifications are made by replanning on a short horizon with only safe controls. By using relative positions of the obstacle to the robot, which is independent of the absolute position of the obstacle, safe controls can be computed offline by solving an invariance control problem, as we have discussed in Section 6.3. In this way, collision-free control strategies are designed by composing the solutions to two independent subproblems: the LTL motion planning based on a static world map and the safety control of the distance between two moving objects.

The replanning scheme is shown in Figure 7.19. The integration of collision avoidance to the safe LTL motion planning is through the module called

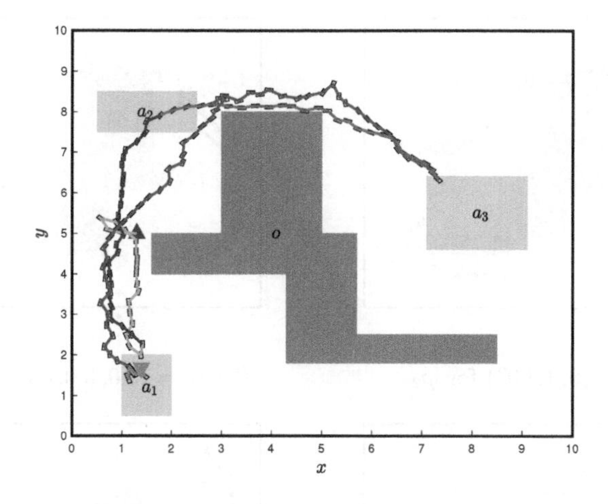

(a) Closed-loop trajectory.

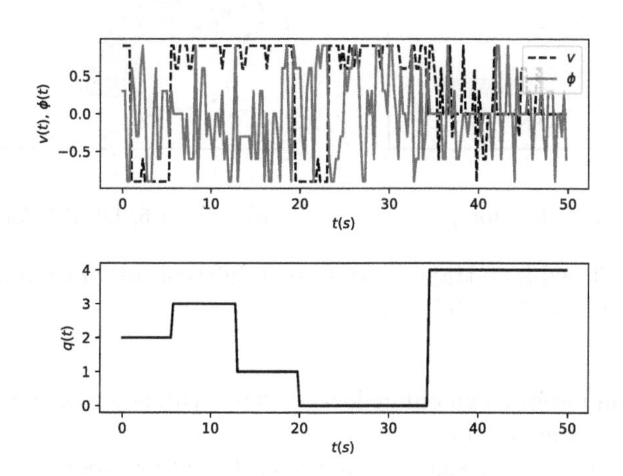

(b) Time history of control inputs and DBA state.

FIGURE 7.15: Closed-loop simulation with $x_0 = (1.3, 5, 135°)$. The upward and downward triangles mark the initial and terminal states, respectively.

`Patcher`. In the absence of obstacles, the output $\tilde{u}$ of the patcher or the control input of the system (7.10) is the same as the output of the static control strategy u, which is generated offline by the module `Static Planner`. When an obstacle is detected, the `Patcher` takes in the information of the *safe control set* $\mathcal{U}_{\text{ca}}(z_r)$ of the current relative state z_r of obstacle to the robot, which is a set of control inputs that can be used to avoid collisions. Such safe control sets are pre-computed offline in the module `Local Collision Avoidance`. The patcher takes a copy of a subgraph of the winning graph around the

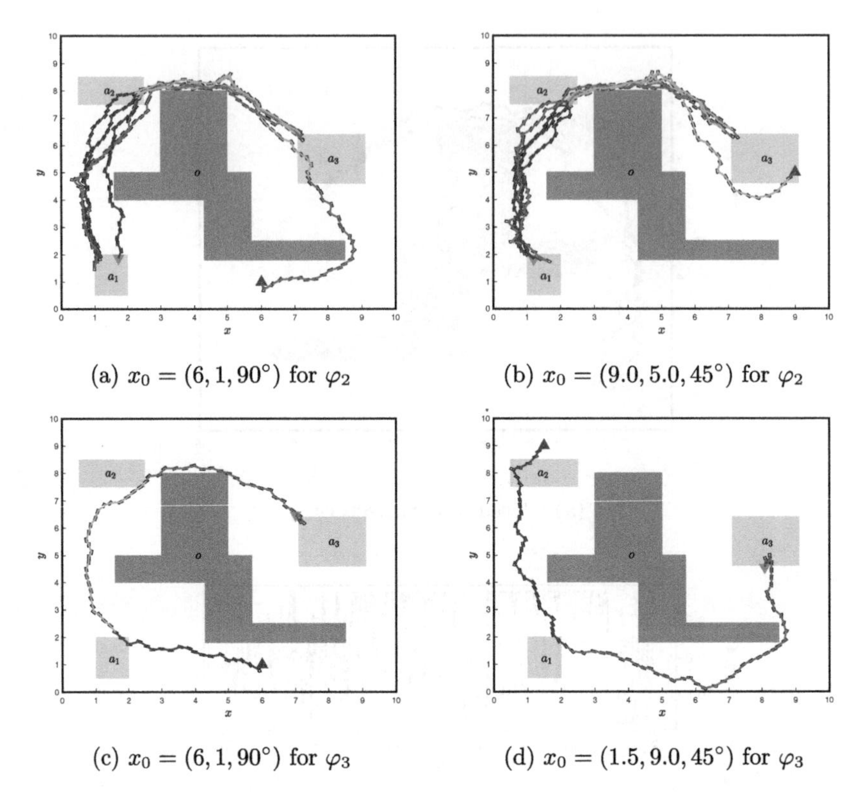

(a) $x_0 = (6, 1, 90°)$ for φ_2

(b) $x_0 = (9.0, 5.0, 45°)$ for φ_2

(c) $x_0 = (6, 1, 90°)$ for φ_3

(d) $x_0 = (1.5, 9.0, 45°)$ for φ_3

FIGURE 7.16: Closed-loop trajectories from 4 different initial conditions that meets φ_2 ((a), (b)) and φ_3 ((c), (d)).

current node and performs an online local control synthesis to generate a new control input $\tilde{u}$ to the system.

The benefit of such a design is that the tools and methods developed for LTL control synthesis in static environments can be reused without modifications, and hence it is more flexible than using the assume-guarantee type of LTL formulas as in [125, 241].

7.5.1 Offline Control Synthesis

Static Planner. For static LTL motion planning, we consider the kinematics model of a mobile robot:

$$\begin{aligned} \dot{x} &= v\cos(\psi), \\ \dot{y} &= v\sin(\psi), \\ \dot{\psi} &= \omega, \end{aligned} \tag{7.10}$$

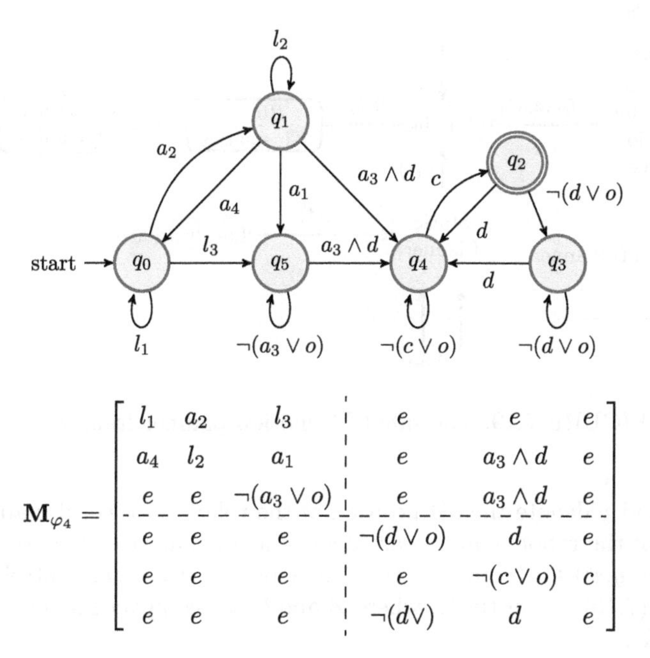

$$\mathbf{M}_{\varphi_4} = \left[\begin{array}{ccc|ccc} l_1 & a_2 & l_3 & e & e & e \\ a_4 & l_2 & a_1 & e & a_3 \wedge d & e \\ e & e & \neg(a_3 \vee o) & e & a_3 \wedge d & e \\ \hline e & e & e & \neg(d \vee o) & d & e \\ e & e & e & e & \neg(c \vee o) & c \\ e & e & e & \neg(d \vee) & d & e \end{array}\right]$$

FIGURE 7.17: The DBA of φ_4 with $l_1 = \neg o \wedge (a_4 \vee (\neg a_1 \wedge \neg a_2))$, $l_2 = \neg(o \vee a_1 \vee a_4 \vee a_3)$ and $a_1 \wedge \neg(a_3 \vee o)$. The topological sort gives the order $q_0 q_1 q_5 q_3 q_4 q_2$ for $\mathbf{M}_{\varphi_4}$.

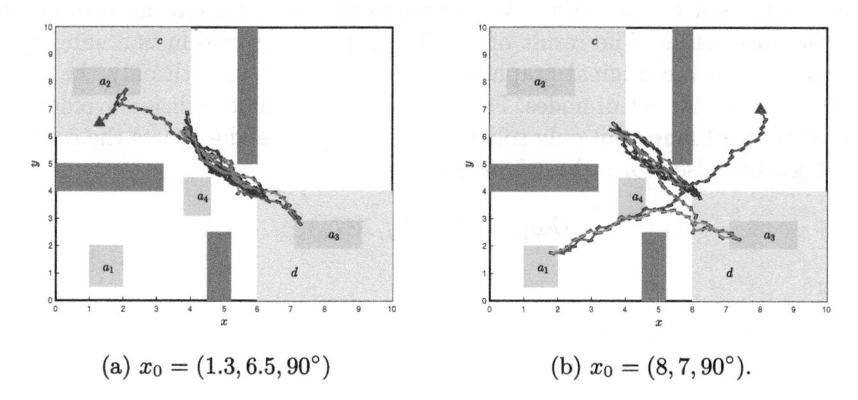

(a) $x_0 = (1.3, 6.5, 90°)$ (b) $x_0 = (8, 7, 90°)$.

FIGURE 7.18: The closed-loop trajectories with two different initial conditions.

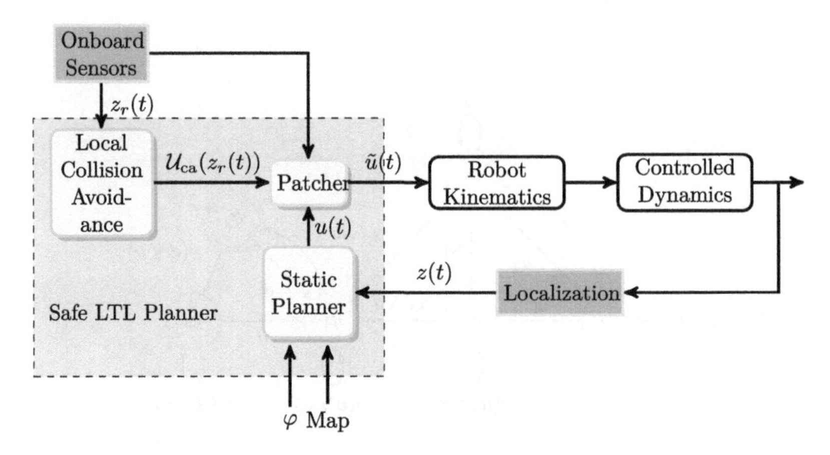

FIGURE 7.19: The safe LTL motion planner framework.

where x and y denote the 2D position of the robot on a world map, ψ is the direction of the robot, v and ω are linear the angular velocities, respectively. Let $z = (x, y, \psi) \in \mathcal{Z}$ and $u = (v, \omega) \in \mathcal{U}$ be the state and control variables of system (7.10), respectively, where $\mathcal{Z}$ and $\mathcal{U}$ are compact sets of $\mathbb{R}^3$ and $\mathbb{R}^2$, respectively.

In `Static Planner`, control synthesis is performed by solving a Rabin/parity game on a graph constructed by taking the product of an abstraction of the system model and an automaton translated by the given LTL formula (See Problem 3.4 in Chapter 3). The product can be treated as an *annotated graph* $\mathcal{G} = (V, E, AN)$, where V and E are sets of nodes and edges, respectively, and $AN = \hat{\mathcal{U}}$ ($\hat{\mathcal{U}}$ is a set of controls of the abstraction) is a set of annotations on the edges. There could be the same annotation for different edges going out of a node $v \in V$ because of the nondeterminism introduced by the abstraction. The result of the Rabin/parity algorithm is a subgraph trimmed from the original graph of the product system with only the winning set W as the set of nodes. For the purpose of online collision avoidance, the `Static Planner` not only solves the Rabin game and extracts the control strategy, but also stores the *winning graph*

$$\mathcal{G}_{\mathbf{Win}} = (V_w, E_w, AN_w),$$

where

$$AN_w = AN, \quad V_w = W,$$
$$E_w = \{(v, u, v') \mid v' \in W, \forall v'(v, u, v') \in E, v \in W, u \in AN\}.$$

It is a subgraph of the original one by trimming off the nodes from which no winning strategy exists and also the edges that cannot be taken in order to meet the winning condition.

Proposition 7.1 *The winning graph $\mathcal{G}_{\mathbf{Win}} \subseteq \mathcal{G}$ of the Rabin game on $\mathcal{G}$ is* closed *and* maximal, *i.e., any run of $\mathcal{G}$ starting in $\mathcal{G}_{\mathbf{Win}}$ can stay inside $\mathcal{G}_{\mathbf{Win}}$ forever and it is the largest possible subgraph of $\mathcal{G}$ such that the winning condition can be satisfied if any edge taken for the current node on $\mathcal{G}_{\mathbf{Win}}$.*

Proof: By definition, $\mathcal{G}_{\mathbf{Win}}$ contains only the winning set W and only the edges connecting nodes in W. Hence, $\mathcal{G}_{\mathbf{Win}}$ is closed. Suppose that $\mathcal{G}_{\mathbf{Win}}$ is not maximal. Since $V_w = W$, it is only possible that there is an edge $(v, u, v') \in E$ but $(v, u, v') \notin E_w$ that can be taken to guarantee that any run stays inside W. Then there exists another edge $(v, u, v'') \in E$ such that $v'' \in V \setminus W$ for the same $u \in AN$. Otherwise, all edges going from v will be included in $\mathcal{G}_{\mathbf{Win}}$ by the definition of E_w. Hence, by taking (v, u, v''), the run can not stay inside W. Therefore, $\mathcal{G}_{\mathbf{Win}}$ is maximal. ■

Local Collision Avoidance. To avoid moving obstacles in local areas, we dynamically modify the corresponding subgraphs of the winning graph by eliminating the transitions that might lead to collisions, and the modification only lives during the period that an obstacle is within a range where collisions are possible. The computation of the safe control set determines those unsafe transitions which can be obtained by invariance control synthesis offline on the two-object system model (7.11).

$$\begin{aligned} \dot{x}_r &= -v + v' \cos(\psi_r) + \omega y_r, \\ \dot{y}_r &= v' \sin(\psi_r) - \omega x_r, \\ \dot{\psi}_r &= \omega' - \omega, \end{aligned} \tag{7.11}$$

where $z_r = (x_r, y_r, \psi_r) \in \mathcal{Z}_r$ is the relative state of the obstacle with respect to the robot, and $u = (v, \omega)$ is the same control inputs as in system (7.10). We consider $d = (v', \omega')$ as a disturbance term, where v' and ω' are the linear and angular velocities of the obstacle with $|v'| \leq \overline{v}'$ and $|\omega'| \leq \overline{\omega}'$, respectively. Let $z' = (x', y', \psi')$ be the state of the obstacle on the world map. Then we have

$$\begin{bmatrix} x_r \\ y_r \end{bmatrix} = \begin{bmatrix} \cos(\psi) & \sin(\psi) \\ -\sin(\psi) & \cos(\psi) \end{bmatrix} \begin{bmatrix} x' - x \\ y' - y \end{bmatrix}.$$

To avoid collision, two objects should maintain at least distance D_{min} for all time, which gives the safety constraint $h(x_r, y_r) = x_r^2 + y_r^2 - D_{\text{min}}^2 \geq 0$. Let

$$\Omega = \left\{ z_r \in \mathbb{R}^3 \mid x_r^2 + y_r^2 - D_{\text{min}}^2 < 0 \right\} \subseteq \mathcal{Z}_r. \tag{7.12}$$

The *safe control set* at a state $z \in \mathbb{R}^3 \setminus \Omega$ for system (7.11) is defined by

$$\begin{aligned} \mathcal{U}_{\text{ca}}(z) = \{u \in \mathcal{U} \mid z(0) = z, u(0) = u, \\ z(t) \in \mathbb{R}^3 \setminus \Omega, u(t) \in \mathcal{U} \text{ satisfies (7.11)} \, \forall t > 0\}. \end{aligned}$$

Then the computation of the safe control set $\mathcal{U}_{\text{ca}}(z_r)$ is equivalent to invariance control in the target set $\mathbb{R}^3 \setminus \Omega$. System (7.11) is only considered when the system state z_r is within $\mathcal{Z}_r$ (where an obstacle can be detected), but the out-of-domain area $\mathbb{R}^3 \setminus \mathcal{Z}_r$ is also considered safe.

Note that systems (7.10) and (7.11) are independent but share the same control inputs of the same robot. When an obstacle is detected, the safe control set defined for (7.11) can provide an explicit collision-free constraint for the LTL motion planning based on (7.10). Moreover, safe control sets can be computed offline since they are independent of the absolute positions of robots on the world map.

7.5.2 Online Replanning

The module `Patcher` modifies the static control strategy online by using the pre-computed safe control sets to avoid collisions with moving obstacles. In the presence of obstacles, the robot may deviate from its original motion plan in order to avoid collisions, but the given LTL specification φ still needs to be realized. As shown in Proposition 7.1, the winning graph $\mathcal{G}_{\mathbf{Win}}$ is the maximal graph to extract a control strategy to realize φ. Hence, the replanning will be based on $\mathcal{G}_{\mathbf{Win}}$.

RH Replanning. This is the main online replanning scheme, which resembles the receding horizon paradigm. By using RH replanning, whose workflow is shown in Figure 7.20, the robot is guaranteed to avoid collisions without failing to realize φ.

In order to react to the moving obstacle in real time, the global state z and the relative state z_r are measured after each sampling time. If no obstacle is detected, then the control strategy κ from Algorithm 4 will be used. If an

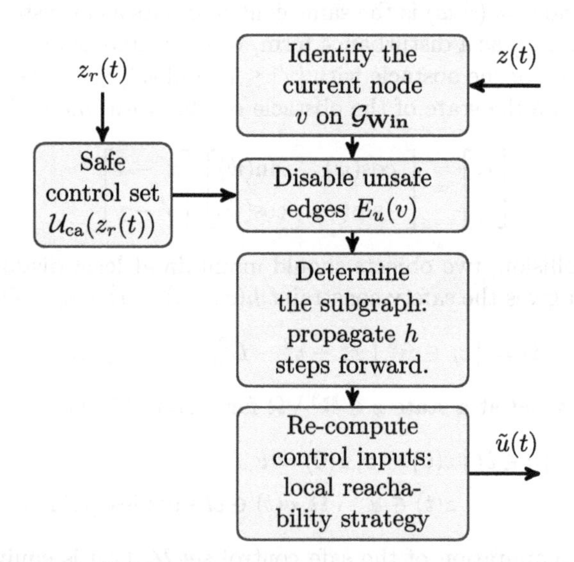

FIGURE 7.20: The workflow of RH replanning.

obstacle appears at time $t = i\tau$ for some $i \in \mathbb{N}$, which means $z_r(t) \in \mathcal{Z}_r$, the control input for the current state $z(t)$ of system (7.10) needs to be recomputed. The same procedure is repeated during each sampling period so that the control input is updated in real time to guarantee safety. Such a recomputation procedure starts with identifying the corresponding node v_i on the winning graph $\mathcal{G}_{\mathbf{Win}}$.

The next step is to determine the unsafe edges of the current node v_i on the winning graph $\mathcal{G}_{\mathbf{Win}}$ by using safe controls in $\mathcal{U}_{\mathcal{G}_r}(v_i)$ and disable them in the winning graph. The set of unsafe edges for a node v is defined as $E_u(v) = \{(v, u, v') \in E_w \mid u \notin \mathcal{U}_{\mathcal{G}_r}(v)\}$, and the modified graph is denoted by $\mathcal{G}_{\mathbf{Win}} \setminus E_u(v_i)$. This prevents the use of unsafe control inputs in the later replanning. Since $\mathcal{U}_{\mathcal{G}_r}(v)$ is pre-computed offline and loaded to memory during the execution of the plan, this information can be accessed immediately.

Note that $\mathcal{G}_{\mathbf{Win}} \setminus E_u(v_i)$ is a subgraph of $\mathcal{G}_{\mathbf{Win}}$. Running Algorithm 4 again on $\mathcal{G}_{\mathbf{Win}} \setminus E_u(v_i)$ yields a new control strategy that can guarantee both the collision avoidance and the satisfaction of φ from the current node v_i, but it is time-consuming. To perform replanning online, we create a copy of subgraph $\mathcal{G}_h(v_i)$ of $\mathcal{G}_{\mathbf{Win}} \setminus E_u(v_i)$ by propagating from v_i forwardly by h ($h \in \mathbb{N}$ is the horizon) steps and including all the traversed nodes and edges. The recomputation of control input at node v_i will be conducted on $\mathcal{G}_h(v_i)$, which is much smaller than $\mathcal{G}_{\mathbf{Win}} \setminus E_u(v_i)$. As a subgraph of $\mathcal{G}_{\mathbf{Win}}$, $\mathcal{G}_h(v_i)$ already contains all the nodes to which the product system state can be controlled in order to still be capable to satisfy φ.

Recall that the control strategy for each node v on $\mathcal{G}_{\mathbf{Win}}$ is essentially the reachability control strategy of $\text{Reach}_{\exists\forall}(\mathcal{T}, \Omega)$. This means that we only need to re-compute a local reachability control strategy on the subgraph $\mathcal{G}_h(v_i)$ for v_i. Technically, any node $v \in \mathcal{G}_h(v_i)$ except for v_i can be included in the local target set B for local backward reachable set computation. Let the *attracting distance* $d_A(v)$ as the remaining number of steps to reach a set $A \subseteq V$ for $v \in V$, and we simply use $d(v)$ as the attracting distance for $v \in V$ to its target set determined by Algorithm 4 in order to realize the LTL specification. Then we can choose the local target set B by

$$B = \left\{ v \in \mathcal{G}_h(v_i) \mid d(v) \in [d_{\min}, \frac{d_{\max} - (1-\gamma)d_{\min}}{\gamma}] \right\},$$

where $\gamma \in (0, 1]$ is a adjustable parameter, and

$$d_{\min} = \min \{d(v) \mid v \in \mathcal{G}_h(v_i), v \neq v_i\},$$
$$d_{\max} = \max \{d(v) \mid v \in \mathcal{G}_h(v_i), v \neq v_i\}.$$

The RH planning is capable to deal with the situation with multiple obstacles by using the union of safe control sets corresponding to different obstacles. Without further assumptions on the number of moving obstacles and how they move, however, a safety stop might be required and deadlock can happen due to the failure in generating any safe control strategy.

Stutter Avoidance. A problem with the RH replanning is that the robot may be blocked and fail to move forward, which we call the *stutter motion*. For example, if the obstacle moves jiggly in front of the robot, the re-computed control inputs by RH replanning may result in a loop in the local subgraph as shown in Figure 7.21: u_2 is applied when the robot state corresponds to v_1 on $\mathcal{G}_{\mathbf{Win}}$ while u_3 is applied when the robot is on v_2 because u_1 is unsafe for both nodes.

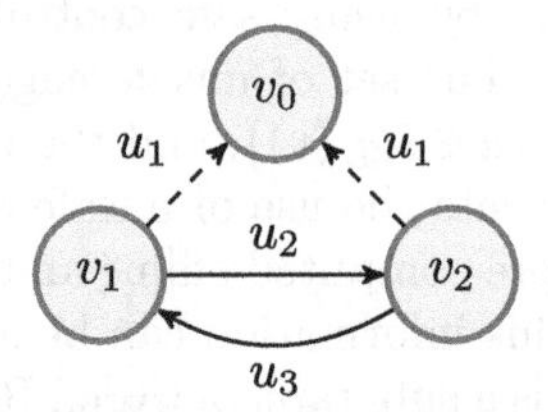

FIGURE 7.21: Disable of the unsafe control u_1 (marked by dashed edges) results in a loop between v_1 and v_2.

To avoid the stutter motion and encourage the robot to move forward, we propose another replanning procedure called *Detouring*. We assume that when the robot is stuttering, the obstacle only moves in a bounded area O near the robot. Under this assumption, the area O is treated as an additional static obstacle on the world map. The *Detouring* procedure is to re-target an obstacle-free area G beside the new obstacle O and generate a control strategy that guides the robot to G. After reaching G, the robot can continue the static plan or using RH to maintain safety again.

We illustrate Detouring in Figure 7.22. Let $(x_o(t), y_o(t))$ be the location of the obstacle on the world map at time t when it is detected. We choose $O = \{(x, y) \mid (x - x_o(t))^2 + (y - y_o(t))^2 \leq r_o^2\}$, where r_o is the radius of O. The target G is determined in the body frame of the robot and projected to the world map. On the x_r-y_r plane, G is a circular area with radius r_g and center

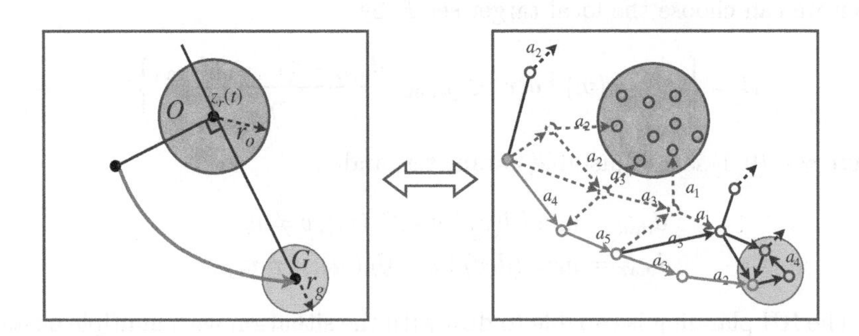

FIGURE 7.22: The Detouring in the body frame of the robot (left) and a subgraph of $\mathcal{G}_{\mathbf{Win}}$ (right). Dash nodes and edges are unsafe and hence deleted, resulting in a detour as shown in the left figure.

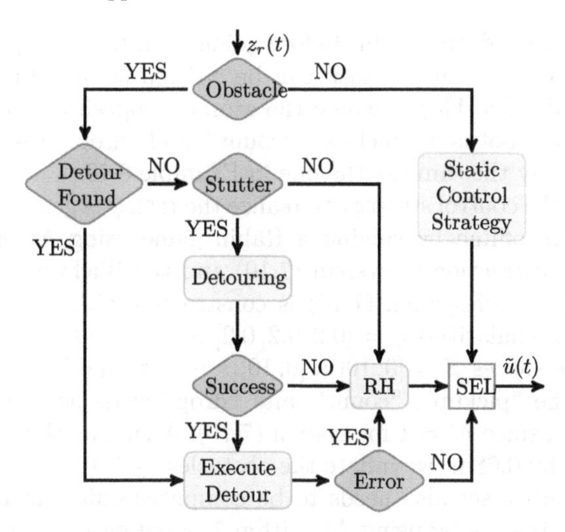

FIGURE 7.23: The logic of the `Patcher`. The "SEL" block only activates one input channel as the output $\tilde{u}(t)$.

on the perpendicular line of the line connecting the center of O and the origin. The distance of the center of G is a tunable parameter $d_g > r_o + r_g$. The sets $\hat{O}$ and $\hat{G}$ of nodes on the $\mathcal{G}_{\mathbf{Win}}$ corresponding to O and G are then identified, respectively. Next we obtain a local subgraph by forward traversing nodes and edges from the current robot node on $\mathcal{G}_{\mathbf{Win}}$ until a certain percentage (e.g., 80%) of nodes in $\hat{G}$ are collected. The nodes, as well as their edges, that cannot avoid reaching $\hat{O}$ will be deleted from the subgraph. Lastly, compute an reachability control strategy of $\hat{G}$ on the subgraph as the output of Detouring.

Overall Workflow. The RH replanning is safety conservative, and the Detouring is more aggressive but cannot always successfully replan a new strategy. For the purpose of safety, we propose the overall workflow of the `Patcher` in Figure 7.23. When an obstacle is at present, the RH replanning is performed if no stuttering occurs, or detour planning is not successful, or there are errors in detour plan execution. If the robot is stuttering, the Detouring procedure will start and be executed (if a detour strategy is generated) until the detour target G is reached.

7.5.3 Simulation

In this section, we demonstrate the effectiveness of the proposed approach by testing in different scenarios where obstacles move in different patterns, which is unknown to the robot. We consider an indoor 10m-by-10m warehouse with walls and blocking areas. The task for the robot is: go to the area "pickup" to pick up a cargo first, move to the "count" area to register the cargo next, then drop off the cargo at "drop" area, and return back to "pickup" to start a new "pick-register-drop" again. Before going back to "pickup", the "count" should not be visited. Such a task can be written as an LTL formula $\varphi = \Diamond(a \wedge \Diamond(c \wedge \Diamond(d \wedge (\neg c)\mathbf{U}a)))$, where the atomic propositions a, c, and d are `True` only if the robot is in "pickup", "count", and "drop", respectively. This formula is actually the same as the one in Example 6.10.

First of all, the control strategy to realize the task φ is generated in module `Static Planner` offline by solving a Rabin game using Algorithm 4 on a product of an abstraction of system (7.10) and the Büchi automaton $\mathcal{A}_\varphi$ of φ. The abstraction of system (7.10) is constructed with the sampling time $\tau = 0.3$s, and granularities $\eta = [0.2, 0.2, 0.2]$ and $\mu = [0.3, 0.3]$ for the state and input state spaces $\mathcal{Z} = [0, 10] \times [0, 10] \times [-\pi, \pi]$ and $\mathcal{U} = [-1, 1] \times [1, 1]$, respectively. The "pickup", "count", and "drop" areas are strengthened by $\varepsilon = \tau/2 = 0.15$, since $M = 1$ for system (7.10). Assuming that the size of the robot is 0.55m by 0.68m, we inflate the obstacles by 0.4m.

The safe control set also needs to be computed offline in module `Local Collision Avoidance` by using Algorithm 7 based on the model (7.11). We consider $\mathcal{Z}_r = [-3, 3] \times [-3, 3] \times [-\pi, \pi]$, $\mathcal{W} = [-0.8, 0.8] \times [-0.8, 0.8]$, and $D_{\min} = 0.8$. The abstraction of system (7.11) is constructed by using the sampling time $\tau_r = 0.1$s, $\eta_r = [0.1, 0.1, 0.1]$, and $\eta_r = \eta$. The safe region is strengthened by $\varepsilon_r = 0.13$.

The stutter motion is detected if the robot position barely changes in, e.g., 3s. To test its effectiveness, we created three obstacle moving patterns: (I) the obstacle enters from $(0, 3)$, heads northeast next and then turns around to exit from the south; (II) the obstacle enters from south and keeps moving in a circle in the middle of the warehouse; (III) the obstacle enters with constant v' from east until around "pickup", then moves back and forth.

We run simulations with time step $\tau_s = 0.1$s on a 2.4 GHz Intel Core i5. By using the horizon $h = 3$, the RH replanning can be finished within 2ms. We choose $r_o = 0.8$ and $r_g = 0.35$ for the Detouring, which can finish in around 20ms. The collision-free trajectories by using the proposed online method are shown in Figure 7.24. For scenario I, the robot waited to let the obstacle pass at the beginning and slightly went back to avoid collision when the obstacle was turning around. Hence the shape of collision-free trajectory is almost the same as the nominal one, but they are different in time frame. For scenario II, the obstacle moved in a circle, hence the robot turned back a little at the north and northeast of the circle, and dodged the obstacle, which was around southwest of the circle, to return to "pickup". In the third scenario, the obstacle moved back and forth blocking the pre-planned route, and as a result, the robot experienced stuttering. The Detouring was then used to search for temporary target areas to circumvent the obstacle. Figure 7.24c shows a successful collision-free path of the robot by using the Detouring.

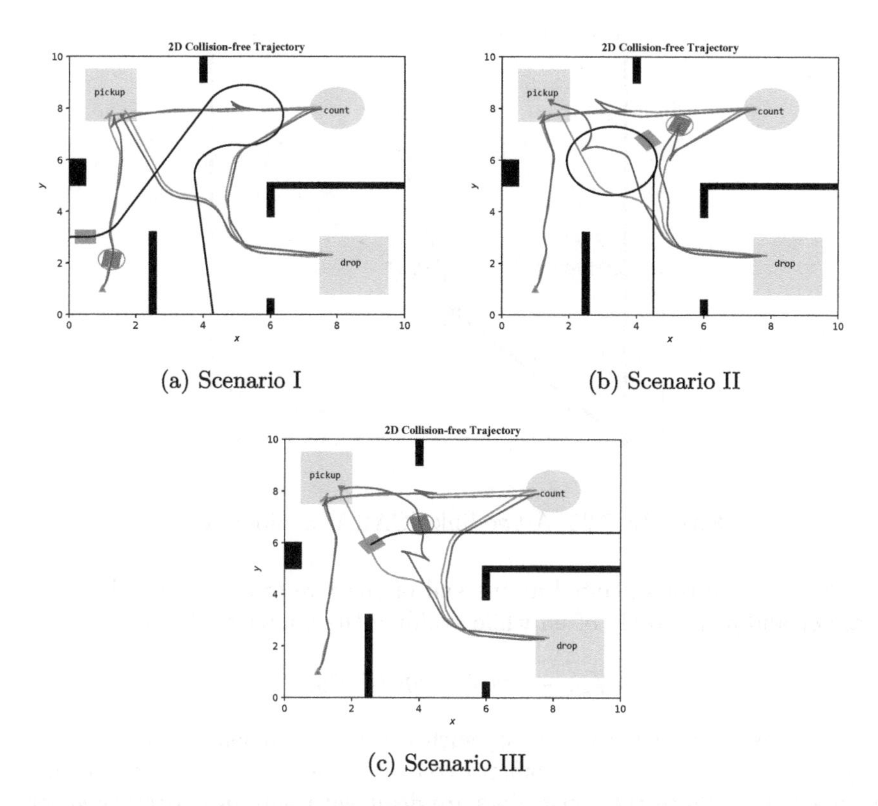

(a) Scenario I

(b) Scenario II

(c) Scenario III

FIGURE 7.24: Online simulation results. The lighter path: the closed-loop trajectory by using the static control strategy; the darker path: the actual collision-free trajectory; the darker path: obstacle trajectory.

7.6 Robotic Manipulator

SCARA (Selective Compliant Articulated Robot for Assembly) is a type of manipulators that are operate on a horizontal plane. They are often used for vertical assembly tasks in industry [214]. In this section, we consider a two-link SCARA manipulator in a workspace shown in Figure 7.25. Its back and fore arms are of equal length and weight ($l_1 = l_2 = 0.15\,\text{m}$, and $m_1 = m_2 = 0.1\,\text{kg}$), and their moment of inertia are $I_1 = I_2 = 1.33 \times 10^{-5}\,\text{kg}\,\text{m}^2$. There are two joints controlling the rotation of the arms, and their angles are denoted by θ_1 and θ_2, respectively. While moving in the workspace, the manipulator has to avoid an obstacle of length $r = 0.5l_1$ at $h = 0.8l_1$ to the origin.

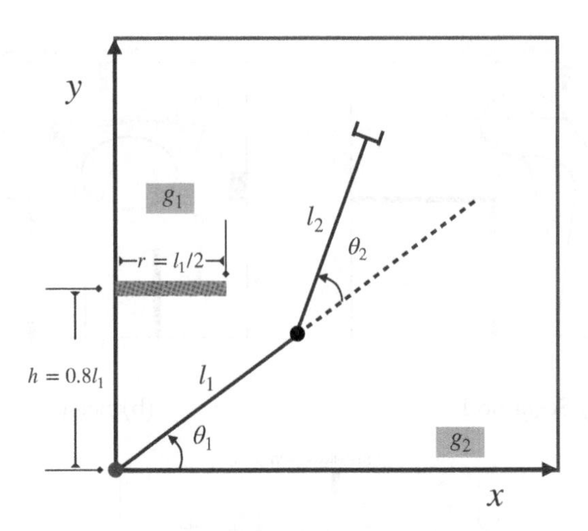

FIGURE 7.25: A two-link SCARA manipulator.

The end-effector mounted at the end of the fore arm is required to visit areas g_1 and g_2 infinitely often while avoiding the horizontal bar o, i.e.,

$$\varphi_{\text{gb}} = \Box\neg o \wedge \Box\Diamond g_1 \wedge \Box\Diamond g_2.$$

The classic approach to solving such a control problem is to design the trajectories that satisfy the geometry constraints and the tracking controller independently. Since the constraints are dealt with only in trajectory generation, there is no collision-free guarantee during tracking. We show here a provably correct control design such that the specification can be satisfied without collision.

Consider the Lagrange dynamics of the manipulator [175]:

$$\begin{bmatrix} z_1 + 2z_2c_2 & z_3 + z_2c_2 \\ z_3 + z_2c_2 & z_3 \end{bmatrix} \begin{bmatrix} \ddot{\theta}_1 \\ \ddot{\theta}_2 \end{bmatrix} + \begin{bmatrix} -z_2s_2\dot{\theta}_2 & -z_2s_2(\dot{\theta}_1 + \dot{\theta}_2) \\ z_2s_2\dot{\theta}_1 & 0 \end{bmatrix} \begin{bmatrix} \dot{\theta}_1 \\ \dot{\theta}_2 \end{bmatrix} = \begin{bmatrix} \tau_1 \\ \tau_2 \end{bmatrix}, \tag{7.13}$$

where the control inputs $\tau_{1,2}$ are the torques at the joints, $r_{1,2} = 0.5l_{1,2}$ are the centers of mass of the arms to the joints, $c_2 = \cos(\theta_2)$, $s_2 = \sin(\theta_2)$, and

$$\begin{aligned} z_1 &= I_1 + I_2 + m_1r_1^2 + m_2(l_1^2 + r_2^2), \\ z_2 &= m_2l_1r_2, \\ z_3 &= I_2 + m_2r_2^2. \end{aligned}$$

Let the state of the system be $[\theta_1, \theta_2, \omega_1, \omega_2]^T$, where ω_i is the angular velocity of the joint i ($i = 1, 2$). Written in the classic first-order form, (7.13) becomes

$$\begin{aligned}
\dot{\theta}_1 &= \omega_1, \\
\dot{\theta}_2 &= \omega_2, \\
\dot{\omega}_1 &= \frac{z_3\tau_1 + z_2z_3s_2(2\omega_1 + \omega_2)\omega_2 + (z_3 + z_2c_2)(z_2\omega_1^2 s_2 - \tau_2)}{\Delta(z_{1,2,3}, c_2)}, \\
\dot{\omega}_2 &= \frac{(z_1 + 2z_2c_2)(\tau_2 - z_2\omega_1^2 s_2) - (z_3 + z_2c_2)(\tau_1 + z_2s_2(2\omega_1 + \omega_2)\omega_2)}{\Delta(z_{1,2,3}, c_2)}.
\end{aligned} \tag{7.14}$$

where $\Delta(z_{1,2,3}, c_2) = z_3(z_1 - z_3) - z_2^2c_2^2$.

Approximating the predecessors based on (7.14) will induce a large approximation error and increase the computational complexity due to the complex expressions. To synthesize a controller within tolerable time, we replace (7.14) by two double integrators:

$$\begin{aligned}
\dot{\theta}_1 &= \omega_1, \quad \dot{\omega}_1 = u_1, \\
\dot{\theta}_2 &= \omega_2, \quad \dot{\omega}_2 = u_2,
\end{aligned} \tag{7.15}$$

where $u_{1,2}$ are virtual control inputs, and (7.13) can be used to convert them to the real inputs $\tau_{1,2}$. As shown in Figure 7.25, the angles for the arms satisfy $\theta_1 \in [0, \pi/2]$, $\theta_2 \in [-\pi, \pi]$.

The end-effector is operating on the Cartesian operational space while the motors at the joints work on the joint space. The inverse kinematics problem consists of the determination of the joint variables corresponding to a given end-effector position and orientation.

$$\begin{aligned}
x &= l_1\cos(\theta_1) + l_2\cos(\theta_1 + \theta_2), \\
y &= l_1\sin(\theta_1) + l_2\sin(\theta_1 + \theta_2).
\end{aligned}$$

For a given point (x, y) in the operational space, we can compute the corresponding angles (θ_1, θ_2) in the joint space for the two motors:

$$\begin{aligned}
\phi_1 &= \arccos\left(\frac{l_1^2 + l_3^2 - l_2^2}{2l_1l_3}\right), \; \phi_2 = \arccos\left(\frac{l_2^2 + l_3^2 - l_1^2}{2l_2l_3}\right), \\
\theta_1 &= \arctan\left(\frac{y}{x}\right) + \phi_1, \quad \theta_2 = -(\phi_1 + \phi_2), \text{ or} \\
\theta_1 &= \arctan\left(\frac{y}{x}\right) - \phi_1, \quad \theta_2 = \phi_1 + \phi_2,
\end{aligned}$$

where $l_3 = \sqrt{x^2 + y^2}$.

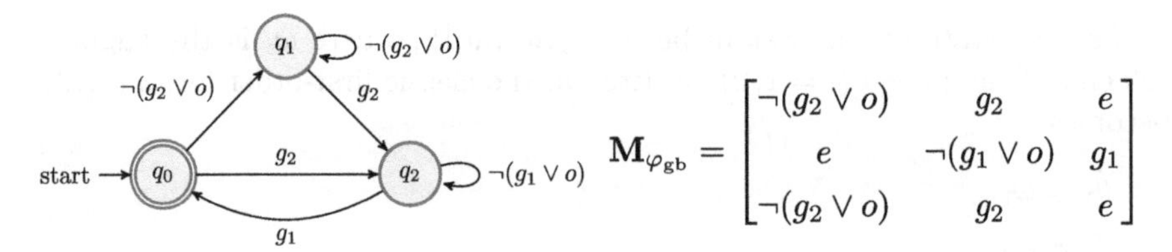

FIGURE 7.26: The DBA of φ_{gb} and the transition matrix with the order q_1, q_2, q_0.

The collision area in the Cartesian operational space is translated into the following inequalities related to the system state:

$$\begin{aligned} \theta_1 &\geq \arctan\left(\frac{h}{r}\right), \\ 0 \leq \theta_1 \leq \arcsin\left(\frac{h}{l_1}\right) &\Rightarrow \\ \theta_1 + \theta_2 &\geq \pi - \arctan\left(\frac{h - l_1 \sin\theta_1}{l_1 \cos\theta_1 - r}\right), \\ \arcsin\left(\frac{h}{l_1}\right) \leq \theta_1 \leq \arctan\left(\frac{h}{r}\right) &\Rightarrow \\ \theta_1 + \theta_2 &\geq \pi + \arctan\left(\frac{l_1 \sin\theta_1 - h}{l_1 \cos\theta_1}\right). \end{aligned} \tag{7.16}$$

With 0.1 s as the sampling time for the discrete-time model of (7.15), the constant $\rho = 1.1$ satisfies Assumption 4.1 and 4.2. A discretization precision of 0.05 is used in Algorithm 10 for $\theta_{1,2}$ and 0.1 for $\omega_{1,2}$. The DBA and transition matrix $\mathbf{M}$ of the specification φ_{gb} is given in Figure 7.26, where propositions are simplified since the target areas are non-overlapping (same in Figure 7.13, 7.14, and 7.17).

As shown in Figure 7.27, the controlled trajectories by using the synthesized feedback controller are collision-free and the DBA state is switching periodically between q_0, q_1, and q_2. The trajectories connecting regions g_1 and g_2 are not exactly the same for different periods. This is because a random valid control input is chosen when there are multiple valid control synthesized by Algorithm 10.

7.7 Bipedal Locomotion

In the field of robotics, planning and control of bipedal locomotion have been one of the attractive topics. Due to the complexity of whole-body

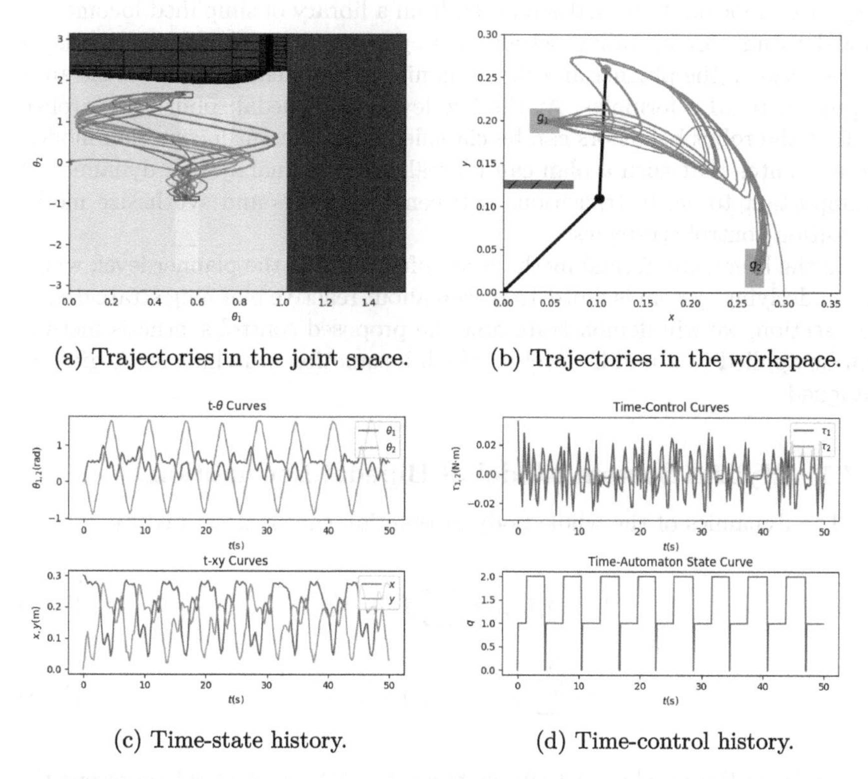

(a) Trajectories in the joint space. (b) Trajectories in the workspace.

(c) Time-state history. (d) Time-control history.

FIGURE 7.27: Closed-loop simulation results. In (a), the gray area is the collision area characterized by (7.12); the target sets $g_{1,2}$ are under-approximated in the joint space by two rectangles; the winning set is projected into shaded region of the θ_1-θ_2 plane.

dynamic locomotion (WBDL) behaviors and the requirements of being reactive to the dynamic environment, planning and control often lives in a hierarchical structure [29]. In this way, the complicated control problem is decomposed into simpler problems which are solved at different levels. The WBDL model, therefore, is simplified accordingly at different levels to serve different control purposes.

In reactive locomotion planning, the bipedal robot is expected to behave in respond to the changes in the environment. Recently, formal methods have gained increasing attention for solving such problems because of the correctness guarantee, and LTL formulas are favored for specifying temporal and reactivity properties [24, 100, 184, 236, 241].

The contact-based decision and planning method, which operates over a set of robotic maneuvers determined by contact points [234], is applied in the design of control hierarchy. At the high level, a motion plan, which is a sequence of locomotion modes (chosen from a library of simplified locomotion models) and corresponding setpoints, is generated by solving a two-player game between the planner and the dynamic environment with the constraints expressed in LTL formulas. At the low level, the bipedal robot is controlled so that the robot behaviors can be classified into different locomotion modes. To guarantee that such a plan can be realized by actual system dynamics, it is important to verify transitions between two modes and synthesize mode-transition control strategies.

In the literature, formal methods are often used at the planner level, where the underlying system is finite, to reason about reactive planning strategies. In this section, we will demonstrate how the proposed control synthesis method can be applied to a middle level in which locomotion switching strategies are designed.

7.7.1 Hybrid System Model of Bipedal Locomotion

The dynamics of the whole-body locomotion can be described by

$$\dot{\boldsymbol{l}} = m\ddot{\boldsymbol{p}}_{\text{com}} = \sum_{i}^{N_c} \boldsymbol{f}_i + m\boldsymbol{g}, \tag{7.17}$$

$$\dot{\boldsymbol{k}} = \sum_{i}^{N_c} (\boldsymbol{p}_i - \boldsymbol{p}_{\text{com}}) \times \boldsymbol{f}_i + \boldsymbol{\tau}_i, \tag{7.18}$$

where N_c is the number of limb contacts, $\boldsymbol{l} \in \mathbb{R}^3$ and $\boldsymbol{k} \in \mathbb{R}^3$ represent the centroidal linear and angular momenta, respectively, $\boldsymbol{f}_i \in \mathbb{R}^3$ is the ith ground reaction force, m is the total mass of the robot, $\boldsymbol{\tau}_i \in \mathbb{R}^3$ is the contact torque of the ith limb, the variables

$$\boldsymbol{g} = \begin{bmatrix} 0 \\ 0 \\ -g \end{bmatrix}, \quad \boldsymbol{p}_i = \begin{bmatrix} p_{i,x} \\ p_{i,y} \\ p_{i,z} \end{bmatrix}, \quad \boldsymbol{p}_{\text{com}} = \begin{bmatrix} x \\ y \\ z \end{bmatrix}$$

correspond to the gravity field, the position of the ith limb contact position, and the center of mass (CoM) position, respectively.

The above general model can be simplified based on different contact modes under the assumptions that are commonly imposed to make the problem tractable [20]. In this WBDL control problem, six locomotion modes are considered to produce various behaviors [248], which are also pictured in Figure 7.28.

The prismatic inverted pendulum mode (PIPM). In a normal environment, the bipedal robot exhibits a normal walking gait: there is a single foot contact with the floor in each walking period. Such walking dynamics can be considered as an inverted pendulum model. Since $N_c = 1$ in this mode, we

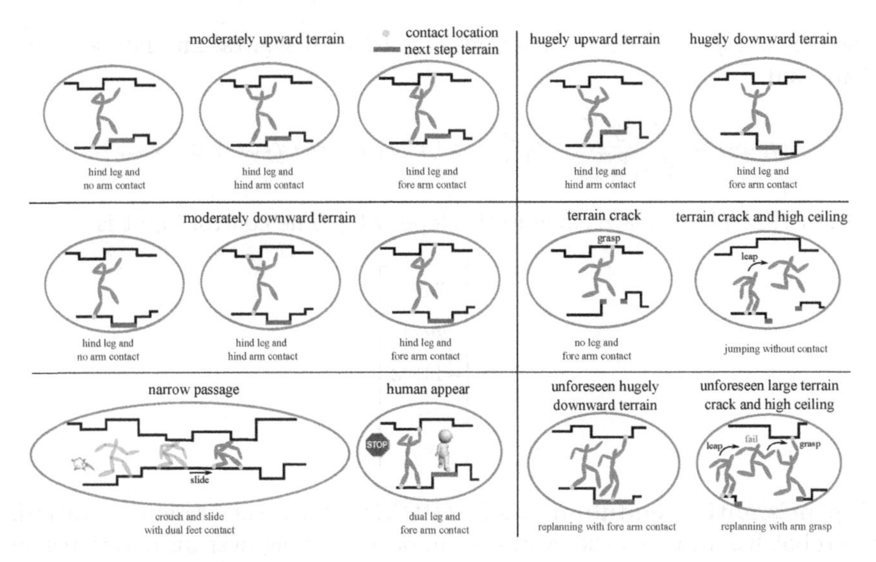

FIGURE 7.28: Contact-based planning strategies for locomotion in rough terrains [249]. Events motivated by ordinary accidents in human daily lives, such as a crack on the terrain and the sudden appearance of a human, are treated as emergency events, and incorporated into the allowable environment.

can simplify (7.18) to

$$(\boldsymbol{p}_{\text{com}} - \boldsymbol{p}_{\text{foot}}) \times (\boldsymbol{f}_{\text{com}} + m\,\boldsymbol{g}) = -\boldsymbol{\tau}_{\text{com}},$$

where $\boldsymbol{p}_{\text{foot}}$ is the position of the contact foot point, and $\boldsymbol{f}_{\text{com}}$ is the vector of CoM inertial forces:

$$\boldsymbol{f}_{\text{com}} = m\ddot{\boldsymbol{p}}_{\text{com}} = m \begin{bmatrix} \ddot{x} \\ \ddot{y} \\ \ddot{z} \end{bmatrix},$$

Assume that the bipedal locomotion follows a piecewise linear CoM path surface

$$\psi_{\text{com}}(x, y, z) = z - ax - by - c = 0, \tag{7.19}$$

where a, b and c are the coefficients of the surface. Thus, the dynamics in the vertical direction are represented by $\ddot{z} = a\ddot{x} + b\ddot{y}$.

Hence, the mathematical model for this mode is

$$\begin{bmatrix} \ddot{x} \\ \ddot{y} \end{bmatrix} = \omega_{\text{PIPM}}^2 \begin{bmatrix} x - x_{\text{foot}} - \frac{\tau_y}{mg} \\ y - y_{\text{foot}} - \frac{\tau_x}{mg} \end{bmatrix}, \tag{7.20}$$

where $\ddot{x}$ and $\ddot{y}$ are CoM accelerations aligned with sagittal and lateral directions, and

$$\omega_{\text{PIPM}} = \sqrt{\frac{g}{z_{\text{PIPM}}^{\text{apex}}}}, \quad z_{\text{PIPM}}^{\text{apex}} = (a \cdot x_{\text{foot}} + b \cdot x_{\text{foot}} + c - z_{\text{foot}})$$

is the PIPM phase-space asymptotic slope [248]. The control input is

$$\boldsymbol{u} = \begin{bmatrix} x_{\text{foot}} \\ y_{\text{foot}} \\ \omega_{\text{PIPM}} \\ \tau_x \\ \tau_y \end{bmatrix}.$$

The prismatic pendulum mode (PPM). When the terrain is cracked, the robot has to grasp the overhead support to swing over an unsafe region using brachiation. The system dynamics can be approximated as a pendulum model. For a single hand contact, we have

$$\begin{bmatrix} \ddot{x} \\ \ddot{y} \end{bmatrix} = -\omega_{\text{PPM}}^2 \begin{bmatrix} x - x_{\text{hand}} - \frac{\tau_y}{mg} \\ y - y_{\text{hand}} - \frac{\tau_x}{mg} \end{bmatrix}, \tag{7.21}$$

where similarly

$$\omega_{\text{PPM}} = \sqrt{\frac{g}{z_{\text{PPM}}^{\text{apex}}}}, \quad z_{\text{PPM}}^{\text{apex}} = (z_{\text{hand}} - a \cdot x_{\text{hand}} - b \cdot x_{\text{hand}} - c)$$

given the same surface (7.19). Similarly, vertical direction dynamics are represented by $\ddot{z} = a\ddot{x} + b\ddot{y}$. A difference between PIPM and PPM lies in that PPM dynamics is inherently stable since the CoM is always attracted to move toward the apex position while the PIPM dynamic is not. This study assumes the robot can firmly grasp the overhead support once receiving the upper limb contact command.

The stop-launch mode (SLM). When a human appears, the robot has to come to a stop, wait until human disappears, and start to move forward. The task in this mode consists of decelerating the CoM motion to zero and accelerating it from zero again. We name this model as a SLM with constant CoM sagittal accelerations:

$$\dot{l}_x = ma_x, \ \dot{l}_y = ma_y, \ \dot{l}_z = ma_z,$$

where a_x, a_y, a_z are the control inputs. The resulting phase space trajectory is a parabolic manifold.

The multi-contact mode (MCM). When the robot maneuvers through unstructured rough terrains, arms and legs in contact can accelerate and decelerate the CoM according to terrain height variations. To make the dynamics tractable, we assume a known constant vertical acceleration a_z in each step and neglect the angular momentum k_z around the z-axis [20], which leads to

$$\sum_{i}^{N_c} f_{i,z} = m(\ddot{z} - g).$$

With multiple point contacts, we let $\boldsymbol{\tau}_i = 0$ for all $i \leq N_c$ in (7.18), and the dynamics can be simplified to

$$\begin{bmatrix} \ddot{x} \\ \ddot{y} \\ \ddot{\varphi} \\ \ddot{\theta} \end{bmatrix} = \begin{bmatrix} \sum_i^{N_c} f_{i,x}/m \\ \sum_i^{N_c} f_{i,y}/m \\ -(\ddot{z} - g) \cdot y + z \cdot \sum_i^{N_c} f_{i,y}/m - \sum_i^{N_c} p_{i,z} \cdot f_{i,x}/m + \sum_i^{N_c} p_{i,z} \cdot f_{i,z}/m \\ (\ddot{z} - g) \cdot x - z \cdot \sum_i^{N_c} f_{i,x}/m + \sum_i^{N_c} p_{i,z} \cdot f_{i,x}/m - \sum_i^{N_c} p_{i,y} \cdot f_{i,y}/m \end{bmatrix},$$

where φ and θ are torso roll and pitch angles aligned with the CoM sagittal and lateral directions as derived from (7.18). The external force vector $(f_{i,x}, f_{i,y}, f_{i,z})$ represents the ith contact force. The vertical position z is a function of x and y defined a priori.

The hopping mode (HM). This model applies when the locomotion model needs to jump over an unsafe region. In this case, the CoM dynamics follows a free-falling ballistic trajectory. We have

$$\ddot{x} = \ddot{y} = 0, \ddot{z} = -g.$$

The trajectory is fully controlled by the initial condition, where a discontinuous jump in the CoM state can occur and be used to generate a desired linear momentum. For instance, when the robot jumps over a cracked terrain, it needs to push the ground as the foot lifts to generate a sufficiently large sagittal linear acceleration.

The sliding mode (SM). This model applies when the robot needs to slide through a constrained region. The CoM dynamics are subject to constant friction. Thus, $\ddot{x}$ is a constant negative value, and we assume $\ddot{y} = 0, \ddot{z} = 0$. The sagittal linear velocity decays at a constant rate.

The considered locomotion modes are selected from the set

$$\mathcal{M} := \{p_{\text{PIPM}}, p_{\text{MCM}}, p_{\text{PPM}}, p_{\text{SLM}}, p_{\text{HM}}, p_{\text{SM}}\}.$$

The switched system representation. Given the continuous locomotion modes above, we formulate the WBDL as a switched system:

$$\dot{\boldsymbol{\xi}}(\zeta) = f_{p(\zeta)}\big(\boldsymbol{\xi}(\zeta), \boldsymbol{u}(\zeta), \boldsymbol{d}(\zeta)\big),\ p(\zeta) \in \mathcal{M}, \tag{7.22}$$

where $\boldsymbol{\xi}(\zeta) \in \Xi \subseteq \mathbb{R}^{12}$ denotes the 12 dimensional CoM position and angular state vector of the robot at $\zeta \geq 0$ on the manifolds of the dynamics, the phase progression variable ζ, analogous to time, represents the current phase progression on a locomotion trajectory, the functions $p(\cdot) : \mathbb{R} \to \mathcal{M}$ and $\boldsymbol{d}(\cdot) \in \mathcal{D} \subseteq \mathbb{R}^d$ $(0 \le d \le 12)$ are the switching control and external disturbance signals, respectively, and f_p denotes the dynamics under mode $p \in \mathcal{M}$. The control input is denoted by

$$\boldsymbol{u} = \begin{bmatrix} \boldsymbol{p}_{\text{contact}} \\ \omega \\ \tau_x \\ \tau_y \\ \tau_z \end{bmatrix} \in \mathcal{U} \subseteq \mathbb{R}^7,$$

where $\boldsymbol{p}_{\text{contact}}$ represents a set of three-dimensional contact position vectors, ω represents the slope of the phase-space asymptote dependent on specific locomotion modes as defined in the above modes, and τ_x, τ_y, and τ_z represent the torso torques along x, y, and z axis, respectively.

The sampled-data system of (7.22) with a constant sampling time $\Delta\zeta \ge 0$ can be written in the form of a transition system (see Definition 3.1)

$$\mathcal{T}_L = (\Xi, \mathcal{M} \times \mathcal{U}, R_L, \Pi, L), \tag{7.23}$$

where $R_L : \Xi \times \mathcal{U} \times \mathcal{M} \to \Xi$ is determined by

$$R_L(\boldsymbol{\xi}, \boldsymbol{u}, p) := \{\boldsymbol{\xi}(\Delta\zeta) \in \Xi \mid \dot{\boldsymbol{\xi}}(\zeta) = f_p(\boldsymbol{\xi}(\zeta), \boldsymbol{u}, \boldsymbol{d}(\zeta)), \boldsymbol{\xi}(0) = \boldsymbol{\xi}, \\ \forall \boldsymbol{d}(\zeta) \in \mathcal{D}, \forall \zeta \in [0, \Delta\zeta]\}.$$

7.7.2 Hierarchical Reactive Planning Strategy

In order to be responsive to any changes in the environment, it is often necessary to predict different scenarios that could happen in the environment and include it in the design of overall motion planning strategy.

Definition 7.1 (Environment System) *The environment can be modeled as a finite transition system:*

$$\mathcal{T}_e := (\mathcal{E}, \mathcal{I}_e, R_e, \Pi_e, L_e), \tag{7.24}$$

where $\mathcal{E}$ is a finite set of environmental states, $R_e : \mathcal{E} \to \mathcal{E}$ is a transition relation, $\mathcal{I}_e = \mathcal{E}_0 \subseteq \mathcal{E}$ is a set of initial states, Π_e is a set of atomic propositions, $L_e : \mathcal{E} \to 2^{\Pi_e}$ is a labeling function mapping the state to an atomic proposition.

The following definition of product system incorporates the external environment $\mathcal{T}_e$ as the model that generates uncontrollable exogenous inputs.

Definition 7.2 (Product System) *The* ***product system*** *of system $\mathcal{T}_L$ and $\mathcal{T}_e$ is a tuple:*

$$\mathcal{T}_{\text{prod}} := (\Xi, \mathcal{M} \times \mathcal{U}, \mathcal{E}, R_{\text{prod}}, \widetilde{\Pi}, \widetilde{L}), \tag{7.25}$$

where Ξ, $\mathcal{M}$, $\mathcal{U}$ are defined in (7.23), and $\mathcal{E}$ is a finite set of uncontrollable environmental actions, which is defined in (7.24) as the set of environmental states, $R_{\text{prod}} : \Xi \times \mathcal{M} \times \mathcal{U} \times \mathcal{E} \to 2^{\Xi}$ is the transition relation, $\widetilde{\Pi}$ is a set of atomic propositions, $\widetilde{L} : \Xi \to 2^{\widetilde{\Pi}}$ is a labeling function mapping the state to an atomic proposition.

Definition 7.3 (Execution of A Product System) *An* execution $\boldsymbol{\varrho}$ *of system $\mathcal{T}_{\text{prod}}$ is an infinite sequence*

$$\boldsymbol{\varrho} = (\boldsymbol{\xi}_0, \boldsymbol{p}_0, \boldsymbol{u}_0, \boldsymbol{e}_0)(\boldsymbol{\xi}_1, \boldsymbol{p}_1, \boldsymbol{u}_1, \boldsymbol{e}_1)(\boldsymbol{\xi}_2, \boldsymbol{p}_2, \boldsymbol{u}_2, \boldsymbol{e}_2) \cdots,$$

where $\boldsymbol{\xi}_i \in \Xi$, $p_i \in \mathcal{M}$, $\boldsymbol{u}_i \in \mathcal{U}$, and $\boldsymbol{e}_i \in \mathcal{E}$ for all $i \in \mathbb{N}$. The word generated from $\boldsymbol{\varrho}$ is $\mathbf{w}_\pi = \widetilde{L}(\boldsymbol{\xi}_0)\widetilde{L}(\boldsymbol{\xi}_1)\widetilde{L}(\boldsymbol{\xi}_2)\cdots$.

The execution $\boldsymbol{\varrho}$ is said to satisfy an LTL formula φ if and only if the word w_γ satisfies φ. If all executions of $\mathcal{T}_{\text{prod}}$ satisfy φ, we say that $\mathcal{T}_{\text{prod}}$ satisfies φ, i.e., $\mathcal{T}_{\text{prod}} \models \varphi$.

Planning and control of a complex robotic system as (7.22), which is high dimensional, is controlled by multiple inputs and subject to environmental constraints. It is often achieved via hierarchical design [249]:

- The high-level planner works on an abstracted state space called keyframe state space $\mathcal{Q}$. A *keyframe state* $q = (\boldsymbol{p}_{\text{contact}}, \dot{x}_{\text{apex}}) \in \mathcal{Q}$ of a locomotion system is in general a pair of contact location $\boldsymbol{p}_{\text{contact}}$ and the apex state $\dot{x}_{\text{apex}}$ when the CoM velocity reaches the local minimal or maximal value. The planner determines the sequence of non-periodic keyframe states and locomotion modes by the planning strategy

$$\kappa_h : \mathcal{M} \times \mathcal{Q} \times \mathcal{E} \to \mathcal{M} \times \mathcal{Q}, \tag{7.26}$$

- The low-level controller within each mode directly controls system dynamics in local region in the state space so that the assumptions for the modeling of different modes can be satisfied. A controller at this level is usually pre-designed and not considered in the planning problem.

- The middle-level mode-transition controller guarantees the feasibility of mode switching required by the planner. It also generate a control strategy κ_l that takes in the command from the planner:

$$\kappa_l : \mathcal{M} \times \mathcal{Q} \times \Xi \to 2^{\mathcal{U}}. \tag{7.27}$$

In this way, the overall control strategy for the locomotion planning problem can be decomposed to a planning strategy (7.26) and a mode-transition

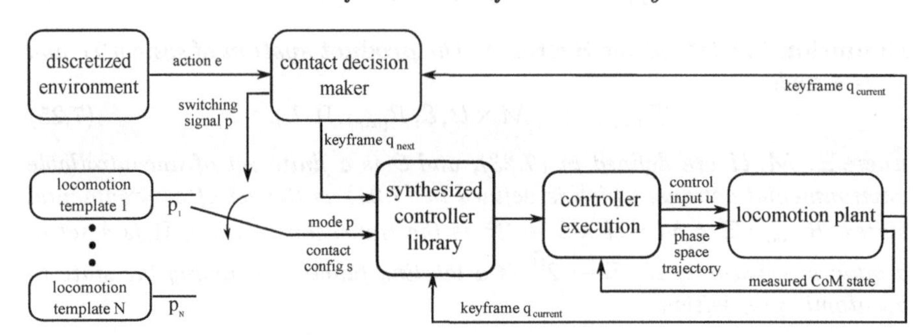

FIGURE 7.29: Hierarchical locomotion planner structure [249].

control strategy (7.27). Figure 7.29 shows the hierarchical framework of locomotion planning described above.

The specifications for the product system that cover the reactivity property is often given in the assume-guarantee form:

$$\varphi = \big(\varphi_e \Rightarrow (\varphi_q \wedge \varphi_s)\big), \tag{7.28}$$

where φ_e and φ_q, φ_s are propositions for the admissible environment actions, the keyframe states, and the correct overall system behavior, respectively. In particular, φ_s specifies the conditions of mode switching.

The formula φ_v $(v \in \{e, q, s\})$ in (7.28) is expressed in the form

$$\varphi_v = \varphi^v_{\text{init}} \bigwedge_{i \in I_{\text{safety}}} \Box \varphi^v_{\text{trans},i} \bigwedge_{i \in I_{\text{goal}}} \Box\Diamond \varphi^v_{\text{goal},i}, \tag{7.29}$$

where φ^v_{init}, $\varphi^v_{\text{trans},i}$, and $\varphi^v_{\text{goal},i}$ are propositional formulas that pose constraints to the initial conditions, transitions, and goals, respectively.

Let the set of states $\mathcal{E}$ for the environment $\mathcal{T}_e$ be

$$\begin{aligned}\mathcal{E} &:= \mathcal{E}_{\text{terrain}} \cup \mathcal{E}_{\text{emergency}} \\ &= \{e_{\text{md}}, e_{\text{hd}}, e_{\text{mu}}, e_{\text{hu}}\} \cup \{e_{\text{tc-nc}}, e_{\text{tc-hc}}, e_{\text{ha}}, e_{\text{np}}\},\end{aligned} \tag{7.30}$$

where the elements in $\mathcal{E}_{\text{terrain}}$ denote different height terrain actions, as illustrated in Figure 7.28. For instance, e_{md} denotes moderatelyDownward terrain. The actions in $\mathcal{E}_{\text{emergency}}$ represent sudden events, i.e., terrainCrack-normalCeiling, terrainCrack-highCeiling, humanAppear, and narrowPassage.

Example 7.2 (Examples of formulas in φ_e) *An example of the initial specification of the environment is*

$$\varphi^e_{\text{init}} = \neg e_{\text{tc-nc}} \wedge \neg e_{\text{tc-hc}} \wedge \neg e_{\text{ha}} \wedge \neg e_{\text{np}},$$

which means the initial environment should not be tough situations as terrain crack in normal or high ceiling, human appear, and narrow passages.

The safety specifications will be given such as "if the current environment action is terrainCrack-highCeiling, *then the next environmental action can not be* terrainCrack-highCeiling, humanAppear, *nor* narrowPassage*" with the equivalent LTL form:*

$$\square\Big(e_{\text{sc-hc}} \Rightarrow \neg(e_{\text{sc-hc}} \wedge e_{\text{ha}} \wedge e_{\text{np}})\Big)$$

To determine the sequence of locomotion modes, the set of robot actions corresponding to different modes is defined as follows:

$$\mathcal{F} := \{s_{\text{li-aj}} \mid \forall (i,j) \in \mathcal{I}_{\text{index}}\}, \tag{7.31}$$

where l and a are short for leg and arm, respectively, and the set of contact limb relative positions is

$$\mathcal{I}_{\text{index}} = \{(\text{h},\text{n}),(\text{h},\text{h}),(\text{h},\text{f}),(\text{d},\text{h}),(\text{d},\text{f}),(\text{d},\text{d}),(\text{d},\text{n}),(\text{n},\text{f}),(\text{n},\text{n})\}$$

with h, f, d and n represent hind, fore, dual and no contacts, respectively.

Example 7.3 (Example of φ_s) *An example for the term φ^s_{trans} of φ_s in response to varying-height terrain $\mathcal{E}_{\text{terrain}}$ is specified as*

$$\square\Big((e_{\text{md}} \vee e_{\text{mu}}) \Rightarrow (p_{\text{PIPM}} \wedge s_{\text{lh-an}}) \vee \big(p_{\text{MCM}} \wedge (s_{\text{lh-ah}} \vee s_{\text{lh-af}})\big)\Big)$$
$$\bigwedge \square(e_{\text{hu}} \Rightarrow p_{\text{MCM}} \wedge s_{\text{lh-ah}}) \bigwedge \square(e_{\text{hd}} \Rightarrow p_{\text{MCM}} \wedge s_{\text{lh-af}}),$$

where $s_{\text{lh-af}}$, for example, means the legHindArmFore *contact configuration in the sense that the robot's hind leg and the fore arm are in contact for that action while the other two limbs are not in contact.*

The keyframe states consist of ordinary and special types:

$$\mathcal{Q} := \mathcal{Q}_{\text{ordinary}} \cup \mathcal{Q}_{\text{special}} = \{q_{i\text{-}j\text{-}k},\ i \in \mathcal{I}_{\text{ordinary-behavior}},\ \forall (j,k) \in \mathcal{I}_{\text{level}} \times \mathcal{I}_{\text{level}}\}$$
$$\cup \{q_{i\text{-}j},\ i \in \mathcal{I}_{\text{special-behavior}},\ \forall j \in \mathcal{I}_{\text{level}}\} \tag{7.32}$$

where

$$\mathcal{I}_{\text{ordinary-behavior}} = \{\text{walk, brachiation}\},$$
$$\mathcal{I}_{\text{special-behavior}} = \{\text{stop, hop, slide}\}$$

are ordinary and special behaviors, respectively. An apex velocity index j and a step length index k refer to the set $\mathcal{Q}_{\text{level}} = \{s, m, l\}$ whose elements are three different keyframe levels: s (Small), m (Medium) and l (Large). For instance, $q_{\text{walk-s-l}}$ represents walkSmallVelocityLargeStep, a walking keyframe with a small apex velocity, and a large step length.

Example 7.4 (Example for φ_q) *One of the formulas for $\varphi^q_{\text{trans},i}$ in φ_q is:*

$$\square(\bigcirc e_{\text{np}} \Rightarrow \bigcirc(q_{\text{slide-s}} \vee q_{\text{slide-m}} \vee q_{\text{slide-l}})),$$

which means that if there is a narrow passage, i.e., e_{np}, then the next key frame state is q_{slide} relying on a specific apex velocity, regardless of the current q.

Details on the full set of specifications that are commonly used can be found in [249, Section 4].

Based on the above definitions, the locomotion planning problem can be described as:

Problem 7.1 (Contact-Based Reactive WBDL Planning) Given bipedal robot $\mathcal{T}_L$ in (7.23) with a set of initial condition $\Xi_0 \subseteq \Xi$, environmental system $\mathcal{T}_e$ in (7.24), and an LTL specification φ in the form of (7.28), synthesize a planning strategy (7.26) and a mode-transition control strategy (7.27) such that the resulting execution π defined in Definition 7.3 satisfies φ in the sense that $\pi \vDash \varphi$ for all initial conditions in Ξ_0.

A two-player game problem can be formulated and analyzed to synthesize a planning strategy over a high-level finite abstract state space $\mathcal{M} \times \mathcal{Q} \times \mathcal{E}$ as illustrated in [249, Section 4], but let us now focus on the synthesis of the mode-transition control strategy for the middle layer.

7.7.3 Robust Switching Between Locomotion Modes

Uncertainty is ubiquitous in the modeling of the WBDL and environment, e.g., sensor noise, model inaccuracy, external disturbances, sudden environmental changes, contact surface geometry uncertainty. As a result, commands from the symbolic task planner are possibly unrealizable for the low-level dynamics. Therefore, we need a middle layer to verify if the transitions between two modes at certain keyframe states can be achieved and construct a mode-transition strategy if possible.

The synthesis of the mode-transition strategy for every single walking step is performed on the robust abstractions for the dynamics of two successive modes with respect to reachability control specifications, which is given by the high-level planner. The finite abstractions are constructed over the phase-space manifolds of the locomotion for the sake of consistency with the dynamics.

Assuming the x and y axes can be decoupled, we define a mapping between the Euclidean and Riemmanian for the dynamics along x axis (the dynamics along y axis is similar):

$$\begin{bmatrix} \zeta \\ \sigma \end{bmatrix} = \mathcal{Z}_p(\boldsymbol{\xi}) = \begin{bmatrix} \mathcal{Z}_{p,\zeta}(x, \dot{x}) \\ \mathcal{Z}_{p,\sigma}(x, \dot{x}) \end{bmatrix} \tag{7.33}$$

where ζ is phase progression variable, σ is the tangent manifold, which can be used to measure deviations from the nominal locomotion trajectory in the phase-space, and $\mathcal{Z}_p(\boldsymbol{\xi})$ is a nonlinear mapping of the CoM state $(x, \dot{x})$ to the Riemannian space states for locomotion mode p. The inverse mapping $\mathcal{Z}_p$ is denoted by $\mathcal{Z}_p^{-1}$.The specific mapping for each of the 6 locomotion modes are given in Appendix D.

Mode transitions usually take place in one walking step. A *one walking step (OWS)* is composed of two consecutive semi-step phase-space trajectories. The first semi-step trajectory starts at the first keyframe state q_1 and ends at the contact switch, which will be determined by the mode-transition control strategy, and the second semi-step trajectory starts at the contact switch and ends at the second keyframe state q_2.

To guarantee that the motion planner yields plans that are robust to disturbances, we introduce ϵ_1 and ϵ_2 as initial and final robustness margins in the one walking step, respectively, so that the neighborhood of nominal initial and final keyframe states q_1 and q_2 can also be considered for mode transition. The formal definition of robustness margin sets is provided below.

Definition 7.4 (Robustness Margin Sets) *Given the initial and final keyframe states q_1 and q_2, let $\zeta_0 = \mathcal{Z}_{p,\zeta}(q_1)$, $0 = \mathcal{Z}_{p,\sigma}(q_1)$, $\zeta_f = \mathcal{Z}_{p,\zeta}(q_2)$, and $0 = \mathcal{Z}_{p,\sigma}(q_2)$, where $\mathcal{Z}_{p,\zeta}(\cdot)$ and $\mathcal{Z}_{p,\sigma}(\cdot)$ are given in (7.33). Also let $\epsilon_1 = [\delta\zeta_{\epsilon_1}, \delta\sigma_{\epsilon_1}]$ and $\epsilon_2 = [\delta\zeta_{\epsilon_2}, \delta\sigma_{\epsilon_2}]$. The robustness margin sets of q_1 and q_2 are*

$$B_{\epsilon_1}(q_1) \triangleq \{\mathcal{Z}_p^{-1}(\zeta, \sigma) \mid \zeta \in [\zeta_0 - \delta\zeta_{\epsilon_1}, \zeta_0 + \delta\zeta_{\epsilon_1}], \sigma \in [-\delta\sigma_{\epsilon_1}, \delta\sigma_{\epsilon_1}]\}, \quad (7.34)$$

and

$$B_{\epsilon_2}(q_2) \triangleq \{\mathcal{Z}_p^{-1}(\zeta, \sigma) \mid \zeta \in [\zeta_f - \delta\zeta_{\epsilon_2}, \zeta_f + \delta\zeta_{\epsilon_2}], \sigma \in [-\delta\sigma_{\epsilon_2}, \delta\sigma_{\epsilon_2}]\}, \quad (7.35)$$

respectively.

The keyframe states q_1 and q_2 and their robustness margin sets $B_{\epsilon_1}(q_1)$ and $B_{\epsilon_2}(q_2)$ are defined in the Euclidean space while the margins ϵ_1 and ϵ_2 are in the phase space. Figure 7.30 gives an intuition of how the robustness margin sets defined in a walking step.

Recall that the high-level planner chooses keyframe states from the set $\mathcal{Q}$ defined in (7.32). These keyframe states represent the cells that are obtained by partitioning the robustness margin sets defined in Definition 7.4. In our case, the ordinary locomotion behaviors (i.e., walk and brachiation) are comprised of 9 keyframe states while the special locomotion behaviors (i.e., stop, hop, and slide) include 3 keyframe states. The goal of mode-transition control synthesis is to determine the possible transitions between these keyframe states. The construction of the set of possible transitions is shown in Figure 7.31.

The goal of mode-transition control synthesis for one walking step is to solve the closed-loop phase-space trajectories starting from the initial robustness margin set $B_{\epsilon_1}(q_1)$ and reaching the final robustness margin set $B_{\epsilon_2}(q_2)$

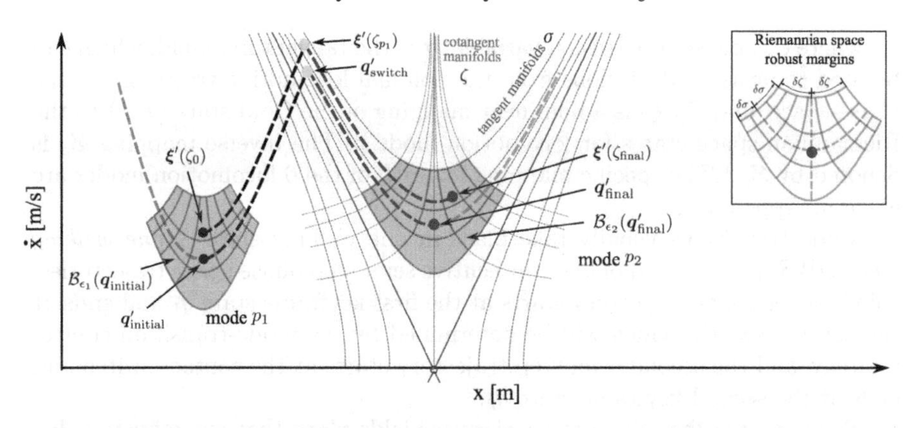

FIGURE 7.30: One walking step with robustness margin sets. The horse shoe shape of the robustness margin sets is the result of mapping from phase space to Euclidean space. The robustness margins are shown in the upper right box. The dot where the state trajectories (dashed lines) of the first and second modes meet is the point where mode switching takes place.

as defined in Definition 7.4, and in the meanwhile, switch from a locomotion mode p_1 to mode p_2. It is fair to consider one walking step as two sequential semi-steps with the first semi-step in mode p_1 and second in mode p_2.

To complete the switching between two locomotion modes in one walking step, as shown in Figure 7.30, the region Ξ_{inter} where the switching happens has to be determined. For the first semi-step, the region Ξ_{inter} can be treated as the target set that robot state is expected to reach in finite time, and the final robustness margin set $B_{\epsilon_2}(q_2)$ is the target set to reach in the second semi-step. Hence, region Ξ_{inter} must lie in the overlap of the winning sets for both semi-steps with respect to reachability specifications.

Additionally, we assume that the function f_p (for all $p \in \mathcal{M}$) in (7.22) is of a particular form:

$$f_p(\boldsymbol{\xi}, \boldsymbol{u}, \boldsymbol{d}) = g_p(\boldsymbol{\xi}) + h_p(\boldsymbol{\xi})\boldsymbol{u} + \boldsymbol{d}, \ \forall p \in \mathcal{M}, \tag{7.36}$$

where g_p is Lipschitz continuous and h_p is bounded on Ξ for all $p \in \mathcal{M}$.

Consider one walking step transiting from keyframe state q_1 to q_2 with robustness margins ϵ_1 and ϵ_2, respectively. The locomotion mode has to switch from p_1 to p_2. Suppose that $\Xi_1 \subseteq \Xi$ and $\Xi_2 \subseteq \Xi$ are two local regions where the first and second semi-step takes place, respectively. Based on the above discussion, we can use the following *two-semistep reachability control synthesis* for a mode transition:

(i) Perform reachability control synthesis with precision ϵ_2 (see Algorithm 9) for the second semi-step (under mode p_2) in the state space Ξ_2. The reachability formula is $\varphi_{r2} = \Diamond B_{\epsilon_2}(q_2)$.

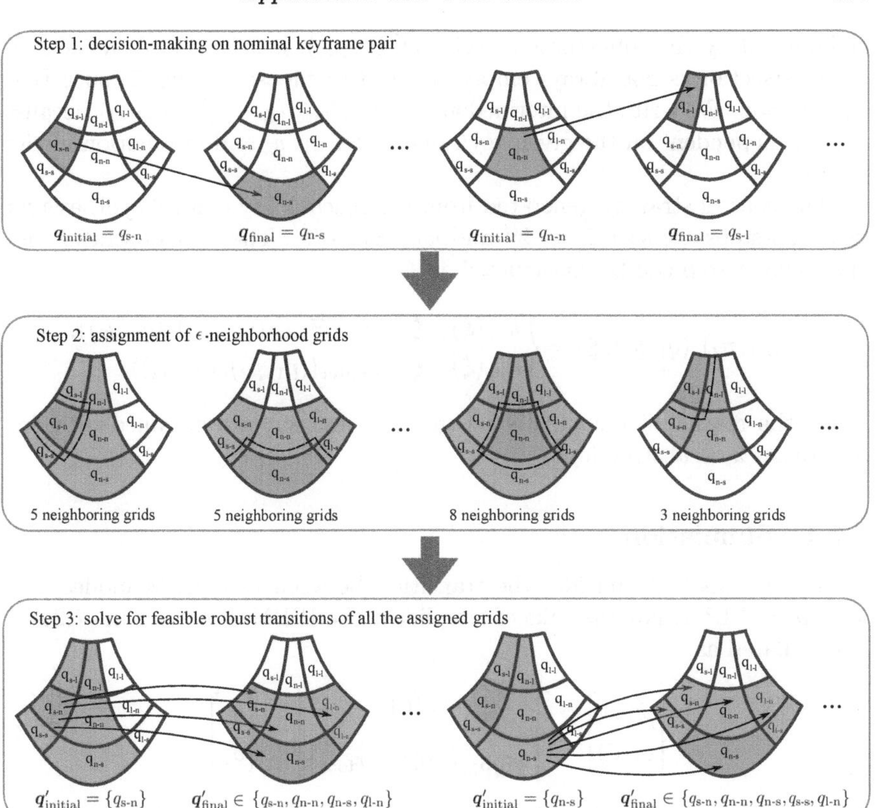

FIGURE 7.31: Construction of a library of possible robust keyframe transitions. The set $\mathcal{Q}$ of keyframe states is obtained by discritizing some neighborhood in the state space Ξ around nominal setpoints. The mode-transition control synthesis verifies the possibility of the transitions between these keyframe states and generate corresponding mode-transition control strategy.

(ii) Determine the intermediate region Ξ_{inter} by

$$\Xi_{\text{inter}} = \left\{ \boldsymbol{\xi} \mid \boldsymbol{\xi} \in \mathbf{Win}_{f_{p_2}}(\varphi_{r2}) \wedge \left\| \mathcal{Z}_{p1,\sigma}(\boldsymbol{\xi}) - \mathcal{Z}_{p1,\sigma}(q_1) \right\|_\infty \leq \delta\sigma_{\epsilon_1} \right\} \tag{7.37}$$

(iii) Perform reachability control synthesis with precision ϵ_1 for the first semi-step (under mode p_1) in the state space Ξ_2. The reachability specification is $\varphi_{r1} = \Diamond\Xi_{\text{inter}}$.

Since the winning set $\mathbf{Win}_{f_{p_1}}(\varphi_{r1})$ of the first semi-step is unknown before the determination of Ξ_{inter} and by definition $\Xi_{\text{inter}} \subseteq \mathbf{Win}_{f_{p_1}}(\varphi_{r1})$, the set Ξ_{inter} can be defined as the intersection of the winning set $\mathbf{Win}_{f_{p_2}}(\varphi_{r2})$ and the tube centered at nominal state trajectory from the initial keyframe state

q_1 bounded by the robustness margin set $B_{\epsilon_1}(q_1)$, i.e., (7.37). To make sure the intersection is not always empty we need to choose Ξ_1 and Ξ_2 such that $\Xi_1 \cap \Xi_2 \neq \emptyset$. We write the intermediate set as $\Xi_{\text{inter}}((p_1, p_2), (q_1, q_2))$ because Ξ_{inter} is dependent on the given keyframe states q_1, q_2 and locomotion modes p_1, p_2.

The control strategy generated from the reachability control synthesis for two semi-steps are $\kappa_1 : \Xi_1 \to 2^{\mathcal{U}}$ and $\kappa_2 : \Xi_2 \to 2^{\mathcal{U}}$. The control strategy for one walking step can be constructed as

$$\kappa((p_1,p_2),(q_1,q_2),\boldsymbol{\xi}) = \begin{cases} \kappa_{p_1}(\boldsymbol{\xi}) & \boldsymbol{\xi} \in \Xi_1 \setminus \Xi_{\text{inter}}((p_1,p_2),(q_1,q_2)), \\ \kappa_{p_2}(\boldsymbol{\xi}) & \boldsymbol{\xi} \in \Xi_{\text{inter}}((p_1,p_2),(q_1,q_2)), \end{cases}$$

where the reachability control strategies κ_{p_1} and κ_{p_2} are obtained by using Algorithm 2 on abstractions.

7.7.4 Simulation

Case I Let us first consider the transition between two PIPM modes, i.e., $p_1 = p_2 =$ PIPM. For the sake of simplicity, the PIPM dynamics in (7.20) is reformulated as

$$\begin{bmatrix} \dot{x}(\zeta) \\ \dot{v}_x(\zeta) \end{bmatrix} = \begin{bmatrix} v_x(\zeta) \\ \omega^2_{\text{PIPM}}(x(\zeta) - x_{\text{foot}}) \end{bmatrix} + \begin{bmatrix} d_1 \\ d_2 \end{bmatrix} \tag{7.38}$$

by assuming $(\tau_x, \tau_y) = \mathbf{0}$ and x_{foot} is a predefined constant. The continuous control input $\omega_{\text{PIPM}} \in [\bar{\omega} - \delta\omega, \bar{\omega} + \delta\omega]$, where $\bar{\omega}$ is the nominal control input and $\delta\omega$ is a pre-defined bound. The disturbance $\boldsymbol{d} = (d_1, d_2)$ satisfies $d_{1,2} \in D_r$, where $D_r \subseteq \mathbb{R}^2$ is a bounded set. Hence (7.38) satisfies (7.36).

Suppose that the high-level planner generates the parameters $x_{\text{foot},1}, x_{\text{foot},2}$ and $\bar{\omega}$ and two nominal keyframe states. Let $\delta\zeta = 2\,\text{ms}$ be the sampling time. The setting of the mode transition problem for the considered one walking step is given in Table 7.5.

TABLE 7.5: Parameters of the PIPM-PIPM mode transition. q_1 and q_2 are the initial and final keyframe states, respectively.

Parameters	Values	Parameters	Values
q_1	$(0\,\text{m}, 0.5\,\text{m/s})$	q_2	$(0.5\,\text{m}, 0.6\,\text{m/s})$
$\delta\zeta_{\epsilon_1}$	0.05	$\delta\sigma_{\epsilon_1}$	0.002
$\delta\zeta_{\epsilon_1}$	0.05	$\delta\sigma_{\epsilon_2}$	0.006
modes	PIPM $\to$ PPM	D_r	$(0.05\,\text{m}, 0.1\,\text{m/s})$
Ξ_p	$[-0.1\,\text{m}, 0.7\,\text{m}] \times [0.1\,\text{m/s}, 1.2\,\text{m/s}]$	$\mathcal{U}_{\text{ows}}$	$[2\,\text{rad/s}, 4\,\text{rad/s}]$

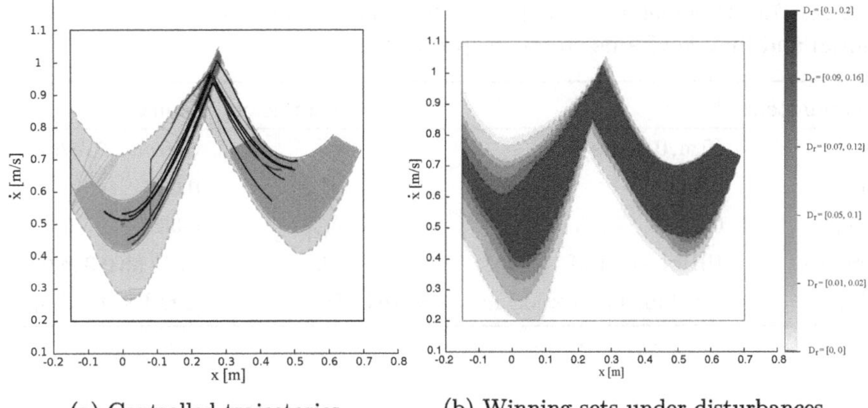

(a) Controlled trajectories. (b) Winning sets under disturbances.

FIGURE 7.32: Control synthesis results for the walking step from PIPM to PPM. (a) The biggest shaded region represents the winning set of this walking step, and the intersection region in the middole is the intermediate robustness margin set. The lines are simulated closed-loop trajectories. (b) Comparison of winning sets under different levels of disturbances.

In this example, we discretize the state space by $\eta = (0.005\,\text{m}, 0.005\,\text{m/s})$ and sample the control space with a granularity $\mu = 0.02\,\text{rad/s}$. Given the setting above, we perform the two-semistep reachability control synthesis. The computed winning sets are shown in Figure 7.32a. As the result shows, the one-walking step reachability is realizable as long as the winning set overlaps (at least partially) the initial and final robustness margin sets. Five simulated trajectories under randomly-sampled bounded disturbances are shown as the lines within the winning sets. The trajectory with a velocity jump in the phase space is a trial under a large disturbance. Figure 7.32b shows the change of the winning set under different disturbance levels. The winning set shrinks as the disturbance set increases because the synthesized controller needs to reach the goal robust set against a larger set of disturbances.

Case II Consider another locomotion mode transition from the PIPM to PPM. Similarly, we can simplify (7.21) to

$$\begin{bmatrix} \dot{x}(\zeta) \\ \dot{v}_x(\zeta) \end{bmatrix} = \begin{bmatrix} v_x(\zeta) \\ -\omega_{\text{PPM}}^2(x(\zeta) - x_{\text{hand}}) \end{bmatrix} + \begin{bmatrix} d_1 \\ d_2 \end{bmatrix} \tag{7.39}$$

with the assumption of $\tau_x = \tau_y = 0$ and a predefined hand contact position x_{hand}. Other parameters are defined in Table 7.6.

To evaluate the performance of the control strategy generated by two-semistep reachability control synthesis, we examine the success rate of reaching the goal robustness margin set through 50 simulation tests under different

TABLE 7.6: Parameters of the PIPM-PPM mode transition. q_1 and q_2 are the initial and final keyframe states, respectively.

Parameters	Values	Parameters	Values
q_1	$(0\,\text{m}, 0.5\,\text{m/s})$	q_2	$(0.6\,\text{m}, 1.7\,\text{m/s})$
$\delta\sigma_{\epsilon_1}$	0.002	$\delta\zeta_{\epsilon_1}$	0.05
$\delta\sigma_{\epsilon_2}$	0.06	$\delta\zeta_{\epsilon_2}$	0.005
modes	PIPM $\rightarrow$ PPM	D_r	$(0.15\,\text{m}, 0.3\,\text{m/s})$
Ξ_p	$[-0.1\,\text{m}, 0.7\,\text{m}] \times [0.1\,\text{m/s}, 1.8\,\text{m/s}]$	$\mathcal{U}_{\text{ows}}$	$[2\,\text{rad/s}, 4\,\text{rad/s}]$

granularities and bounded disturbances. Each trial is run for one walking step with PIPM to PPM mode transition. The exerted disturbances in the simulation is the same as those used in the controller synthesis process. All the trials reach the final robustness margin set successfully.

We evaluate the effect of the control synthesis precision and the magnitude of disturbances used in the controller synthesis process as shown in Figure 7.33. Figure 7.33 shows 4 sets of simulation results for different control precisions ranging from 0.002 to 0.005. For each set of simulations, the success rate increases as the modeled disturbance in the controller synthesis increases, and it reaches 100 % when the modeled disturbance matches the actual disturbance D_r used in the simulation. If we compare the results for different control synthesis precisions under a same disturbance D_i $(i = 0, 1, 2, 3, 4)$, the success rate almost remains the same. This is because we use the same disturbance for simulation and control analysis. In addition, it can be observed that the success

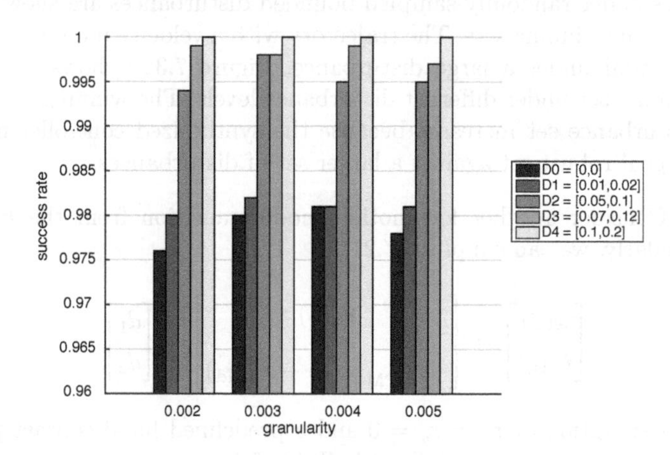

FIGURE 7.33: Performance evaluation result for the control of PIPM to PPM walking step. 1000 trials are run for each case with a specific precision and a bounded disturbance $D_r = (0.1\,\text{m}, 0.2\,\text{m/s})$.

rates for all the synthesized controllers are greater than 97 %, even in the case no disturbance is considered in the controller synthesis. Moreover, under the same disturbance D_r, the nominal phase-space planner with a fixed open-loop control input only achieves a success rate of 29 %. This huge discrepancy in success rate clearly shows the advantage of using the proposed method in the middle-layer of control synthesis within the planning framework.

Multi-Step Locomotion Transition Now we simulate the closed-loop multi-step given the mode switching sequence generated by the high-level planner:

$$\text{PIPM} \rightarrow \text{PIPM} \rightarrow \text{PPM} \rightarrow \text{PIPM} \rightarrow \text{MCM} \rightarrow \text{PIPM} \rightarrow \text{PIPM}$$

To enable the initial and final keyframe robustness margin sets to cover a sufficiently larger phase space, we extend the default 3×3 keyframe grid to a 5×5 keyframe grid for each mode. This allows the mode-transition control strategy to be applicable to a larger set of keyframe states. For each locomotion mode transition, we synthesize all the possible control strategies that reach the final keyframe robustness margin set under bounded disturbances. We enumerate all the combinations of the allowable locomotion mode pairs and generate all the reachability control policies offline. These controllers are saved as a control library and are executed at runtime according to the high-level decision and measured states under bounded disturbances.

The controller synthesis and execution process use the same disturbance bound $D_r = (0.05\,\text{m}, 0.1\,\text{m/s})$. The full state space is $\Xi_{\text{full}} = [-0.2\,\text{m}, 3.8\,\text{m}] \times [0.2\,\text{m/s}, 1.9\,\text{m/s}]$. The local state space of each walking step is chosen so that it is sufficiently large to cover the space around the two keyframe states. A time step $\delta\zeta = 0.02\,\text{m s}$ is used for the abstraction construction of each walking step. The control inputs for PIPM, PPM and MCM satisfy $\omega_{\text{PIPM}} \in [2, 4]$, $\omega_{\text{PPM}} \in [2, 4]$ and $\omega_{\text{MCM}} \in [1, 3]$. We obtain the sets of sampled control values by a granularity of 0.02. The robustness margins of the phase space manifolds are $\delta\sigma_{\text{PIPM}} = 0.002$, $\delta\zeta_{\text{PIPM}} = 0.002$; $\delta\sigma_{\text{PPM}} = 0.04$, $\delta\zeta_{\text{PPM}} = 0.003$; $\delta\sigma_{\text{MCM}} = 0.15$, $\delta\zeta_{\text{MCM}} = 0.9 \times 10^{-5}$.

We perform the two-semistep reachability control synthesis for each walking step with the precision $(0.003\,\text{m}, 0.003\,\text{m/s})$. The computational time is around 30 s by average for synthesizing a reachability controller corresponding to each keyframe pair. Since we run 625 (i.e., 25×25) times of such reachability control synthesis for each walking step, the time of generating all the controller policies is approximately 90 min for each walking step. In the simulation of these six consecutive walking steps, all the local reachability control strategies are patched together to cover the overall state space. The time for simulating a single closed-loop walking trajectory is around 2s. As the results show in Figure 7.34, we simulate six different trials with different initial conditions, i.e., starting from different initial robustness margin sets. Each locomotion trajectory is guaranteed to reach one of the robustness

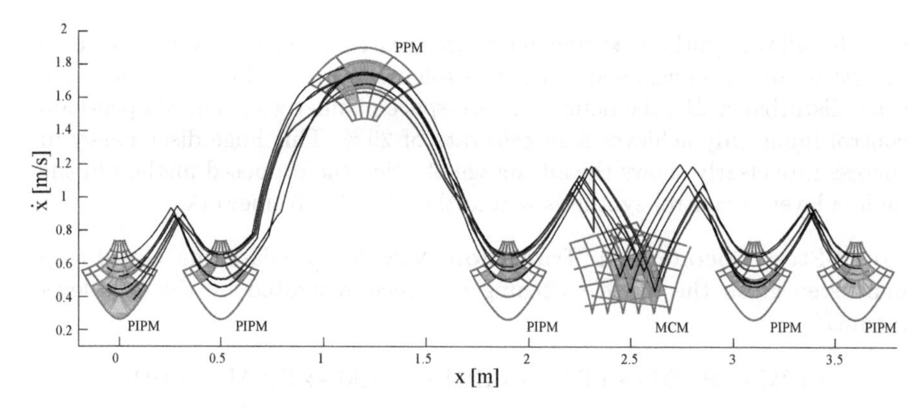

FIGURE 7.34: The state trajectories of multi-step mode transition under bounded disturbances.

margin sets at the next walking step by using the reachability controller from the control library.

7.8 Summary

This chapter is a collection of use cases of formal methods for solving control synthesis problems with respect to linear temporal logic specifications, taken from [139, 142, 143, 147, 165, 249]. Although most of the chapter has been dedicated to robotic applications ranging from various motion planning scenarios for mobile robots and manipulators to bipedal robots, Robotics is not the only field where formal methods apply. The potential of controlling power electronics devices such as DC-DC boost converters, jet engines, and computing domains-of-attraction for general dynamical systems were also seen from the first three examples.

It is also interesting to compare the time and space efficiencies of the abstraction-based and specification-guided methods we have discussed in Chapters 5 and 6. Table 7.7 lists the performance data collected after running some of the examples in this chapter on the high performance computing (HPC) cluster Béluga[2]. All the jobs are run sequentially on a node with an Intel Xeon(R) Gold 6148 processor @2.4 GHz. Each node is allocated Maximum 750 GB of memory. The run time for an abstraction-based method is the sum of the time for abstraction and synthesis. The specification-guided outperforms abstraction-based methods in terms of memory efficiency in call cases while the abstraction-based methods are more time efficient in most of

[2]`https://docs.alliancecan.ca/wiki/B%C3%A9luga/en`

TABLE 7.7: Performance results."OFM"=out of memory.

System	$\|X\|$	$\|U\|$	Spec	Abstraction-based				Specification-guided		
				#Partitions	#Trans	Time(s)	Mem(GB)	#Partitions	Time(s)	Mem(GB)
DC-DC	2	1	$\square\Omega$	160400	1037282	0.384	N/A	5624	0.085	N/A
			$\Diamond(\square\Omega)$	1002001	7464930	7.843	N/A	192340	6.29	N/A
Engine	2	2	φ_{rs}	1236544	1670780774	5142.08	117.58	837599	21883	0.357
Mobile robot	4	2	φ_1	93636	26888839	12.569	3.01	1113312	347.207	0.435
			φ_2			11.712	2.97	836009	556.5	0.33
			φ_3			21.837	4.08	1923704	576.296	0.727
			φ_4			18.373	3.28	241380	852.76	0.556
SCARA	4	2	φ_{gb}	N/A	N/A	N/A	OFM	7921889	41405.7	5.45

the cases, especially for the mobile robot motion planning examples. For the cases that require high partition precision, the specification-guided method shows more advantage since memory consumption becomes a bottleneck for abstraction-based method and the time consumption for both methods are close. For example, in SCARA motion planning, we need more than 750 GB memory to be able to use the abstraction-based methods.

Appendix A

Basic Theory of Ordinary Differential Equations

This chapter reviews some basic theory of ordinary differential equations and proves the existence and uniqueness theorem stated in Chapter 1.

A.1 Initial Value Problem and Carathéodory Solutions

Recall the following initial problem:

$$x'(t) = f(t, x(t), u(t), w(t)), \quad x(t_0) = x_0, \tag{A.1}$$

where $J \subseteq \mathbb{R}$ is an interval, $D \subseteq \mathbb{R}^n$ is an open set, and $f : J \times D \times U \times W \to \mathbb{R}^n$ is a function that is assumed to satisfy:

(C1) $f(\cdot, x, u, w) : J \to \mathbb{R}^n$ is measurable for each $(x, u, w) \in D \times U \times W$; and

(C2) $f(t, \cdot, \cdot, \cdot) : D \times U \times W \to \mathbb{R}^n$ is continuous for each $t \in J$.

In the statement of the existence and uniqueness theorem, two additional conditions are assumed:

(C3) (**local integrable**) For every compact set $K \subseteq J \times D \times U \times W$, there exists a Lebesgue integrable function $m_K(t)$ such that $\|f(t, x, u, w)\| \leq m_K(t)$ for all $(t, x, u, w) \in K$.

(C4) (**local Lipschitz**) For every compact set $K \subseteq J \times D \times U \times W$, there exists a Lebesgue integrable function $l_K(t)$ such that

$$\|f(t, x, u, w) - f(t, y, u, w)\| \leq l_K(t) \|x - y\|,$$

for all $(t, x, u, w) \in K$.

The inputs $u : J \to U \subseteq \mathbb{R}^m$ and $w : J \to M \subseteq \mathbb{R}^p$ are assumed to be locally essentially bounded. A **solution** to the IVP (A.1) on an interval

DOI: 10.1201/9780429270253-A

$I \subseteq J$ is a locally absolutely continuous function $x : I \to \mathbb{R}^n$ that satisfies the integral equation

$$x(t) = x_0 + \int_{t_0}^{t} f(s, x(s), u(s), w(s))ds, \quad \text{for all } t \in I. \qquad \text{(A.2)}$$

An absolute continuous function is almost everywhere differentiable and the fundamental theorem of calculus also holds for absolutely continuous functions (see, e.g., [209, p. 126, Theorem 14, Chapter 6]). Hence (A.2) is equivalent to

$$x'(t) = f(t, x(t), u(t), w(t)), \quad \text{for almost all } t \in J. \qquad \text{(A.3)}$$

Such a solution is often called a **Carathéodory solution.**

A.2 Existence and Uniqueness Theorem

Given $u(t)$ and $w(t)$, define

$$F(t, x) = f(t, x, u(t), w(t)), \quad (t, x) \in J \times D.$$

Under the assumption that f satisfies conditions (**C1**)–(**C4**), we can verify that F satisfies the following conditions:

(**D1**) $F(\cdot, x) : J \to \mathbb{R}^n$ is measurable for each $x \in D$;

(**D2**) $F(t, \cdot) : D \to \mathbb{R}^n$ is continuous for each $t \in J$;

(**D3**) (**local integrable**) For every compact set $K \subseteq J \times D$, there exists a Lebesgue integrable function $m_K(t)$ such that $\|F(t, x)\| \le m_K(t)$ for all $(t, x) \in K$.

(**D4**) (**local Lipschitz**) For every compact set $K \subseteq J \times D$, there exists a Lebesgue integrable function $l_K(t)$ such that

$$\|F(t, x) - F(t, y)\| \le l_K(t) \|x - y\|,$$

for all $(t, x) \in K$.

The proof of (**D2**) from (**C2**) is obvious. To verify (**D1**), note that, by (**C1**) and (**C2**), $f(t, x, u, w)$ is measurable in t and continuous in (x, u, w). Such a function is sometimes called a *Carathéodory function.* Accordingly to [2, p. 153, Lemma 4.51, Chapter 4], $f(t, x, u, w)$ is *jointly measurable*, or more precisely, $(\mathcal{L}_J \times \mathcal{B}_D \times \mathcal{B}_U \times \mathcal{B}_W, \mathcal{B}_{\mathbb{R}^n})$-measurable, where $\mathcal{L}_J \times \mathcal{B}_D \times \mathcal{B}_U \times \mathcal{B}_W$ denotes a joint σ-algebra, $\mathcal{L}_J$ is the Lebesgue σ-algebra on J, and $\mathcal{B}_D$, $\mathcal{B}_U$, $\mathcal{B}_W$, and $\mathcal{B}_{\mathbb{R}^n}$ are the the Borel σ-algebras on D, U, W, and $\mathbb{R}^n$, respectively. Hence, for a fixed x, $f(\cdot, x, \cdot, \cdot)$ is $(\mathcal{L}_J \times \mathcal{B}_U \times \mathcal{B}_W, \mathcal{B}_{\mathbb{R}^n})$-measurable. Consider the function $t \mapsto (t, u(t), w(t))$. According to Lemma 4.49 in [2], this function is $(\mathcal{L}_J, \mathcal{L}_J \times \mathcal{B}_U \times \mathcal{B}_W)$-measurable, because the identity function $t \mapsto t$ is trivially $(\mathcal{L}_J, \mathcal{L}_J)$-measurable, and $u(t)$ and $w(t)$ are assumed to be Lebesgue measurable. Hence, the composition function $t \mapsto f(t, x, u(t), w(t))$ is $(\mathcal{L}_J, \mathcal{B}_{\mathbb{R}^n})$, i.e., Lebesgue measurable. To verify (**D3**), note that the compact set K projected to J gives a compact set $J_K \subseteq J$. Since $u(t)$ and $w(t)$ are locally essentially bounded, there exists a compact set $B \subseteq U \times W$ such that $u(t) \in B$ and $w(t) \in B$ for almost all $t \in J_K$. Now consider the compact set $K \times B$. By condition (**C3**), there exists an integrable function $m_{K\times B}(t)$ such that $\|f(t, x, u, w)\| \leq m_{K\times B}(t)$ for all $(t, x, u, w) \in K \times B$. This implies $\|F(t, x)\| \leq m_{K\times B}(t)$ for all $(t, x) \in K$. Condition (**D3**) is verified. Similarly, condition (**D4**) can be checked.

We have now reduced the initial value problem to that for a more standard Carathéodory equation:

$$x' = F(t, x), \quad x(t_0) = x_0, \tag{A.4}$$

where $F : J \times D \to \mathbb{R}^n$ satisfies the **Carathéodory conditions** (**D1**)–(**D3**). Given an interval $I \subseteq J$, a solution to the IVP (A.4) on I is a locally absolutely continuous function $x : I \to D$ such that

$$x(t) = x_0 + \int_{t_0}^{t} F(s, x(s))ds \tag{A.5}$$

for all $t \in I$.

The existence and uniqueness results for Carathéodory equations can be found, e.g., in [51, Chapter 2], [99, Section I.5], and [71, Chapter 1]. We include them here for completeness.

Proposition A.1 (Local existence) *Let $J \subseteq \mathbb{R}$ be an interval and $D \subseteq \mathbb{R}^n$ be an open set. Suppose that $F : J \times D \to \mathbb{R}^n$ satisfies conditions (**D1**)–(**D3**). Then, for any $(t_0, x_0) \in J \times D$, there exists some $c > 0$ such that $x(t)$ is a solution to the IVP (A.4) on the interval $[t_0 - c, t_0 + c] \cap J$.*

Proof: Without loss of generality, suppose that t_0 is not a right endpoint of J. Choose $a > 0$ such that $[t_0, t_0 + a] \subseteq J$. Choose $b > 0$ such that the set $B_b(x_0) = \{x \in \mathbb{R}^n : \|x - x_0\| \leq b\}$ is contained in D. Let $K = [t_0, t_0 + a] \times B_b(x_0)$. Consider the function $m_K(t)$ guaranteed by (**D3**). Choose $c > 0$ such that $\int_{t_0}^{t_0+c} m_K(t)dt \leq b$. This is always possible because $m_K(t)$ is integrable.

Define a sequence of functions $\{x_n(t)\}_{n=1}^{\infty}$ by

$$x_n(t) = \begin{cases} x_0, & t \in [t_0 - c, t_0], \\ x_0 + \displaystyle\int_{t_0}^{t} F(s, x_n(s - \frac{c}{n})), & t \in [t_0, t_0 + c]. \end{cases} \tag{A.6}$$

Note that $x_n(t)$ is well-defined on $[t_0, t_0 + \frac{c}{n}]$, because $x_n(t)$ is well-defined on $[t_0 - \frac{c}{n}, t_0]$ and the integral in (A.6) is meaningful in view of (**D1**) (cf. the argument for showing (**D1**)). Repeating this argument, we can conclude that $x_n(t)$ is well-defined on $[t_0, t_0 + c]$.

Clearly, for $t \in [t_0 - c, t_0]$ and every n,

$$\|x_n(t) - x_0\| \leq b. \tag{A.7}$$

Then, for $t \in [t_0, t_0 + \frac{c}{n}]$ and every n, by (**D3**) we have

$$\begin{aligned}\|x_n(t) - x_0\| &= \left\| \int_{t_0}^{t} F(s, x_n(s - \frac{c}{n}))ds \right\| \\ &\leq \int_{t_0}^{t} \left\| F(s, x_n(s - \frac{c}{n})) \right\| ds \leq \int_{t_0}^{t_0+c} m_K(s)ds \leq b.\end{aligned}$$

By this argument, we can show that (A.7) holds for all $t \in [t_0 - c, t_0 + c]$ and every n. This also shows that the sequence $\{x_n(t)\}_{n=1}^{\infty}$ is *uniformly bounded* on $[t_0, t_0 + c]$.

We verify that the sequence is also *equicontinuous*. Note that the function $\int_{t_0}^{t} m_K(s)ds$ is uniformly continuous on $[t_0, t_0 + c]$ (in fact absolutely continuous, because it is an indefinite integral; see [209, p. 125, Theorem 11]). For any $\varepsilon > 0$, choose $\delta > 0$ such that $t_1, t_2 \in [t_0, t_0 + c]$ and $0 \leq t_2 - t_1 \leq \delta$ imply $\int_{t_1}^{t_2} m_K(s)ds \leq \varepsilon$. Then, for $t_1, t_2 \in [t_0, t_0 + c]$ such that $0 \leq t_2 - t_1 \leq \delta$, we have

$$\begin{aligned}|x_n(t_2) - x_n(t_1)| &= \left\| \int_{t_1}^{t_2} F(s, x_n(s - \frac{c}{n}))ds \right\| \\ &\leq \int_{t_1}^{t_2} \left\| F(s, x_n(s - \frac{c}{n})) \right\| ds \leq \int_{t_1}^{t_2} m_K(s)ds \leq \varepsilon.\end{aligned}$$

Hence $\{x_n(t)\}_{n=1}^{\infty}$ is equicontinuous.

The Arzelà–Ascoli theorem [209, p. 208] states that a uniformly bounded and equicontinuous sequence of real-valued functions defined on a compact interval must have a subsequence that converges uniformly to a continuous function on this interval. Hence there exists a subsequence $\{x_{n_k}(t)\}$ that converges uniformly to a continuous function $x(t)$ on $[t_0, t_0 + c]$.

We show that $x(t)$ is a solution to the IVP (A.4) on $[t_0, t_0 + c]$ by showing that (A.5) holds on this interval. To do so, we only need to verify that

$$\lim_{k\to\infty} \int_{t_0}^{t} F(s, x_{n_k}(s - \frac{c}{n_k}))ds = \int_{t_0}^{t} F(s, x(s))ds, \quad t \in [t_0, t_0 + c]. \tag{A.8}$$

Fix any $t \in [t_0, t_0 + c]$. By (**D3**), we have

$$\left\| F(s, x_{n_k}(s - \frac{c}{n_k})) \right\| \leq m_K(s), \quad s \in [t_0, t],$$

where the function $m_K(s)$ is integrable on $[t_0, t]$. Hence (A.8) follows from Lebesgue's dominated convergence theorem [209, p. 88]. Passing the limit in (A.8) requires (**D2**). The fact that $x(t)$ is an integral of the form $x_0 + \int_{t_0}^t F(s, x(s))ds$ shows that $x(t)$ is absolutely continuous on $[t_0, t_0 + c]$ [209, p. 125, Theorem 11]. Finally, if t_0 is not a left endpoint of J, we can similarly construct a solution on $[t_0 - c, t_0]$. ∎

A local solution can be extended to a solution that is defined on a maximal interval of existence, as shown by the following results.

Let $x(t)$ be a solution of the IVP (A.4) on an interval I. We say that $y(t)$ is a **continuation** of $x(t)$, if $y(t)$ is a solution of the IVP (A.4) on an interval $\hat{I} \supsetneq I$ and $y(t) = x(t)$ on I. If no such continuation exists, we say that $x(t)$ is **noncontinuable** and I is a **maximal interval of existence**.

Lemma A.1 *Let $J \subseteq \mathbb{R}$ be an interval and $D \subseteq \mathbb{R}^n$ be an open set. Suppose that $F : J \times D \to \mathbb{R}^n$ satisfies conditions (**D1**)–(**D3**). Let $x(t)$ be a solution to the IVP* (A.4) *on a bounded interval $(a, b) \subseteq J$. Then the following holds:*

1. *if $x(t)$ is bounded on (a, b), then the limits $x(b^-) = \lim_{t\to b^-} x(t)$ and $x(a^+) = \lim_{t\to a^+} x(t)$ exist;*

2. *if a is not a left endpoint of J and $x(a^+) \in D$, then $x(t)$ can be continued to the left past $t = a$; and*

3. *if b is not a right endpoint of J and $x(b^-) \in D$, then $x(t)$ can be continued to the right past $t = b$.*

Proof: We prove the case for the left endpoint $t = a$ and the case for the right endpoint is similar. Let $K \subseteq J \times D$ be a compact set that contains $(t, x(t))$ for all $t \in (a, b)$. Let $\{t_n\}$ be any monotonically decreasing sequence approaching a. We have, for $m \geq n$,

$$\|x(t_m) - x(t_n)\| = \left\| \int_{t_m}^{t_n} F(s, x(s))ds \right\| \leq \int_{t_m}^{t_n} \|F(s, x(s))\|\, ds \leq \int_{t_m}^{t_n} m_K(s)ds.$$

Since $\int_{t_0}^t m_K(s)ds$ is uniformly continuous, the above inequality shows that $\{x(t_n)\}$ is a Cauchy sequence for any such sequence $\{t_n\}$. Hence, $x(a^+) = \lim_{t\to a^+} x(t)$ exists.

Now suppose that a is in the interior of J and $x(a^+)) \in D$. Consider the IVP:

$$\begin{cases} y' = F(t, y), \\ y(a) = x(a^+). \end{cases} \tag{A.9}$$

Then by local existence, there exists a solution to this IVP on $[a - c, a]$ for some $c > 0$. Define

$$y(t) = x(t), \quad t \in (a, b).$$

Then y is continuous on $[a-c,b)$ and satisfies

$$y(t) = x(a^+) + \int_a^t F(s, y(s))ds, \quad t \in [a-c, a],$$

$$y(t) = x_0 + \int_{t_0}^t F(s, y(s))ds, \quad t \in (a, b).$$

Note that

$$x(a^+) = \lim_{t \to a^+} x(t) = \lim_{t \to a^+} \left[x_0 + \int_{t_0}^t F(s, y(s))ds \right] = x_0 + \int_{t_0}^a F(s, y(s))ds.$$

This shows that

$$y(t) = x_0 + \int_{t_0}^t F(s, y(s))ds, \quad t \in [a-c, b),$$

which is a continuation of $x(t)$ to the left past the left endpoint $t = a$. ■

Proposition A.2 (Continuation) *Let $J \subseteq \mathbb{R}$ be an interval and $D \subseteq \mathbb{R}^n$ be an open set. Suppose that $F : J \times D \to \mathbb{R}^n$ satisfies conditions **(D1)**–**(D3)**. Then any solution $x(t)$ to the IVP* (A.4) *can be continued to a maximal solution $x^*(t)$ defined on a maximal interval of existence J^*. Moreover, the following holds:*

1. *J^* is open relative to J;*
2. *if the right endpoint β of J^* is in the interior of J, then $\limsup_{t \to \beta^-} \|x^*(t)\| = \infty$ or $x^*(t) \to \partial D$ as $t \to \beta^-$;*
3. *if the left endpoint α of J^* is in the interior of J, then $\limsup_{t \to \alpha^+} \|x^*(t)\| = \infty$ or $x^*(t) \to \partial D$ as $t \to \alpha^+$.*

Proof: If x is not a noncontinuable solution, let P denote the set of all continuations of x. Define a partial order on P by $x_1 \leq x_2$ if and only if x_2 is a continuation of x_1. For every chain (totally ordered subset) $\{x_i\}$ in P. Denote by J_i the interval on which x_i is defined. Let $\hat{J} = \bigcup_i J_i$. Then $\hat{J}$ is also an interval. Define y on $\hat{J}$ by $y(t) = x_i(t)$ when $t \in J_i$. Then y is well-defined and is a continuation of every element in the chain, i.e., y is an upper bound of the chain. By Zorn's Lemma, the set P has a maximal element. We denote this by x^* and let its domain be J^*. Clearly, x^* is noncontinuable and J^* is a maximal interval of existence.

The interval J^* must be open relative to J. Otherwise, by Lemma A.1, x^* can be extended past the endpoint at which J^* is closed.

We only prove the case for the right endpoint β (the proof for α is similar). Suppose that β is not a right endpoint of J and $\limsup_{t \to \beta^-} \|x^*(t)\| < \infty$. Then $x^*(t)$ is bounded on $[c, \beta)$ for any $c \in (\alpha, \beta)$. Fix any $c \in (\alpha, \beta)$. Suppose that there does not exist a sequence $t_n \to \beta^-$ as $n \to \infty$ such that

$x^*(t_n) \to \partial D$ as $n \to \infty$. Let Q be a bounded open set such that $x^*(t) \in Q$ and $\overline{Q} \subseteq D$ for all $t \in [c, \beta)$. Hence $x^*(t)$ is bounded on $[c, \beta)$. By Lemma A.1, $\lim_{t \to \beta^-} x^*(t)$ exists. Moreover, $x(\beta^-) \in \overline{Q} \in D$ and, by Lemma A.1, $x^*(t)$ can be extended past $t = \beta$, which contradicts the fact that $x^*(t)$ is noncontinuable. Hence there exists a sequence $\{t_n\} \to \beta^-$ as $n \to \infty$ such that $x^*(t_n) \to \partial D$ as $n \to \infty$.

We further show that $x^*(t) \to \partial D$ as $t \to \beta^-$. Suppose this is not the case. Since $x^*(t)$ is bounded on $[c, \beta)$, there exists a sequence $\tau_n \to \beta^-$ as $n \to \infty$ such that $x^*(\tau_n) \to p$ for some $p \in D$. We now derive a contradiction. Choose $\varepsilon > 0$ such that $B_\varepsilon(p) = \{x \in \mathbb{R}^n : \|x - p\| \leq \varepsilon\}$ is contained in D. Consider the compact set $K = [c, \beta] \times B_\varepsilon(p)$ and let $m_K(t)$ be an integrable function on $[c, \beta]$ given by (**D3**). Choose some N sufficiently large such that $\tau_N \in [c, \beta)$, $\|x^*(\tau_N) - p\| < \frac{\varepsilon}{2}$, and $\int_{t_N}^{\beta} m_K(s)ds < \frac{\varepsilon}{2}$. Since $x^*(t_n) \to \partial D$ as $n \to \infty$, $x^*(t)$ has to exit $B_\varepsilon(p)$ after τ_N. Let $\bar{t}$ be the first exit time, i.e., $\bar{t} = \inf\{t \in [\tau_N, \beta) : \|x^*(t) - p\| > \varepsilon\}$. Then $\|x^*(t) - p\| \leq \varepsilon$ and hence $(t, x^*(t)) \in K$ for all $t \in [\tau_N, \bar{t}]$. It follows that

$$\begin{aligned} \left\|x^*(\bar{t}) - x^*(\tau_N)\right\| &= \left\| \int_{\tau_N}^{\bar{t}} F(s, x^*(s))ds \right\| \\ &\leq \int_{\tau_N}^{\bar{t}} \|F(s, x^*(s))\| \, ds \leq \int_{\tau_N}^{\beta} m_K(s)ds < \frac{\varepsilon}{2}, \end{aligned}$$

which implies that $\left\|x^*(\bar{t}) - p\right\| \leq \left\|x^*(\bar{t}) - x^*(\tau_N)\right\| + \|x^*(\tau_N) - p\| < \frac{\varepsilon}{2} + \frac{\varepsilon}{2} = \varepsilon$. This contradicts the definition of $\bar{t}$ by continuity of $x^*(t)$. ■

If both the Carathéodory conditions (**D1**)–(**D3**) and the local Lipschitz condition (**D4**) are satisfied, then the IVP (A.4) has a unique solution defined on the maximal interval of existence. Before proving the uniqueness result, we introduce a widely used inequality called Gronwall's inequality, which can be written in the form of a differential or integral inequality and is a useful tool for estimating solutions of differential equations.

Lemma A.2 (Gronwall's Inequality) *Let C be a constant. Let m and v be real-valued functions defined on an interval $[a, b]$. Suppose that $v(t)$ is integrable on $[a, b]$, $m(t)$ is continuous on $[a, b]$, and $v(t) \geq 0$ for all $t \in [a, b]$ and*

$$m(t) \leq C + \int_a^t v(s)m(s)ds, \quad t \in [a, b], \tag{A.10}$$

then

$$m(t) \leq Ce^{\int_a^t v(s)ds}, \quad t \in [a, b]. \tag{A.11}$$

Proof: Let

$$u(t) = C + \int_a^t v(s)m(s)ds, \quad t \in [a, b].$$

Then, because $m(t) \leq u(t)$ and $v(t) \geq 0$, we have

$$u'(t) = v(t)m(t) \leq v(t)u(t), \quad \text{for almost all } t \in [a,b].$$

Multiplying $e^{-\int_a^t v(s)ds}$ to both sides of the previous inequality gives

$$u'(t)e^{-\int_a^t v(s)ds} - v(t)u(t)e^{-\int_a^t v(s)ds} \leq 0, \quad \text{for almost all } t \in [a,b],$$

which is

$$\frac{d}{dt}\left[u(t)e^{-\int_a^t v(s)ds}\right] \leq 0, \quad \text{for almost all } t \in [a,b].$$

This implies that $u(t)e^{-\int_a^t v(s)ds}$ is non-increasing on $[a,b]$. Hence

$$u(t)e^{-\int_a^t v(s)ds} \leq u(a)e^{-\int_a^a v(s)ds} = u(a) = C, \quad t \in [a,b].$$

which leads to (A.11). ■

Proposition A.3 (Uniqueness) *Let $J \subseteq \mathbb{R}$ be an interval and $D \subseteq \mathbb{R}^n$ be an open set. Suppose that $F : J \times D \to \mathbb{R}^n$ satisfies conditions (**D1**)-(**D4**). Then the IVP (A.4) has a unique solution defined on J^*, where J^* is the maximal interval of existence.*

Proof: Suppose that $y(t)$ and $x(t)$ are both solutions to the IVP (A.4) on the interval $I = [t_0 - c, t_0 + c] \cap J$. Let K be a compact set that contains the set $\{(t,z) : t \in I,\ z = x(t) \text{ or } y(t)\}$. Then, for $t \in [t_0, t_0 + c] \cap J$, we have

$$\begin{aligned}\|y(t) - x(t)\| &\leq \int_{t_0}^t \|F(s, y(s)) - F(s, x(s))\|\, ds \\ &\leq \int_{t_0}^t l_K(s)\, \|y(x) - x(s)\|\, ds,\end{aligned}$$

for all $t \in [t_0, t_0 + c] \cap J$. Applying Gronwalls' inequality (Lemma A.2 with $C = 0$ and $v(t) = l_K(t)$), we obtain

$$\|y(t) - x(t)\| \leq 0e^{\int_{t_0}^t l_K(s)ds} = 0. \tag{A.12}$$

Hence, $y(t) = x(t)$ for all $t \in [t_0, t_0 + c] \cap J$. The proof for $t \in [t_0 - c, t_0] \cap J$ is similar and left to the reader.

Now suppose that the two solutions $y(t)$ and $x(t)$ are extended to their maximal intervals of existence J_1^* and J_2^*, respectively. We show that $J_1^* = J_2^*$ and the two solutions are identical on this interval. Suppose without loss of generality that $y(t)$ is maximally extended to the right on $[t_0, \beta_1) \cap J$ and $x(t)$ on $[t_0, \beta_2) \cap J$ with $\beta_2 > \beta_1$. Clearly, β_1 is not a right endpoint of J. Then the argument in the previous paragraph shows that $y(t) = x(t)$ on any interval of the form $[t_0, c]$, where $c \in (0, \beta_1)$. Hence, $y(t) = x(t)$ on $[t_0, \beta_1)$. Since $x(t)$ is defined at $t = \beta_1$, we can define $y(\beta_1) = x(\beta_1)$ and extend the solution past $t = \beta_1$. This contradicts that J_1^* is a maximal interval of existence. We can conclude that the IVP (A.4) has a unique solution defined on the maximal interval of existence. ■

The same analysis in the uniqueness theorem can be easily applied to obtain a result on continuous dependence of solutions on initial conditions.

Proposition A.4 (Continuous Dependence) *Let $J \subseteq \mathbb{R}$ be a compact interval containing t_0 and $D \subseteq \mathbb{R}^n$ be an open set. Suppose that $F : J \times D \to \mathbb{R}^n$ satisfies conditions **(D1)**–**(D4)**. Let $x(t)$ and $y(t)$ be two solutions to the IVP (A.4) with initial conditions $x(t_0) = x_0$ and $y(t_0) = y_0$, respectively. Then there exists a constant C such that*

$$\|y(t) - x(t)\| \leq C \|y_0 - x_0\|, \quad \forall t \in J. \tag{A.13}$$

*In particular, if $l_K(t)$ in the Carathéodory condition **(D4)** can be taken as a constant L (that depends on K), then we have*

$$\|y(t) - x(t)\| \leq \|y_0 - x_0\| e^{L|t-t_0|}, \quad \forall t \in J. \tag{A.14}$$

Proof: Following the same argument as in the proof of Proposition A.3, let K be a compact set that contains the set $\{(t, z) : t \in J,\ z = x(t) \text{ or } y(t)\}$. Instead of (A.12), we obtain

$$\|y(t) - x(t)\| \leq \|y_0 - x_0\| e^{\int_{t_0}^{t} l_K(s) ds}, \tag{A.15}$$

for $t \geq t_0$. We obtain a similar bound for $t \leq t_0$. Since l_K is integrable, we obtain (A.13). If there is a Lipschitz constant L, we get (A.13). ■

A.3 Global Existence

Based on the continuation result Proposition A.2, several corollaries on global existence of solutions can be stated.

Corollary A.1 (Local boundedness of solutions implies global existence) *Suppose that $F : \mathbb{R} \times \mathbb{R}^n \to \mathbb{R}^n$ satisfies conditions **(D1)**–**(D3)** and $x(t)$ is a solution of initial value problem (A.4) on a maximal interval of existence $J = (\alpha, \beta)$.*

- *If, for any $c > t_0$, $x(t)$ is bounded on $J \cap [t_0, c)$, then $\beta = \infty$.*
- *If, for any $c < t_0$, $x(t)$ is bounded on $J \cap (c, t_0]$, then $\alpha = -\infty$.*

- *If, for any $c > 0$, $x(t)$ is bounded on $J \cap (-c, c)$, then $J = (-\infty, \infty)$.*

Proof: We only prove the first case. If $\beta < \infty$, we take $c = \beta$. By Proposition A.2, we have $\limsup_{t \to c^-} |x(t)| = \infty$. This contradicts that $x(t)$ is bounded on $J \cap [t_0, c)$. ■

Corollary A.2 (Linear growth of vector fields implies global existence) *Suppose that $F : \mathbb{R} \times \mathbb{R}^n \to \mathbb{R}^n$ satisfies conditions (**D1**)-(**D2**) and*

$$\|F(t, x)\| \leq a(t) \|x\| + b(t), \quad (t, x) \in \mathbb{R} \times \mathbb{R}^n,$$

for some locally integrable[1] functions $a : \mathbb{R} \to [0, \infty)$ and $b : \mathbb{R} \to [0, \infty)$. Then, for every $(t_0, x_0) \in \mathbb{R} \times \mathbb{R}^n$, every solution of the IVP (A.1) *can be continued to a global solution defined on $\mathbb{R}$.*

Proof: Given any solution, let $x(t)$ be its continuation defined on a maximal interval of existence $J = (\alpha, \beta)$ (as guaranteed by Proposition A.2). We show that $\beta = \infty$. The proof for $\alpha = -\infty$ is similar. Pick any $c > t_0$. We have, for $t \in J \cap [t_0, c)$,

$$\begin{aligned} \|x(t)\| &\leq \|x_0\| + \int_{t_0}^{t} \|F(s, x(s))\| \, ds \\ &\leq |x_0| + \int_{t_0}^{t} a(s) \|x(s)\| \, ds + \int_{t_0}^{t} b(s) ds. \end{aligned}$$

Let $B = \int_{t_0}^{c} b(s) ds$. Then

$$\|x(t)\| \leq (|x_0| + B) + \int_{t_0}^{t} a(s) \|x(s)\| \, ds,$$

for all $t \in [t_0, c) \cap J$. By Gronwall's inequality (Lemma A.2),

$$\|x(t)\| \leq (\|x_0\| + B) e^{\int_{t_0}^{t} a(s) ds} \leq (\|x_0\| + B) e^{\int_{t_0}^{c} a(s) ds}$$

for all $t \in [t_0, c) \cap J$. Hence $x(t)$ is bounded on $[t_0, c) \cap J$. By Corollary A.1, $\beta = \infty$. ■

Note that the linear growth condition implies the local integrability condition (**D3**); hence we do not need (**D3**). The following are special cases of systems that satisfy the linear growth condition.

Corollary A.3 (Boundedness of vector fields implies global existence) *Suppose that $F : \mathbb{R} \times \mathbb{R}^n \to \mathbb{R}^n$ satisfies conditions (**D1**)-(**D2**) and is bounded on $\mathbb{R} \times \mathbb{R}^n$. Then, for every $(t_0, x_0) \in \mathbb{R} \times \mathbb{R}^n$, every solution of the IVP* (A.1) *can be continued to a global solution defined on $\mathbb{R}$.*

[1] A function is locally integrable if it is integrable on any compact interval.

Corollary A.4 (Global Lipschitz vector fields implies global existence) *Suppose that $F : \mathbb{R} \times \mathbb{R}^n \to \mathbb{R}^n$ satisfies conditions **(D1)**-**(D3)** and*

(D4') ***(global Lipschitz)*** *there exists a locally integrable function $l(t)$ such that*

$$\|F(t,x) - F(t,y)\| \le l(t)\,\|x-y\|,$$

for all $(t,x,y) \in \mathbb{R} \times \mathbb{R}^n \times \mathbb{R}^n$.

Then, for every $(t_0,x_0) \in \mathbb{R} \times \mathbb{R}^n$, the IVP (A.1) *has a unique solution on $\mathbb{R}$.*

Proof: Pick $y = 0$. We obtain

$$\|F(t,x)\| \le l(t)\,\|x\| + \|F(t,0)\|.$$

By **(D3)**, $\|F(t,0)\|$ is locally integrable. Hence F satisfies the linear growth condition in Corollary A.2, from which the conclusion follows. ■

We stated all the global existence results for solutions defined on $\mathbb{R}$. It is clear from the proofs that if the respective conditions are only satisfied for $t \in [t_0,\infty)$, then solutions exist on $[t_0,\infty)$; similarly for $(-\infty,t_0] \times \mathbb{R}^n$. In practical applications, we mostly care about solutions defined in positive time.

A.4 Differential Inequalities and a Comparison Theorem

In this section, we will introduce a comparison theorem that can be used to estimate solutions of ordinary differential equations. We have introduced a simple comparison theorem, i.e., Gronwall's inequality (Lemma A.2), in the form of an integral inequality. The more general comparison theorem presented in this section are in the form of differential inequalities. The differential inequalities we state and prove here involve a locally Lipschitz function. For differential and integral inequalities in more general settings, see [133, 221, 237].

We are interested in comparison for vector-valued functions. For $x, y \in \mathbb{R}^n$, we write $x \le y$ if $x_i \le y_i$ for all $i \in \{1,2,\cdots,n\}$; $x < y$, $x \ge y$, and $x > y$ are similarly defined.

Definition A.1 *A function $f : \mathbb{R}^n \to \mathbb{R}^n$ is said to satisfy the quasi-monotonicity property*[2] *if, for each $i \in \{1,2,\cdots,n\}$, we have $f_i(x) \le f_i(y)$ whenever $x_i = y_i$ and $x \le y$.*

[2] Also called the type K or Kamke condition, named after Kamke [111]

Proposition A.5 *Let $J = [t_0, T] \subseteq \mathbb{R}$ be a compact interval and $D \subseteq \mathbb{R}^n$ be an open set. Suppose that $f : J \times D \to \mathbb{R}^n$ is locally Lipschitz in the sense that, for each compact set $K \subseteq D$, there exists a constant L such that*

$$\|f(t,x) - f(t,y)\| \le L \|x - y\|_\infty, \quad \forall t \in J, \forall x \in K. \tag{A.16}$$

Furthermore, $f(t,x)$ satisfies the quasi-monotonicity property in x on D for each $t \in J$. Let $r : J \to \mathbb{R}^n$ and $m : J \to \mathbb{R}^n$ be two absolutely continuous functions that satisfy

1. *$r(t_0) \le m(t_0)$;*

2. *$r'(t) \le f(t, r(t))$ and $m'(t) \ge f(t, m(t)))$ for almost all $t \in J$.*

Suppose that D contains the smallest interval enclosure[3] of the images of r and m on J. Then $r(t) \le m(t)$ for all $t \in J$. Furthermore, if we have $r(t_0) < m(t_0)$, then $r(t) < m(t)$ for all $t \in J$.

Proof: (1) We first prove the strict inequality version. Suppose that this is not true. Then there exists a first $\bar{t} > t_0$ such that $r_i(\bar{t}) \ge m_i(\bar{t})$ for at least one $i \in \{1, 2, \cdots, n\}$. Indeed, we can define

$$\bar{t} = \inf \{t \in J \mid r_i(t) \ge m_i(t) \text{ for some } i\}.$$

By continuity, $r_i(\bar{t}) = m_i(\bar{t})$ for some i and $r_j(t) \le m_j(t)$ for all $t \in [t_0, \bar{t}]$ and all $j \in \{1, 2, \cdots, n\}$. Let L be the Lipschitz constant for $f(t,x)$ in x on $J \times K$, where K is some compact interval containing the imagine of r and m on J. For almost all $t \in [t_0, \bar{t}]$, we have

$$\begin{aligned}
r_i'(t) &\le f_i(t, r(t)) = f_i(t, m(t)) + f_i(t, r(t)) - f_i(t, m(t)) \\
&\le m_i'(t) + f_i(t, r_1(t), \ldots, r_i(t), \ldots, r_n(t)) - f_i(t, m(t)) \\
&\le m_i'(t) + f_i(t, m_1(t), \ldots, r_i(t), \ldots, m_n(t)) - f_i(t, m(t)) \\
&\le m_i'(t) + L |r_i(t) - m_i(t))| \\
&= m_i'(t) - L(r_i(t) - m_i(t)),
\end{aligned} \tag{A.17}$$

where we used the fact that $r_j(t) \le m_j(t)$ for all $t \in [t_0, \bar{t}]$ and all $j \in \{1, 2, \cdots, n\}$ and the quasi-monotonicity of f. Let

$$v(t) = -(r_i(\bar{t} - t) - m_i(\bar{t} - t))$$

on $[0, \bar{t} - t_0]$. Then, by (A.17), $v(t)$ is nonnegative on $[0, \bar{t} - t_0]$ and

$$v'(t) = r_i'(\bar{t} - t) - m_i'(\bar{t} - t) \le -L(r_i'(\bar{t} - t) - m_i'(\bar{t} - t)) = Lv(t),$$

[3] Using interval analysis notation, this means

$$[\underline{x}, \overline{x}] \subseteq D,$$

where $\underline{x}_i = \min(\min_{t \in J} r_i(t), \min_{t \in J} m_i(t))$ and $\overline{x}_i = \max(\max_{t \in J} r_i(t), \max_{t \in J} m_i(t))$.

for almost all $t \in [0, \bar{t} - t_0]$. Note also that $v(0) = 0$. By Gronwall's inequality (Lemma A.2), $v(t) \equiv 0$ on $[0, \bar{t} - t_0]$. In particular, $v(\bar{t} - t_0) = 0$ implies $r_i(t_0) = m_i(t_0)$, which contradicts $r(t_0) < m(t_0)$.

(2) We now prove the non-strict inequality version. Let K be a compact interval containing an ε_0-neighborhood of the imagine of r and m on J, where $\varepsilon_0 > 0$ is sufficiently small such that $K \subseteq D$. Let L be the Lipschitz constant for $f(t, x)$ in x on $J \times K$. Choose $\varepsilon > 0$ such that $\rho_i(t) := \varepsilon e^{L(t-t_0)} \leq \varepsilon_0$ for all $t \in J$ and $i \in \{1, 2, \cdots, n\}$.

Consider $\hat{m}(t) := m(t) + \rho(t)$. Then

$$\begin{aligned}
\hat{m}'(t) &= m'(t) + \rho'(t) \\
&\geq f(t, m(t)) + L\rho(t) \\
&= f(t, \hat{m}(t)) + f(t, m(t)) - f(t, \hat{m}(t)) + L\rho(t) \\
&\geq f(t, \hat{m}(t)) - L \left\| m(t) - \hat{m}(t) \right\|_\infty + L\rho(t) \\
&= f(t, \hat{m}(t)) - L\rho(t) + L\rho(t) = f(t, \hat{m}(t)),
\end{aligned} \tag{A.18}$$

for almost all $t \in J$, where we used the assumption on m and the Lipschitz continuity of f in x. Clearly, $\hat{m}(t_0) = m(t_0) + \varepsilon \geq r(t_0) + \varepsilon > r(t_0)$. We have verified that $\hat{m}$ and r satisfy the assumptions of the strict inequality version of the proposition proved in part (1). Hence, $\hat{m}(t) > r(t)$ for all $t \in J$. Since $\varepsilon > 0$ can be taken to be arbitrarily small. We have $m(t) \geq r(t)$ for all $t \in J$ by letting $\varepsilon \to 0$. ∎

Appendix B

Interval Analysis

B.1 Proof of Convergence for Taylor Inclusions (Proposition 4.6)

Suppose that $f : \mathbb{R} \to \mathbb{R}$ is k-times continuously differentiable on an interval X_0. Recall that the kth-order **Taylor inclusion function** of f is given by

$$[f]_{\mathrm{t}}([x]) := f(m) + f'(m) \cdot ([x] - m) + \cdots + f^{(k-1)}(m) \cdot \frac{([x] - m)^{k-1}}{(k-1)!} + [f^{(k)}]([x]) \cdot \frac{([x] - m)^k}{k!}, \quad [x] \subseteq X_0.$$

where $m = \mathrm{mid}([x])$ and $[f^{(k)}]$ is an inclusion function of the kth derivative function $f^{(k)}(x)$.

The proof of Proposition 4.6 requires the following technical lemma on properties of intervals.

Lemma B.1 *The following properties hold for intervals A and B in $\mathbb{IR}$:*

(i) if $A \subseteq B$, then $d(A, B) \leq \mathrm{w}(B) - \mathrm{w}(A)$;

(ii) $\mathrm{w}(A + B) = \mathrm{w}(A) + \mathrm{w}(B)$;

(iii) $\mathrm{w}(\alpha A) = |\alpha| \,\mathrm{w}(A)$ for $\alpha \in \mathbb{R}$;

(iv) if $0 \in A$, then $\mathrm{w}(A^k) \leq \mathrm{w}(A)^k$ for integers $k \geq 1$;

(v) if $a \in A$, then $|A| \leq \mathrm{w}(A) + |a|$;

(vi) if $A = -A$, then $BA^k = |B|\, A^k$ for integers $k \geq 1$;

(vii) if $\mathrm{w}(A \cdot B) \leq |A| \,\mathrm{w}(B) + |B| \,\mathrm{w}(A)$.

Proof: One can easily verify properties (i)–(iii), (v), and (vii) by definition. To prove (*iv*), let $A = [-a, b]$, where $b \geq 0$ and $a \geq 0$. Assume $b \geq a$. Then

$$A^2 = [-ab, b^2], \quad A^3 = [-ab^2, b^3], \quad \cdots.$$

DOI: 10.1201/9780429270253-B

It can be shown by induction that, for $k \geq 1$,

$$A^k = [-ab^{k-1}, b^k].$$

Hence

$$\mathrm{w}(A^k) = b^k + ab^{k-1} = b^{k-1}(a+b) \leq (a+b)^k = \mathrm{w}(A)^k,$$

since $0 \leq b \leq a+b$. If $a \geq b$, we apply the above argument to $-A$ and obtain

$$\mathrm{w}((A)^k) = \mathrm{w}((-1)^k(-A)^k) = \mathrm{w}((-A)^k) \leq \mathrm{w}(-A)^k = \mathrm{w}(A)^k.$$

For property (vi), let $A = [-a, a]$, where $a \geq 0$. Then

$$BA = [-a\,|B|\,, a\,|B|] = |B|\,A.$$

It follows that $BA^k = |B|\,A^k$ for $k \geq 1$. ■

B.1.1 Proof of Proposition 4.6

Note that the Taylor inclusion with $k = 1$ gives the mean-value inclusion. So we consider two cases: 1) $k > 1$ and 2) $k = 1$. Let $[x] \subseteq X_0$, on which f is k-times continuously differentiable.

Case $k > 1$: By Lemma B.1 (ii), we have

$$\begin{aligned}\mathrm{w}([f]_{\mathrm{t}}([x])) = \mathrm{w}(f'(m)([x]-m)) + \cdots + \mathrm{w}\left(f^{(k-1)}(m)\cdot\frac{([x]-m)^{k-1}}{(k-1)!}\right) \\ + \mathrm{w}\left([f^{(k)}]([x])\cdot\frac{([x]-m)^k}{k!}\right).\end{aligned} \tag{B.1}$$

By Lemma B.1 (iii) and (iv),

$$\begin{aligned}\mathrm{w}\left(f^{(k-1)}(m)\cdot\frac{([x]-m)^{k-1}}{(k-1)!}\right) &\leq \frac{\left|f^{(k-1)}(m)\right|}{(k-1)!}\mathrm{w}\left(([x]-m)^{k-1}\right) \\ &\leq \frac{\left|f^{(k-1)}(m)\right|}{(k-1)!}\mathrm{w}([x])^{k-1}, \quad k \geq 2,\end{aligned} \tag{B.2}$$

where we used Lemma B.1 (iv) with $A = [x]-m$ and the fact that $\mathrm{w}([x]-m) = \mathrm{w}([x])$. Since f is continuously differentiable on X_0, there exists a constant C_1 (dependent on X_0, but not $[x]$) such that

$$\sum_{i=2}^{k-1} \mathrm{w}\left(f^{(i)}(m)\cdot\frac{([x]-m)^i}{(i)!}\right) \leq C_1\mathrm{w}([x])^2. \tag{B.3}$$

By Lemma B.1 (vi) and (iv),

$$\mathrm{w}\left([f^{(k)}]([x])\cdot\frac{([x]-m)^k}{k!}\right) \leq \frac{\left|[f^{(k)}]([x])\right|}{k!}\mathrm{w}([x])^k.$$

By the assumption on $[f^{(k)}]$, we there exists a constant c_1 (dependent on X_0, but not $[x]$) such that $\mathrm{w}([f^{(k)}]([x])) \leq c_1 \mathrm{w}([x])$. Furthermore, $f^{(k)}(x)$ is bounded on X_0 by continuity of $f^{(k)}$. We can choose any $\xi \in X_0$ and apply Lemma B.1 (v) to obtain

$$\left|[f^{(k)}]([x])\right| \leq \left|f^{(k)}(\xi)\right| + \mathrm{w}([f^{(k)}]([x])) \leq c_2(1 + \mathrm{w}([x])),$$

which implies that there exists a constant C_2 (dependent on X_0, but not $[x]$) such that

$$\mathrm{w}\left([f^{(k)}]([x]) \cdot \frac{([x] - m)^k}{k!}\right) \leq C_2 \mathrm{w}([x])^2, \quad k \geq 2. \tag{B.4}$$

Combining (B.1), (B.2), (B.3), and (B.4) gives

$$\mathrm{w}([f]_{\mathrm{t}}([x])) \leq |f'(m)|\,\mathrm{w}([x]) + (C_1 + C_2)\mathrm{w}([x])^2. \tag{B.5}$$

Let $[x] = [\underline{x}, \overline{x}]$. By the mean value theorem for f,

$$\mathrm{w}([f]^*([x])) \geq |f(\overline{x}) - f(\underline{x})| = |f'(\zeta)|\,\mathrm{w}([x]), \quad \zeta \in [x].$$

By the mean value theorem for f',

$$|f'(m)| - |f'(\zeta)| \leq |f'(m) - f'(\zeta)| = |f''(\theta)|\,|m - \zeta| \leq |f''(\theta)|\,\mathrm{w}([x]).$$

The last two estimates, combined with Lemma B.1 (i) and (B.5), imply

$$\begin{aligned} d([f]_{\mathrm{t}}([x]), [f]^*([x])) &\leq \mathrm{w}([f]_{\mathrm{t}}([x])) - \mathrm{w}([f]^*([x])) \\ &\leq (|f'(m)| - |f'(\zeta)|)\mathrm{w}([x]) + (C_1 + C_2)\mathrm{w}([x])^2 \\ &\leq |f''(\theta)|\,\mathrm{w}([x])^2 + (C_1 + C_2)\mathrm{w}([x])^2. \end{aligned}$$

By boundedness of f'' on X_0, there exists a constant C (dependent on X_0, but not $[x]$) such that

$$d([f]_{\mathrm{t}}([x]), [f]^*([x])) \leq C\mathrm{w}([x])^2.$$

Case $k = 1$: In this case, the proof above can be modified to yield

$$\begin{aligned} d([f]_{\mathrm{m}}([x]), [f]^*([x])) &\leq \mathrm{w}([f]_{\mathrm{m}}([x])) - \mathrm{w}([f]^*([x])) \\ &\leq (|[f']([x])| - |f'(\zeta)|)\mathrm{w}([x]) \\ &\leq \mathrm{w}([f']([x]))\mathrm{w}([x]) \\ &\leq C\mathrm{w}([x])^2, \end{aligned}$$

where the first inequality follows from Lemma B.1 (i), the second from Lemma B.1 (iii) and the mean value theorem, the third from Lemma B.1 (v), and the last from the assumption on $[f']$. □

Appendix C

Basic Set Theory

C.1 Set Convergence

In approximating winning sets, we rely on the following definitions and results on set limits and convergence.

Definition C.1 *The **limit inferior** of a sequence $\{x_i\}$ is defined by*

$$\liminf_{i\to\infty} x_i = \lim_{i\to\infty}\left(\inf_{j\geq i} x_j\right).$$

*Similarly, the **limit superior** of $\{x_n\}$ is defined by*

$$\limsup_{i\to\infty} x_i = \lim_{i\to\infty}\left(\sup_{j\geq i} x_j\right).$$

Definition C.2 (Painlevé-Kuratowski Convergence) *For a sequence $\{A_i\}_{i=1}^{\infty}$ of subsets of $\mathbb{R}^n$. The **outer limit** of $\{A_i\}_{i=1}^{\infty}$ is defined by*

$$\limsup_{i\to\infty} A_i = \left\{x \in \mathbb{R}^n \mid \liminf_{i\to\infty} d(x, A_i) = 0\right\}.$$

*The **inner limit** of $\{A_i\}_{i=1}^{\infty}$ is defined by*

$$\liminf_{i\to\infty} A_i = \left\{x \in \mathbb{R}^n \mid \limsup_{i\to\infty} d(x, A_i) = 0\right\}.$$

*The **(set) limit** of $\{A_i\}_{i=1}^{\infty}$ exists if and only if the outer and inner limit sets are equal:*

$$\lim_{i\to\infty} A_i = \limsup_{i\to\infty} A_i = \liminf_{i\to\infty} A_i.$$

Both inner and outer limits of any sequence of subsets on $\mathbb{R}^n$ by definition are closed [206]. Specifically for any *monotone* sequence $\{A_i\}_{i=1}^{\infty}$, i.e., either $A_i \subseteq A_{i+1}$ or $A_i \supseteq A_{i+1}$ for all $i \in \mathbb{Z}_{>0}$, the set limit always exists.

Proposition C.1 ([206]) *Consider a sequence of sets $\{A_i\}_{i=1}^{\infty}$. Then*

(i) $\lim_{i\to\infty} A_i = \overline{\bigcup_{i=1}^{\infty} A_i}$ *whenever* $A_i \subseteq A_{i+1}$ *for all* $i \in \mathbb{Z}_{>0}$.

(ii) $\lim_{i\to\infty} A_i = \bigcap_{i=1}^{\infty} \overline{A_i}$ *whenever* $A_i \supseteq A_{i+1}$ *for all* $i \in \mathbb{Z}_{>0}$.

DOI: 10.1201/9780429270253-C

C.2 Heine-Borel Theorem

Definition C.3 *An* ***open cover*** *of a set E in a metric space X is a collection $\{G_i\}$ of open subsets of X such that $E \subseteq \cup_i G_i$.*

There are different forms to state the Heine-Borel theorem. Here we follow the version in [210], and the theorem provides three alternative definitions (or necessary and sufficient conditions) of a compact set in the $\mathbb{R}^n$ space.

Definition C.4 *A subset E of a metric space X is said to be* ***compact*** *if every open cover of E contains a finite sub-cover.*

Theorem C.1 (Heine-Borel Theorem) *Let E be a subset in $\mathbb{R}^n$. Then the following three statements are equivalent:*

(i) E is closed and bounded.

(ii) Every open cover of E contains a finite sub-cover.

(iii) Every infinite subset of E has a limit point in E.

C.3 Minkowski Sum and Minkowski Difference

Given two sets $A, B \subseteq \mathbb{R}^n$, the Minkowski sum $A+B$ and difference[1] $A-B$ are defined by

$$A + B = \{x + y \mid x \in A,\ y \in B\}, \tag{C.1}$$

$$A - B = \{x \mid x + y \in A,\ \forall y \in B\}, \tag{C.2}$$

respectively. The following properties of Minkowski sum and difference among sets will be used.

Proposition C.2 *Let $A, B \subseteq \mathbb{R}^n$ and assume that $A - B \neq \emptyset$. Then the following properties hold.*

(i) $A - B \subseteq A$ if $\mathbf{0}^n \in B$.

(ii) $A - B + B \subseteq A \subseteq A + B - B$.

(iii) $A - B = (A_1 - B) \cap (A_2 - B), \quad A = A_1 \cap A_2, A_1, A_2 \subseteq \mathbb{R}^n$.

(iv) $A - B = \bigcap_{i=1}^{\infty}(A_i - B)$, where $A = \bigcap_{i=1}^{\infty} A_i$.

[1] Also known as the Pontryagin difference.

(v) $A - B$ is closed (compact) if A is closed (compact).

Proof: Property (i) is straightforward since for all $a \in A - B$, $\mathbf{0}^n \in B$ implies that $a + \mathbf{0}^n \in A$.

To show property (ii), let $z \in A - B + B$. Then we can find $y \in A - B$ and $b \in B$ such that $z = y + b$ by (C.1). By (C.2), $y + b \in A$, which gives that $z \in A$. Hence $A - B + B \subseteq A$. Let $a \in A$ be arbitrary. Then $a + b \in A + B$ for all $b \in B$. It follows that $a \in A + B - B$ by (C.2).

We now show (iii). By (C.2), we have

$$\begin{aligned} A - B = (A_1 \cap A_2) - B &= \{x \in A_1 \cap A_2 \mid x + b \in A_1 \cap A_2, \forall b \in B\} \\ &= \{x \in A_1 \cap A_2 \mid (x + b \in A_1) \wedge (x + b \in A_2), \forall b \in B\}, \\ (A_1 - B) \cap (A_2 - B)& \\ &= \{y \in A_1 \cap A_2 \mid (y + b_1 \in A_1) \wedge (y + b_2 \in A_2), \forall b_1, b_2 \in B\}. \end{aligned}$$

Then $(A_1 - B) \cap (A_2 - B) \subseteq A - B$ clearly. For any $y \notin (A_1 - B) \cap (A_2 - B)$, there exists $b' \in B$ so that $(y + b' \notin A_1) \vee (y + b' \notin A_2)$, which indicates that $y \notin A - B$. Hence $A - B \subseteq (A_1 - B) \cap (A_2 - B)$, and iii holds.

To show (iv), we prove both $(\bigcap_{i=1}^{\infty} A_i) - B \subseteq \bigcap_{i=1}^{\infty}(A_i - B)$ and $\bigcap_{i=1}^{\infty}(A_i \cap B) \subseteq (\bigcap_{i=1}^{\infty} A_i) - B$. Let $x \in (\bigcap_{i=1}^{\infty} A_i) - B$. Then $x + b \in \bigcap_{i=1}^{\infty} A_i$ for all $b \in B$. That is to say $x + b \in A_i$ for all $b \in B$ and $i \in \mathbb{Z}_{>0}$. It then implies that $x \in A_i - B$ for all $i \in \mathbb{Z}_{>0}$, i.e., $x \in \bigcap_{i=1}^{\infty}(A_i - B)$. For the other direction, let $x \notin \bigcap_{i=1}^{\infty} A_i - B$. Then there exists $b \in B$ such that $x + b \notin \bigcap_{i=1}^{\infty} A_i$, i.e., there exists $j \in \mathbb{Z}_{>0}$ such that $x + b \notin A_j$. It follows that $x \notin A_j - B$ and hence $x \notin \bigcap_{i=1}^{\infty}(A_i - B)$.

For (v), we first show the closedness property. By (C.2), $A - B = \bigcap_{b \in B}(A \setminus \{b\})$. If A is closed, then $A \setminus \{b\}$ is closed for all $b \in B$. It follows that $\bigcap_{b \in B}(A \setminus \{b\})$ is also closed. By ii, we have $A - B + B \subseteq A$. If additionally A is bounded, then $A - B$ is also bounded. ■

Part of the proof can also be found in [120].

C.4 Proof of Proposition 6.2

Proof: (i) For any subset $S \subseteq A \cap B$, $S \subseteq A$ and $S \subseteq B$. Hence, $h(S) \subseteq h(A)$ and $h(S) \subseteq h(B)$, which implies $h(S) \subseteq h(A) \cap h(B)$. Next we provide an example where $h(A \cap B)$ is strictly inside $h(A) \cap h(B)$. Consider a function

$$h(x) = \begin{cases} 0 & x \in [1, 2] \cup [4, 5], \\ x & \textit{otherwise}. \end{cases}$$

If $A = [1.3, 3]$ and $B = [[2, 5]$, then, for the induced function h of sets of real numbers, $h(A) = 0 \cup [2, 3]$ and $h(B) = 0 \cup [2, 3]$ but $h(A \cup B) = h([2, 3]) = [2, 3] \subseteq h(A) \cup h(B)$.

(ii) For any $A, B \subseteq Y$, $h(A) \subseteq h(A \cup B)$ and $h(B) \subseteq h(A \cup B)$, and hence $h(A) \cup h(B) \subseteq h(A \cup B)$. To see why we don't have $h(A) \cup h(B) = h(A \cup B)$, let us consider the function $h(S) = S - B_r$ for any $S \subseteq \mathbb{R}^2$ and $B_r \subseteq \mathbb{R}^2$ is a ball with radius r ($r > 0$). Let $A = \{x \in \mathbb{R}^2 \mid 3r < |x| \leq 5r\}$ and $B = \{x \in \mathbb{R}^2 \mid r < |x| \leq 3r\}$. Then $h(A) = h(B) = \emptyset$ but $h(A \cup B) = B \supseteq h(A) \cup h(B)$. ■

C.5 Proof of Proposition 6.3

The proofs of Proposition 6.3 (i) for nonlinear disturbed systems under similar assumptions to Assumption 6.1 can also be found in [199, Theorem 2], and [32, Theorem 5.2].

Proof: Given that Ω is closed (compact), $\Omega - W = \Omega - \mathbb{B}_\delta$ is also closed (compact) by Proposition C.2 (v). The conclusion trivially hold if $\Omega - W = \emptyset$, because $\mathrm{Pre}^\delta(\Omega) = \emptyset$, which is compact. Hence we assume that $\Omega - W \neq \emptyset$. By Proposition 6.1 (iii), we can simplify the proof by considering Pre only.

Let $\{x_k\}_{k=0}^\infty$ be a convergent sequence in the set $\mathrm{Pre}(\Omega)$ with the limit x^*, i.e., $\lim_{k\to\infty} x_k = x^*$. By (4.29), for all k, there exists $u_k \in U$ such that $f(x_k, u_k) = \widetilde{x}_k \in \Omega$. We aim to show that $x^* \in \Omega$.

Under Assumption 6.1, the input space U is closed (compact). Then there exists a sub-sequence $\{u_{k_i}\}_{i=0}^\infty$ of $\{u_k\}_{k=0}^\infty$ ($0 \leq k_i \leq k$) that converges to a point $u^* \in U$. Let $\{x_{k_i}\}_{i=0}^\infty$ be the corresponding sub-sequence of $\{x_k\}_{k=0}^\infty$. By the continuity of f with respect to both arguments, we have

$$\lim_{i\to\infty} \widetilde{x}_{k_i} = \lim_{i\to\infty} f(x_{k_i}, u_{k_i}) = f(\lim_{i\to\infty} x_{k_i}, \lim_{i\to\infty} u_{k_i}) = f(x^*, u^*),$$

which means that $\{\widetilde{x}_{k_i}\}_{i=0}^\infty$ converges to some point $\widetilde{x}^* = f(x^*, u^*)$. Since Ω is closed (compact), $f(x^*, u^*) = \widetilde{x}^* \in \Omega$, which implies $x^* \in \mathrm{Pre}(\Omega)$.

Under Assumption 6.2, U is finite. Let $\{x_{k_i}\}_{i=0}^\infty$ be the sub-sequence of $\{x_k\}_{k=0}^\infty$ that belong to Pre by using a common $u \in U$, and $\lim_{i\to\infty} x_{k_i} = x^*$. By the continuity of $f(x, u)$ with respect to x for a fixed u, we have

$$\widetilde{x}^* = \lim_{i\to\infty} \widetilde{x}_{k_i} = \lim_{i\to\infty} f(x_{k_i}, u) = f(\lim_{i\to\infty} x_{k_i}, u) = f(x^*, u). \tag{C.3}$$

Similarly, (C.3) implies $x^* \in \Omega$.

To show ii, we consider the complement $\left(\mathrm{Pre}^\delta(\Omega)\right)^c$ of $\mathrm{Pre}^\delta(\Omega)$:

$$\left(\mathrm{Pre}^\delta(\Omega)\right)^c = \{x \in X \mid \forall u \in U, \exists w \in W, \text{ s.t. } f(x, u) + w \in \Omega^c\},$$

where $\Omega^c = X \setminus \Omega$ is closed with respect to X since $\Omega \subseteq X$ is open.

Let $\{x_i\}_{i=0}^\infty$ be a convergent sequence in $\left(\mathrm{Pre}^\delta(\Omega)\right)^c$ with $x = \lim_{i\to\infty} x_i$.

Then for any given $u \in U$, there exists $w_i \in W$ such that $f(x_i, u) + w_i = y_i \in \Omega^c$ for all $i \in \mathbb{N}$. Since W is compact, there exists a convergent sub-sequence $\{w_{i_j}\}_{j=0}^{\infty}$ of $\{w_i\}_{i=0}^{\infty}$ with $w = \lim_{j\to\infty} w_{i_j} \in W$. Then with the continuity of f with respect to x under Assumption 6.1 or 6.2 we have

$$\begin{aligned}\lim_{j\to\infty} (f(x_{i_j}, u) + w_{i_j}) &= f(\lim_{j\to\infty} x_{i_j}, u) + \lim_{j\to\infty} w_{i_j} \\ &= f(x, u) + w = \lim_{j\to\infty} y_{i_j} = y \in \Omega^c\end{aligned}$$

if Ω is open. It follows that $x \in \left(\mathrm{Pre}^{\delta}(\Omega)\right)^c$ (i.e., for any $u \in U$ there exists $w \in W$ such that $f(x, u) + w \in \Omega^c$). Hence, $\mathrm{Pre}^{\delta}(\Omega)$ is open with respect to X since $x \in \left(\mathrm{Pre}^{\delta}(\Omega)\right)^c$ is closed. ■

C.6 Proof of Proposition 6.4

Proof: The set $K_{\Omega}^{\delta}(x)$ is trivially compact if U is finite under Assumption 6.2. Suppose that Assumption 6.1 holds. Let $x \in \mathrm{Pre}^{\delta}(\Omega)$ and $\{u_i\}_{i=0}^{\infty} \subseteq K_{\Omega}^{\delta}(x)$ be a convergent sequence with $u = \lim_{i\to\infty} u_i$. Then $f(x, u_i) \in \Omega - W$ for all $i \in \mathbb{N}$. Since $\Omega - W$ is compact, we can find a convergent sub-sequence $\{f(x, u_{i_k})\}_{k=0}^{\infty}$ with $\lim_{k\to\infty} f(x, u_{i_k}) \in \Omega - W$. By the continuity of f with respect to u, we have

$$\lim_{k\to\infty} f(x, u_{i_k}) = f(x, \lim_{k\to\infty} u_{i_k}) = f(x, u) \in \Omega - W,$$

which means $u \in K_{\Omega}^{\delta}(x)$. Therefore, the set $K_{\Omega}^{\delta}(x)$ is closed and hence compact for all $x \in \mathrm{Pre}^{\delta}(x)$. ■

Appendix D

Mappings of Locomotion Modes

Closed-form solutions of the phase-space manifolds are required to define the robustness margin sets in Definition 7.4. The followings are the closed-form solutions the locomotion modes presented in Section 7.7.1. A detailed derivation can be found in [248].

Proposition D.1 (PIPM phase-space tangent manifold) *Given the PIPM mode defined in (7.20) with initial conditions* $(x_0, \dot{x}_0) = (x_{\text{foot}}, \dot{x}_{\text{apex}})$ *and known foot placement* x_{foot}*, the phase-space tangent manifold is characterized by the states* $(x, \dot{x}, x_{\text{foot}}, \dot{x}_{\text{apex}})$ *such that*

$$\sigma(x, \dot{x}, x_{\text{foot}}, \dot{x}_{\text{apex}}) = \frac{\dot{x}^2_{\text{apex}}}{\omega^2_{\text{PIPM}}}\left(\dot{x}^2 - \dot{x}^2_{\text{apex}} - \omega^2_{\text{PIPM}}(x - x_{\text{foot}})^2\right), \quad \text{(D.1)}$$

where σ *denotes the Riemannian distance to the nominal phase-space manifold i.e.,* $\sigma = 0$*).*

Proposition D.2 (PIPM phase-space cotangent manifold) *Let* ζ_0 *be a nonnegative scaling value representing the initial phase of a cotangent manifold. Given the PIPM in (7.20) and a specific initial state* $(x_0, \dot{x}_0)$ *different from the keyframe* $(x_{\text{foot}}, \dot{x}_{\text{apex}})$*, the cotangent manifold is characterized by the states* $(x, \dot{x}, x_0, \dot{x}_0)$ *such that*

$$\zeta(x, \dot{x}, x_0, \dot{x}_0) = \zeta_0\left(\frac{\dot{x}}{\dot{x}_0}\right)^{\omega^2_{\text{PIPM}}} \frac{x - x_{\text{foot}}}{x_0 - x_{\text{foot}}}, \quad \text{(D.2)}$$

where ζ_0 *is chosen as the phase progression value at the keyframe state in this study.*

This cotangent manifold represents the arc length along the tangent manifold σ in Eq. (D.1). We use this cotangent manifold to quantify the length of a phase-space robustness margin. Detailed derivations of these two closed-form solutions above, i.e., $\sigma(x, \dot{x}, x_{\text{foot}}, \dot{x}_{\text{apex}}) = 0$ and $\zeta(x, \dot{x}, x_0, \dot{x}_0) = 0$, are provided in [248].

Proposition D.3 (PPM phase-space tangent manifold) *Given the PPM in (7.39) with initial conditions* $(x_0, \dot{x}_0) = (x_{\text{foot}}, \dot{x}_{\text{apex}})$ *and known arm placement* x_{foot}*, the PPM phase-space tangent manifold is defined as*

$$\sigma(x, \dot{x}, \dot{x}_{\text{apex}}, x_{\text{foot}}) = \frac{\dot{x}^2_{\text{apex}}}{-\omega^2_{\text{PPM}}}\left(\dot{x}^2 - \dot{x}^2_{\text{apex}} + \omega^2_{\text{PPM}}(x - x_{\text{foot}})^2\right), \quad \text{(D.3)}$$

DOI: 10.1201/9780429270253-D

Compared to the PIPM tangent manifold in Proposition D.1, the PPM tangent manifold has a negative asymptote slope square, i.e., $-\omega_{\text{PPM}}^2$. Thus, the tangent manifold with $\sigma > 0$ locates beneath the nominal $\sigma = 0$ tangent manifold. This property is in contrast to that of the PIPM tangent manifold.

Proposition D.4 (PPM phase-space cotangent manifold) *Given the PPM in (7.39), the PPM cotangent manifold is*

$$\zeta = \zeta_0\left(\frac{\dot{x}}{\dot{x}_0}\right)^{-\omega_{\text{PPM}}^2} \frac{x - x_{\text{foot}}}{x_0 - x_{\text{foot}}}, \tag{D.4}$$

Proposition D.5 (MCM phase-space tangent manifold) *Given the MCM with a constant acceleration ω_{MCM} (i.e., the control input), an initial condition $(x_0, \dot{x}_0) = (x_{\text{foot}}, \dot{x}_{\text{apex}})$, and a known foot placement x_{foot}, the MCM phase-space tangent manifold is*

$$\sigma(x, \dot{x}, x_{\text{foot}}, \dot{x}_{\text{apex}}) = 2\omega_{\text{MCM}}(x - x_{\text{apex}}) - (\dot{x}^2 - \dot{x}_{\text{apex}}^2), \tag{D.5}$$

where $\sigma = 0$ represents the nominal phase-space tangent manifold.

Proposition D.6 (MCM phase-space cotangent manifold) *Given the MCM with a constant acceleration and initial conditions $(x_0, \dot{x}_0) = (x_{\text{foot}}, \dot{x}_{\text{apex}})$ and known foot placement x_{foot}, the phase-space cotangent manifold is*

$$\zeta(x, \dot{x}, x_{\text{foot}}, \dot{x}_{\text{apex}}) = \omega_{\text{MCM}} \cdot \ln\left(\frac{\dot{x}}{\dot{x}_{\text{apex}}}\right) - (x - x_{\text{foot}}), \tag{D.6}$$

The phase-space manifolds of the hopping model are trivial since its tangent phase-space manifold is a horizontal line. The stop-launch model and sliding model have similar phase-space manifolds (i.e., parabolic trajectories) as those of the multi-contact model since all of them has a constant sagittal acceleration. Their derivations are omitted for brevity.

Bibliography

[1] Gotz Alefeld and Jurgen Herzberger. *Introduction to Interval Computation.* Academic Press, 1983.

[2] Charalambos D Aliprantis and Kim C Border. *Infinite Dimensional Analysis: A Hitchhiker's Guide.* Springer Science & Business Media, 2006.

[3] Rajeev Alur, Costas Courcoubetis, Nicolas Halbwachs, Thomas A Henzinger, P-H Ho, Xavier Nicollin, Alfredo Olivero, Joseph Sifakis, and Sergio Yovine. The algorithmic analysis of hybrid systems. *Theoretical Computer Science*, 138(1):3–34, 1995.

[4] Rajeev Alur, Thao Dang, and Franjo Ivančić. Reachability analysis of hybrid systems via predicate abstraction. In *International Workshop on Hybrid Systems: Computation and Control*, pages 35–48. Springer, 2002.

[5] Rajeev Alur, Radu Grosu, Yerang Hur, Vijay Kumar, and Insup Lee. Modular specification of hybrid systems in CHARON. In *International Workshop on Hybrid Systems: Computation and Control*, pages 6–19. Springer, 2000.

[6] Rajeev Alur, Thomas A Henzinger, Gerardo Lafferriere, and George J Pappas. Discrete abstractions of hybrid systems. *Proceedings of the IEEE*, 88(7):971–984, 2000.

[7] Rajeev Alur, Thomas A Henzinger, and Eduardo D Sontag, editors. *Hybrid Systems III.* Springer, 1996.

[8] Aaron D Ames, Samuel Coogan, Magnus Egerstedt, Gennaro Notomista, Koushil Sreenath, and Paulo Tabuada. Control barrier functions: Theory and applications. In *European Control Conference*, pages 3420–3431. IEEE, 2019.

[9] Aaron D Ames, Xiangru Xu, Jessy W Grizzle, and Paulo Tabuada. Control barrier function based quadratic programs for safety critical systems. *IEEE Transactions on Automatic Control*, 62(8):3861–3876, 2016.

[10] David Angeli. A Lyapunov approach to incremental stability properties. *IEEE Transactions on Automatic Control*, 47(3):410–421, 2002.

[11] Marco Antoniotti and Bud Mishra. Discrete event models + temporal logic = supervisory controller: Automatic synthesis of locomotion controllers. In *IEEE International Conference on Robotics and Automation*, pages 1441–1446. IEEE, 1995.

[12] Panos J Antsaklis, Wolf Kohn, Anil Nerode, and Shankar Sastry, editors. *Hybrid Systems II*. Springer, 1995.

[13] Zvi Artstein. Stabilization with relaxed controls. *Nonlinear Analysis: Theory, Methods & Applications*, 7(11):1163–1173, 1983.

[14] Eugene Asarin and Ahmed Bouajjani. Perturbed turing machines and hybrid systems. In *IEEE Symposium on Logic in Computer Science*, pages 269–278. IEEE, 2001.

[15] Eugene Asarin, Olivier Bournez, Thao Dang, Oded Maler, and Amir Pnueli. Effective synthesis of switching controllers for linear systems. *Proceedings of the IEEE*, 88(7):1011–1025, 2000.

[16] Eugene Asarin, Thao Dang, and Oded Maler. The d/dt tool for verification of hybrid systems. In *International Conference on Computer Aided Verification*, pages 365–370. Springer, 2002.

[17] Eugene Asarin, Oded Maler, and Amir Pnueli. Reachability analysis of dynamical systems having piecewise-constant derivatives. *Theoretical Computer Science*, 138(1):35–65, 1995.

[18] Eugene Asarin, Gerardo Schneider, and Sergio Yovine. Algorithmic analysis of polygonal hybrid systems, part I: Reachability. *Theoretical Computer Science*, 379(1-2):231–265, 2007.

[19] Karl Johan Astrom and Richard M Murray. *Feedback Systems: An Introduction for Scientists and Engineers*. Princeton, 2008.

[20] Hervé Audren, Joris Vaillant, Abderrahmane Kheddar, Adrien Escande, Kunihiko Kaneko, and Erika Yoshida. Model preview control in multi-contact motion-application to a humanoid robot. In *IEEE/RSJ International Conference on Intelligent Robots and Systems*, pages 4030–4035, 2014.

[21] Tomáš Babiak, Mojmír Křetínský, Vojtěch Řehák, and Jan Strejček. LTL to Büchi automata translation: Fast and more deterministic. In *International Conference on Tools and Algorithms for the Construction and Analysis of Systems*, pages 95–109. Springer, 2012.

[22] O O Badmus, S Chowdhury, and Carl N Nett. Nonlinear control of surge in axial compression systems. *Automatica*, 32(1):59–70, 1996.

[23] Christel Baier and Joost-Pieter Katoen. *Principles of Model Checking*. MIT Press, 2008.

[24] Calin Belta, Antonio Bicchi, Magnus Egerstedt, Emilio Frazzoli, Eric Klavins, and George Pappas. Symbolic planning and control of robot motion. *IEEE Robotics & Automation Magazine*, 14(1):61–70, 2007.

[25] Calin Belta, Boyan Yordanov, and Ebru Aydin Gol. *Formal Methods for Discrete-Time Dynamical Systems*. Springer, 2017.

[26] Alberto Bemporad, Giancarlo Ferrari-Trecate, and Manfred Morari. Observability and controllability of piecewise affine and hybrid systems. *IEEE Transactions on Automatic Control*, 45(10):1864–1876, 2000.

[27] Dimitri P Bertsekas. Infinite-time reachability of state-space regions by using feedback control. *IEEE Transactions on Automatic Control*, 17(5):604–613, 1972.

[28] Dimitri P Bertsekas and Ian B Rhodes. On the minimax reachability of target sets and target tubes. *Automatica*, 7(2):233–247, 1971.

[29] Amit Bhatia, Matthew R Maly, Lydia E Kavraki, and Moshe Y Vardi. Motion planning with complex goals. *IEEE Robotics & Automation Magazine*, 18(3):55–64, 2011.

[30] Franco Blanchini. Minimum-time control for uncertain discrete-time linear systems. In *IEEE Conference on Decision and Control*, pages 2629–2634, 1992.

[31] Franco Blanchini. Ultimate boundedness control for uncertain discrete-time systems via set-induced lyapunov functions. *IEEE Transactions on Automatic Control*, 39(2):428–433, 1994.

[32] Franco Blanchini. Set invariance in control. *Automatica*, 35(11):1747–1767, 1999.

[33] Roderick Bloem, Barbara Jobstmann, Nir Piterman, Amir Pnueli, and Yaniv Sa'ar. Synthesis of reactive (1) designs. *Journal of Computer and System Sciences*, 78(3):911–938, 2012.

[34] Jean-Michel Bony. Principe du maximum, inégalité de harnack et unicité du probleme de cauchy pour les opérateurs elliptiques dégénérés. *Annales de l'institut Fourier*, 19(1):277–304, 1969.

[35] Francesco Borrelli. *Constrained Optimal Control for Hybrid Systems*. Springer, 2003.

[36] Stephen Boyd, Laurent El Ghaoui, Eric Feron, and Venkataramanan Balakrishnan. *Linear Matrix Inequalities in System and Control Theory*. SIAM, 1994.

[37] Michael S Branicky, Vivek S Borkar, and Sanjoy K Mitter. A unified framework for hybrid control: Model and optimal control theory. *IEEE Transactions on Automatic Control*, 43(1):31–45, 1998.

[38] Haïm Brezis. On a characterization of flow-invariant sets. *Communications on Pure and Applied Mathematics*, 23(2):261–263, 1970.

[39] Mireille E Broucke. Reach control on simplices by continuous state feedback. *SIAM Journal on Control and Optimization*, 48(5):3482–3500, 2010.

[40] Mireille E Broucke and Marcus Ganness. Reach control on simplices by piecewise affine feedback. *SIAM Journal on Control and Optimization*, 52(5):3261–3286, 2014.

[41] L P Burton and William M Whyburn. Minimax solutions of ordinary differential systems. *Proceedings of the American Mathematical Society*, 3(5):794–803, 1952.

[42] Peter E Caines and Yuan-Jun Wei. Hierarchical hybrid control systems. In *Control Using Logic-Based Switching*, pages 39–48. Springer, 1997.

[43] Peter E Caines and Yuan-Jun Wei. Hierarchical hybrid control systems: A lattice theoretic formulation. *IEEE Transactions on Automatic Control*, 43(4):501–508, 1998.

[44] Cristian S Calude, Sanjay Jain, Bakhadyr Khoussainov, Wei Li, and Frank Stephan. Deciding parity games in quasi-polynomial time. *SIAM Journal on Computing*, 51(2):STOC17–152–STOC17–188, 2022.

[45] Christos G Cassandras and Stéphane Lafortune. *Introduction to Discrete Event Systems*. Springer, 2008.

[46] Graziano Chesi. Estimating the domain of attraction for non-polynomial systems via LMI optimizations. *Automatica*, 45(6):1536–1541, 2009.

[47] Alessandro Cimatti, Edmund Clarke, Enrico Giunchiglia, Fausto Giunchiglia, Marco Pistore, Marco Roveri, Roberto Sebastiani, and Armando Tacchella. Nusmv 2: An opensource tool for symbolic model checking. In *International Conference on Computer Aided Verification*, pages 359–364. Springer, 2002.

[48] Edmund M Clarke, E Allen Emerson, and Joseph Sifakis. Model checking: Algorithmic verification and debugging. *Communications of the ACM*, 52(11):74–84, 2009.

[49] Edmund M Clarke, Thomas A Henzinger, Helmut Veith, and Roderick Bloem. *Handbook of Model Checking*. Springer, 2018.

[50] Edmund M Clarke Jr, Orna Grumberg, Daniel Kroening, Doron Peled, and Helmut Veith. *Model Checking*. MIT Press, 2018.

[51] Earl A Coddington and Norman Levinson. *Theory of Ordinary Differential Equations*. Tata McGraw-Hill Education, 1955.

[52] Pieter Collins. Continuity and computability of reachable sets. *Theoretical Computer Science*, 341(1-3):162–195, 2005.

[53] Pieter Collins. On the computability of reachable and invariant sets. In *IEEE Conference on Decision and Control*, pages 4187–4192. IEEE, 2005.

[54] Charles C Conley. *Isolated Invariant Sets and the Morse Index*. American Mathematical Society, 1978.

[55] Samuel Coogan. Mixed monotonicity for reachability and safety in dynamical systems. In *IEEE Conference on Decision and Control*, pages 5074–5085. IEEE, 2020.

[56] Samuel Coogan and Murat Arcak. Efficient finite abstraction of mixed monotone systems. In *International Conference on Hybrid Systems: Computation and Control*, pages 58–67, 2015.

[57] José ER Cury, Bruce H Krogh, and Toshihiko Niinomi. Synthesis of supervisory controllers for hybrid systems based on approximating automata. *IEEE Transactions on Automatic Control*, 43(4):564–568, 1998.

[58] Thi Xuan Thao Dang. *Verification and Synthesis of Hybrid Systems*. PhD thesis, Institut National Polytechnique de Grenoble-INPG, 2000.

[59] Sumanth Dathathri and Richard M Murray. Decomposing GR(1) games with singleton liveness guarantees for efficient synthesis. In *IEEE Conference on Decision and Control*, pages 911–917, 2017.

[60] Jennifer M Davoren and Anil Nerode. Logics for hybrid systems. *Proceedings of the IEEE*, 88(7):985–1010, 2000.

[61] Alexandre Duret-Lutz. LTL translation improvements in Spot 1.0. *International Journal of Critical Computer-Based Systems 5*, 5(1-2):31–54, 2014.

[62] Alexandre Duret-Lutz, Alexandre Lewkowicz, Amaury Fauchille, Thibaud Michaud, Étienne Renault, and Laurent Xu. Spot 2.0 –a framework for LTL and ω-automata manipulation. In *Automated Technology for Verification and Analysis*, pages 122–129, 2016.

[63] Rüdiger Ehlers, Stéphane Lafortune, Stavros Tripakis, and Moshe Y Vardi. Supervisory control and reactive synthesis: A comparative introduction. *Discrete Event Dynamic Systems*, 27(2):209–260, 2017.

[64] Javier Esparza, Jan Křetínskỳ, Jean-François Raskin, and Salomon Sickert. From LTL and limit-deterministic Büchi automata to deterministic parity automata. In *International Conference on Tools and Algorithms for the Construction and Analysis of Systems*, pages 426–442. Springer, 2017.

[65] Javier Esparza, Jan Křetínský, and Salomon Sickert. A unified translation of linear temporal logic to ω-automata. *Journal of the ACM*, 67(6):1–61, 2020.

[66] Georgios E Fainekos, Antoine Girard, Hadas Kress-Gazit, and George J Pappas. Temporal logic motion planning for dynamic robots. *Automatica*, 45(2):343–352, 2009.

[67] Georgios E Fainekos, Hadas Kress-Gazit, and George J Pappas. Hybrid controllers for path planning: A temporal logic approach. In *IEEE Conference on Decision and Control*, pages 4885–4890. IEEE, 2005.

[68] Georgios E Fainekos, Hadas Kress-Gazit, and George J Pappas. Temporal logic motion planning for mobile robots. In *IEEE International Conference on Robotics and Automation*, pages 2020–2025. IEEE, 2005.

[69] Georgios E Fainekos and George J Pappas. Robustness of temporal logic specifications for continuous-time signals. *Theoretical Computer Science*, 410(42):4262–4291, 2009.

[70] Timm Faulwasser, Benjamin Kern, and Rolf Findeisen. Model predictive path-following for constrained nonlinear systems. In *IEEE Conference on Decision and Control held jointly with Chinese Control Conference*, pages 8642–8647, 2009.

[71] Aleksei Fedorovich Filippov. *Differential Equations with Discontinuous Righthand Sides*. Springer Science & Business Media, 1988.

[72] Martin Fränzle. Analysis of hybrid systems: An ounce of realism can save an infinity of states. In *International Workshop on Computer Science Logic*, pages 126–139. Springer, 1999.

[73] Martin Fränzle. What will be eventually true of polynomial hybrid automata? In *International Symposium on Theoretical Aspects of Computer Software*, pages 340–359. Springer, 2001.

[74] Emilio Frazzoli, Munther A Dahleh, Emilio Frazzoli, Munther A Dahleh, and Eric Feron. Maneuver-based motion planning for nonlinear systems with symmetries. *IEEE Transactions on Robotics*, 21(6):1077–1091, 2005.

[75] Randy Freeman and Petar V Kokotovic. *Robust Nonlinear Control Design: State-Space and Lyapunov Techniques*. Springer Science & Business Media, 2008.

[76] Goran Frehse. PHAVer: Algorithmic verification of hybrid systems past HyTech. *International Journal on Software Tools for Technology Transfer*, 10(3):263–279, 2008.

[77] Goran Frehse, Colas Le Guernic, Alexandre Donzé, Scott Cotton, Rajarshi Ray, Olivier Lebeltel, Rodolfo Ripado, Antoine Girard, Thao Dang, and Oded Maler. SpaceEx: Scalable verification of hybrid systems. In *International Conference on Computer Aided Verification*, pages 379–395. Springer, 2011.

[78] Laurent Fribourg and Romain Soulat. *Control of Switching Systems by Invariance Analysis: Application to Power Electronics*. Wiley-ISTE, 2013.

[79] Andreas Gaiser, Jan Křetínský, and Javier Esparza. Rabinizer: Small deterministic automata for LTL (F, G). In *International Symposium on Automated Technology for Verification and Analysis*, pages 72–76. Springer, 2012.

[80] Sicun Gao, Jeremy Avigad, and Edmund M Clarke. δ-Complete decision procedures for satisfiability over the reals. In *International Joint Conference on Automated Reasoning*, pages 286–300. Springer, 2012.

[81] Paul Gastin and D Oddoux. LTL 2 BA: Fast translation from LTL formulae to Büchi automata, 2001.

[82] Giuseppe De Giacomo and Moshe Y Vardi. Automata-theoretic approach to planning for temporally extended goals. In *European Conference on Planning*, pages 226–238. Springer, 1999.

[83] Antoine Girard. Controller synthesis for safety and reachability via approximate bisimulation. *Automatica*, 48(5):947–953, 2012.

[84] Antoine Girard, A Agung Julius, and George J Pappas. Approximate simulation relations for hybrid systems. *Discrete event dynamic systems*, 18(2):163–179, 2008.

[85] Antoine Girard and George J Pappas. Approximation metrics for discrete and continuous systems. *IEEE Transactions on Automatic Control*, 52(5):782–798, 2007.

[86] Antoine Girard, Giordano Pola, and Paulo Tabuada. Approximately bisimilar symbolic models for incrementally stable switched systems. *IEEE Transactions on Automatic Control*, 55(1):116–126, 2010.

[87] Antoine Girard, Giordano Pola, and Paulo Tabuada. Approximately bisimilar symbolic models for incrementally stable switched systems. *IEEE Transactions on Automatic Control*, 55(1):116–126, 2010.

[88] Fausto Giunchiglia and Paolo Traverso. Planning as model checking. In *European Conference on Planning*, pages 1–20. Springer, 1999.

[89] Rafal Goebel, Ricardo G Sanfelice, and Andrew R Teel. *Hybrid Dynamical Systems: Modeling, Stability, and Robustness.* Princeton University Press, 2012.

[90] Eric Goubault and Sylvie Putot. Forward inner-approximated reachability of non-linear continuous systems. In *International Conference on Hybrid Systems: Computation and Control*, pages 1–10, 2017.

[91] Nicolas Gourdain, Frédéric Sicot, Florent Duchaine, and Laurent Gicquel. Large eddy simulation of flows in industrial compressors: A path from 2015 to 2035. *Philosophical Transactions of the Royal Society A*, 372(2022):20130323, 2014.

[92] Erich Gradel and Wolfgang Thomas. *Automata, Logics, and Infinite Games: A Guide to Current Research.* Springer Science & Business Media, 2002.

[93] Robert L Grossman, Anil Nerode, Anders P Ravn, and Hans Rischel, editors. *Hybrid Systems.* Springer, 1994.

[94] Colas Le Guernic and Antoine Girard. Reachability analysis of hybrid systems using support functions. In *International Conference on Computer Aided Verification*, pages 540–554. Springer, 2009.

[95] Sumit Gulwani and Ashish Tiwari. Constraint-based approach for analysis of hybrid systems. In *International Conference on Computer Aided Verification*, pages 190–203. Springer, 2008.

[96] Per-Olof Gutman and Michael Cwikel. Admissible sets and feedback control for discrete-time linear dynamical systems with bounded controls and states. *IEEE Transactions on Automatic Control*, 31(4):373–376, 1986.

[97] Luc CGJM Habets, Pieter J Collins, and Jan H van Schuppen. Reachability and control synthesis for piecewise-affine hybrid systems on simplices. *IEEE Transactions on Automatic Control*, 51(6):938–948, 2006.

[98] Wassim M Haddad and VijaySekhar Chellaboina. *Nonlinear Dynamical Systems and Control.* Princeton University Press, 2011.

[99] Jack K Hale. *Ordinary Differential Equations.* John Wiley and Sons, 1969.

[100] Keliang He, Morteza Lahijanian, Lydia E Kavraki, and Moshe Y Vardi. Towards manipulation planning with temporal logic specifications. In *IEEE International Conference on Robotics and Automation*, pages 346–352, 2015.

[101] Thomas A Henzinger, Pei-Hsin Ho, and Howard Wong-Toi. HyTech: A model checker for hybrid systems. In *International Conference on Computer Aided Verification*, pages 460–463. Springer, 1997.

[102] Thomas A Henzinger, Pei-Hsin Ho, and Howard Wong-Toi. Algorithmic analysis of nonlinear hybrid systems. *IEEE Transactions on Automatic Control*, 43(4):540–554, 1998.

[103] Thomas A Henzinger, Benjamin Horowitz, Rupak Majumdar, and Howard Wong-Toi. Beyond HyTech: Hybrid systems analysis using interval numerical methods. In *International Workshop on Hybrid Systems: Computation and Control*, pages 130–144. Springer, 2000.

[104] Thomas A Henzinger, Peter W Kopke, Anuj Puri, and Pravin Varaiya. What's decidable about hybrid automata? *Journal of Computer and System Sciences*, 57(1):94–124, 1998.

[105] Gerard J Holzmann. The model checker SPIN. *IEEE Transactions on Software Engineering*, 23(5):279–295, 1997.

[106] Florian Horn. Streett games on finite graphs. In *Games in Design Verification*, 2005.

[107] Kyle Hsu, Rupak Majumdar, Kaushik Mallik, and Anne-Kathrin Schmuck. Multi-layered abstraction-based controller synthesis for continuous-time systems. In *International Conference on Hybrid Systems: Computation and Control*, pages 120–129, 2018.

[108] Luc Jaulin. *Applied Interval Analysis: with Examples in Parameter and State Estimation, Robust Control and Robotics.* Springer, 2001.

[109] Shengbing Jiang and Ratnesh Kumar. Supervisory control of discrete event systems with CTL* temporal logic specifications. *SIAM Journal on Control and Optimization*, 44(6):2079–2103, 2006.

[110] A Agung Julius, Georgios E Fainekos, Madhukar Anand, Insup Lee, and George J Pappas. Robust test generation and coverage for hybrid systems. In *International Workshop on Hybrid Systems: Computation and Control*, pages 329–342. Springer, 2007.

[111] Erich Kamke. Zur theorie der systeme gewöhnlicher differentialgleichungen. ii. *Acta Mathematica*, 58(1):57–85, 1932.

[112] Eric C Kerrigan. *Robust Constraint Satisfaction: Invariant Sets and Predictive Control.* PhD thesis, Department of Engineering, University of Cambridge, 2000.

[113] Mahmoud Khaled, Eric S Kim, Murat Arcak, and Majid Zamani. Synthesis of symbolic controllers: A parallelized and sparsity-aware approach. In *Tools and Algorithms for the Construction and Analysis of Systems*, page 265–281, 2019.

[114] Hassan K Khalil. *Nonlinear Systems.* Patience Hall, 2002.

[115] Eric S Kim, Murat Arcak, and Majid Zamani. Constructing control system abstractions from modular components. In *International Conference on Hybrid Systems: Computation and Control*, pages 137–146, 2018.

[116] Marius Kloetzer and Calin Belta. Temporal logic planning and control of robotic swarms by hierarchical abstractions. *IEEE Transactions on Robotics*, 23(2):320–330, 2007.

[117] Marius Kloetzer and Calin Belta. A fully automated framework for control of linear systems from temporal logic specifications. *IEEE Transactions on Automatic Control*, 53(1):287–297, 2008.

[118] Marius Kloetzer and Calin Belta. Automatic deployment of distributed teams of robots from temporal logic motion specifications. *IEEE Transactions on Robotics*, 26(1):48–61, 2009.

[119] Pascal Koiran, Michel Cosnard, and Max Garzon. Computability with low-dimensional dynamical systems. *Theoretical Computer Science*, 132(1-2):113–128, 1994.

[120] Ilya Kolmanovsky and Elmer G Gilbert. Theory and computation of disturbance invariant sets for discrete-time linear systems. *Mathematical Problems in Engineering*, 4(4):317–367, 1998.

[121] Zuzana Komárková and Jan Křetínskỳ. Rabinizer 3: Safraless translation of LTL to small deterministic automata. In *International Symposium on Automated Technology for Verification and Analysis*, pages 235–241. Springer, 2014.

[122] Soonho Kong, Sicun Gao, Wei Chen, and Edmund Clarke. dReach: δ-Reachability analysis for hybrid systems. In *International Conference on Tools and Algorithms for the Construction and Analysis of Systems*, pages 200–205. Springer, 2015.

[123] Hadas Kress-Gazit. *Transforming High Level Tasks to Low Level Controllers.* PhD thesis, University of Pennsylvania, 2008.

[124] Hadas Kress-Gazit, Georgios E Fainekos, and George J Pappas. Where's waldo? sensor-based temporal logic motion planning. In *IEEE International Conference on Robotics and Automation*, pages 3116–3121. IEEE, 2007.

[125] Hadas Kress-Gazit, Georgios E Fainekos, and George J Pappas. Temporal-logic-based reactive mission and motion planning. *IEEE Transactions on Robotics*, 25(6):1370–1381, 2009.

[126] Hadas Kress-Gazit, Morteza Lahijanian, and Vasumathi Raman. Synthesis for robots: Guarantees and feedback for robot behavior. *Annual Review of Control, Robotics, and Autonomous Systems*, 1:211–236, 2018.

[127] Hadas Kress-Gazit, Tichakorn Wongpiromsarn, and Ufuk Topcu. Correct, reactive, high-level robot control. *IEEE Robotics & Automation Magazine*, 18(3):65–74, 2011.

[128] Jan Křetínský and Ruslan Ledesma Garza. Rabinizer 2: Small deterministic automata for LTL\ GU. In *Automated Technology for Verification and Analysis*, pages 446–450. Springer, 2013.

[129] Jan Křetínský, Tobias Meggendorfer, Salomon Sickert, and Christopher Ziegler. Rabinizer 4: From LTL to your favourite deterministic automaton. In *International Conference on Computer Aided Verification*, pages 567–577. Springer, 2018.

[130] Orna Kupferman and Mohse Y Vardi. Freedom, weakness, and determinism: From linear-time to branching-time. In *IEEE Symposium on Logic in Computer Science*, pages 81–92. IEEE, 1998.

[131] Gerardo Lafferriere, George J Pappas, and Shankar Sastry. O-minimal hybrid systems. *Mathematics of Control, Signals and Systems*, 13(1):1–21, 2000.

[132] Gerardo Lafferriere, George J Pappas, and Sergio Yovine. A new class of decidable hybrid systems. In *International Workshop on Hybrid Systems: Computation and Control*, pages 137–151. Springer, 1999.

[133] V Lakshmikantham and S Leela. *Differential and Integral Inequalities: Theory and Applications: Volume I: Ordinary Differential Equations*. Academic Press, 1969.

[134] Shengbo Li, Keqiang Li, Rajesh Rajamani, and Jianqiang Wang. Model predictive multi-objective vehicular adaptive cruise control. *IEEE Transactions on Control Systems Technology*, 19(3):556–566, 2011.

[135] Yinan Li. *Robustly Complete Temporal Logic Control Synthesis for Nonlinear Systems*. PhD thesis, University of Waterloo, 2019.

[136] Yinan Li and Jun Liu. Switching control of differential-algebraic equations with temporal logic specifications. In *American Control Conference*, pages 1941–1946. IEEE, 2015.

[137] Yinan Li and Jun Liu. Computing maximal invariant sets for switched nonlinear systems. In *IEEE Conference on Computer Aided Control System Design*, pages 862–867. IEEE, 2016.

[138] Yinan Li and Jun Liu. An interval analysis approach to invariance control synthesis for discrete-time switched systems. In *IEEE Conference on Decision and Control*, pages 6388–6394. IEEE, 2016.

[139] Yinan Li and Jun Liu. Invariance control synthesis for switched nonlinear systems: An interval analysis approach. *IEEE Transactions on Automatic Control*, 63(7):2206–2211, 2018.

[140] Yinan Li and Jun Liu. Robustly complete reach-and-stay control synthesis for switched systems via interval analysis. In *American Control Conference*, pages 2350–2355. IEEE, 2018.

[141] Yinan Li and Jun Liu. ROCS: A robustly complete control synthesis tool for nonlinear dynamical systems. In *International Conference on Hybrid Systems: Computation and Control*, pages 130–135, 2018.

[142] Yinan Li and Jun Liu. Robustly complete synthesis of memoryless controllers for nonlinear systems with reach-and-stay specifications. *IEEE Transactions on Automatic Control*, 66(3):1199–1206, 2021.

[143] Yinan Li and Jun Liu. Robustly complete synthesis of sampled-data control for continuous-time nonlinear systems with reach-and-stay objectives. *Nonlinear Analysis: Hybrid Systems*, 44:101170, 2022.

[144] Yinan Li, Jun Liu, and Necmiye Ozay. Computing finite abstractions with robustness margins via local reachable set over-approximation. *IFAC-PapersOnLine*, 48(27):1–6, 2015.

[145] Yinan Li, Ebrahim Moradi Shahrivar, and Jun Liu. Safe linear temporal logic motion planning in dynamic environments. In *IEEE/RSJ International Conference on Intelligent Robots and Systems*, pages 9818–9825. IEEE, 2021.

[146] Yinan Li, Zhibing Sun, and Jun Liu. ROCS 2.0: An integrated temporal logic control synthesis tool for nonlinear dynamical systems. *IFAC-PapersOnLine*, 54(5):31–36, 2021.

[147] Yinan Li, Zhibing Sun, and Jun Liu. A specification-guided framework for temporal logic control of nonlinear systems. *IEEE Transactions on Automatic Control*, 2022, in press.

[148] Daniel Liberzon. *Switching in Systems and Control*, volume 190. Springer, 2003.

[149] Feng Lin. Analysis and synthesis of discrete event systems using temporal logic. In *IEEE International Symposium on Intelligent Control*, pages 140–145. IEEE, 1991.

[150] Hai Lin. Mission accomplished: An introduction to formal methods in mobile robot motion planning and control. *Unmanned Systems*, 2(02):201–216, 2014.

[151] Hai Lin and Panos J Antsaklis. Hybrid dynamical systems: An introduction to control and verification. *Foundations and Trends® in Systems and Control*, 1(1):1–172, 2014.

[152] Jun Liu. Robust abstractions for control synthesis: Completeness via robustness for linear-time properties. In *International Conference on Hybrid Systems: Computation and Control*, pages 101–110, 2017.

[153] Jun Liu. Closing the gap between discrete abstractions and continuous control: Completeness via robustness and controllability. In *International Conference on Formal Modeling and Analysis of Timed Systems*, pages 67–83. Springer, 2021.

[154] Jun Liu and Necmiye Ozay. Abstraction, discretization, and robustness in temporal logic control of dynamical systems. In *International Conference on Hybrid Systems: Computation and Control*, pages 293–302, 2014.

[155] Jun Liu and Necmiye Ozay. Finite abstractions with robustness margins for temporal logic-based control synthesis. *Nonlinear Analysis: Hybrid Systems*, 22:1–15, 2016.

[156] Jun Liu, Necmiye Ozay, Ufuk Topcu, and Richard M Murray. Synthesis of reactive switching protocols from temporal logic specifications. *IEEE Transactions on Automatic Control*, 58(7):1771–1785, 2013.

[157] Jun Liu and Pavithra Prabhakar. Switching control of dynamical systems from metric temporal logic specifications. In *IEEE International Conference on Robotics and Automation*, pages 5333–5338. IEEE, 2014.

[158] Jun Liu, Ufuk Topcu, Necmiye Ozay, and Richard M Murray. Reactive controllers for differentially flat systems with temporal logic constraints. In *IEEE Conference on Decision and Control*, pages 7664–7670. IEEE, 2012.

[159] Savvas G Loizou and Kostas J Kyriakopoulos. Automatic synthesis of multi-agent motion tasks based on ltl specifications. In *IEEE Conference on Decision and Control*, pages 153–158. IEEE, 2004.

[160] Matt Luckcuck, Marie Farrell, Louise A Dennis, Clare Dixon, and Michael Fisher. Formal specification and verification of autonomous robotic systems: A survey. *ACM Computing Surveys*, 52(5):1–41, 2019.

[161] Jan Lunze and Françoise Lamnabhi-Lagarrigue. *Handbook of Hybrid Systems Control: Theory, Tools, Applications*. Cambridge University Press, 2009.

[162] John Lygeros, Claire Tomlin, and Shankar Sastry. Controllers for reachability specifications for hybrid systems. *Automatica*, 35(3):349–370, 1999.

[163] Oded Maler, Amir Pnueli, and Joseph Sifakis. On the synthesis of discrete controllers for timed systems. In *Annual Symposium on Theoretical Aspects of Computer Science*, pages 229–242. Springer, 1995.

[164] Yiming Meng, Yinan Li, Maxwell Fitzsimmons, and Jun Liu. Smooth converse Lyapunov-barrier theorems for asymptotic stability with safety constraints and reach-avoid-stay specifications. *Automatica*, 2022, in press.

[165] Yiming Meng, Yinan Li, and Jun Liu. Control of nonlinear systems with reach-avoid-stay specifications: A Lyapunov-barrier approach with an application to the moore-greizer model. In *American Control Conference*, pages 2284–2291. IEEE, 2021.

[166] Pierre-Jean Meyer, Alex Devonport, and Murat Arcak. TIRA: Toolbox for interval reachability analysis. In *ACM International Conference on Hybrid Systems: Computation and Control*, pages 224–229, 2019.

[167] Pierre-Jean Meyer and Dimos V Dimarogonas. Hierarchical decomposition of LTL synthesis problem for nonlinear control systems. *IEEE Transactions on Automatic Control*, 64(11):4676–4683, 2019.

[168] Ian Mitchell and Claire J Tomlin. Level set methods for computation in hybrid systems. In *International Workshop on Hybrid Systems: Computation and Control*, pages 310–323. Springer, 2000.

[169] Felipe J Montana, Jun Liu, and Tony J Dodd. Sampling-based stochastic optimal control with metric interval temporal logic specifications. In *IEEE Conference on Control Applications*, pages 767–773. IEEE, 2016.

[170] Felipe J Montana, Jun Liu, and Tony J Dodd. Sampling-based path planning for multi-robot systems with co-safe linear temporal logic specifications. In *Critical Systems: Formal Methods and Automated Verification*, pages 150–164. Springer, 2017.

[171] Felipe J Montana, Jun Liu, and Tony J Dodd. Sampling-based reactive motion planning with temporal logic constraints and imperfect state information. In *Critical Systems: Formal Methods and Automated Verification*, pages 134–149. Springer, 2017.

[172] Cristopher Moore. Unpredictability and undecidability in dynamical systems. *Physical Review Letters*, 64(20):2354, 1990.

[173] Ramon E Moore. *Interval Analysis*. Prentice-Hall, 1966.

[174] Ramon E Moore. *Methods and Applications of Interval Analysis.* SIAM, 1979.

[175] Richard M Murray, S Shankar Sastry, and Li Zexiang. *A Mathematical Introduction to Robotic Manipulation.* CRC Press, 1994.

[176] Mitio Nagumo. Über die lage der integralkurven gewöhnlicher differentialgleichungen. *Proceedings of the Physico-Mathematical Society of Japan. 3rd Series*, 24:551–559, 1942.

[177] Nedialko S Nedialkov, Kenneth R Jackson, and John D Pryce. An effective high-order interval method for validating existence and uniqueness of the solution of an ivp for an ode. *Reliable Computing*, 7(6):449–465, 2001.

[178] Petter Nilsson, Omar Hussien, Ayca Balkan, Yuxiao Chen, Aaron D Ames, Jessy W Grizzle, Necmiye Ozay, Huei Peng, and Paulo Tabuada. Correct-by-construction adaptive cruise control: Two approaches. *IEEE Transactions on Control System Technology*, 24(4):1294–1307, 2016.

[179] Petter Nilsson, Necmiye Ozay, and Jun Liu. Augmented finite transition systems as abstractions for control synthesis. *Discrete Event Dynamic Systems*, 27(2):301–340, 2017.

[180] Melkior Ornik, Miad Moarref, and Mireille E Broucke. An automated parallel parking strategy using reach control theory. *IFAC-PapersOnLine*, 50(1):9089–9094, 2017.

[181] Necmiye Ozay, Jun Liu, Pavithra Prabhakar, and Richard M Murray. Computing augmented finite transition systems to synthesize switching protocols for polynomial switched systems. In *American Control Conference*, pages 6237–6244. IEEE, 2013.

[182] Nir Peterman and Amir Pnueli. Faster solutions of Rabin and Streett games. In *IEEE Symposium on Logic in Computer Science*, pages 275–284, 2006.

[183] Nir Piterman, Amir Pnueli, and Yaniv Sa'ar. Synthesis of reactive (1) designs. In *International Workshop on Verification, Model Checking, and Abstract Interpretation*, pages 364–380. Springer, 2006.

[184] Erion Plaku and Sertac Karaman. Motion planning with temporal-logic specifications: Progress and challenges. *AI communications*, 29(1):151–162, 2016.

[185] Amir Pnueli. The temporal logic of programs. In *Annual Symposium on Foundations of Computer Science*, pages 46–57. IEEE, 1977.

[186] Amir Pnueli and Roni Rosner. On the synthesis of a reactive module. In *ACM Symposium on Principles of Programming Languages*, pages 179–190, 1989.

[187] Giordano Pola and Maria Domenica Di Benedetto. Control of cyber-physical-systems with logic specifications: A formal methods approach. *Annual Reviews in Control*, 47:178–192, 2019.

[188] Giordano Pola, Antoine Girard, and Paulo Tabuada. Approximately bisimilar symbolic models for nonlinear control systems. *Automatica*, 44(10):2508–2516, 2008.

[189] Pavithra Prabhakar, Parasara Sridhar Duggirala, Sayan Mitra, and Mahesh Viswanathan. Hybrid automata-based cegar for rectangular hybrid systems. *Formal Methods in System Design*, 46(2):105–134, 2015.

[190] Pavithra Prabhakar, Geir Dullerud, and Mahesh Viswanathan. Pre-orders for reasoning about stability. In *ACM International Conference on Hybrid Systems: Computation and Control*, pages 197–206, 2012.

[191] Pavithra Prabhakar and Miriam Garcia Soto. Abstraction based model-checking of stability of hybrid systems. In *International Conference on Computer Aided Verification*, pages 280–295. Springer, 2013.

[192] Pavithra Prabhakar, Jun Liu, and Richard M Murray. Pre-orders for reasoning about stability properties with respect to input of hybrid systems. In *International Conference on Embedded Software*, pages 1–10. IEEE, 2013.

[193] Stephen Prajna. Barrier certificates for nonlinear model validation. *Automatica*, 42(1):117–126, 2006.

[194] Stephen Prajna and Ali Jadbabaie. Safety verification of hybrid systems using barrier certificates. In *International Workshop on Hybrid Systems: Computation and Control*, pages 477–492. Springer, 2004.

[195] Stephen Prajna, Antonis Papachristodoulou, and Pablo A Parrilo. Introducing sostools: A general purpose sum of squares programming solver. In *IEEE Conference on Decision and Control*, pages 741–746. IEEE, 2002.

[196] Michael Melholt Quottrup, Thomas Bak, and RI Zamanabadi. Multi-robot planning: A timed automata approach. In *IEEE International Conference on Robotics and Automation*, pages 4417–4422. IEEE, 2004.

[197] Jörg Raisch and Siu O'Young. A totally ordered set of discrete abstractions for a given hybrid or continuous system. In *Hybrid Systems IV*, pages 342–360. Springer, 1996.

[198] Jörg Raisch and Siu D O'Young. Discrete approximation and supervisory control of continuous systems. *IEEE Transactions on Automatic Control*, 43(4):569–573, 1998.

[199] Saša V Rakovic, Eric C Kerrigan, David Q Mayne, and John Lygeros. Reachability analysis of discrete-time systems with disturbances. *IEEE Transactions on Automatic Control*, 51(4):546–561, 2006.

[200] Peter J Ramadge and W Murray Wonham. Supervisory control of a class of discrete event processes. *SIAM Journal on Control and Optimization*, 25(1):206–230, 1987.

[201] R M Redheffer. The theorems of Bony and Brezis on flow-invariant sets. *The American Mathematical Monthly*, 79(7):740–747, 1972.

[202] Gunther Reißig. Computing abstractions of nonlinear systems. *IEEE Transactions on Automatic Control*, 56(11):2583–2598, 2011.

[203] Gunther Reissig and Matthias Rungger. Feedback refinement relations for symbolic controller synthesis. In *IEEE Conference on Decision and Control*, pages 88–94. IEEE, 2014.

[204] Gunther Reissig, Alexander Weber, and Matthias Rungger. Feedback refinement relations for the synthesis of symbolic controllers. *IEEE Transactions on Automatic Control*, 62(4):1781–1796, 2016.

[205] Michael Rinehart, Munther A Dahleh, Dennis Reed, and Ilya Kolmanovsky. Suboptimal control of switched systems with an application to the disc engine. *IEEE Transactions on Control System Technology*, 16(2):189–201, 2008.

[206] R Tyrrell Rockafellar and Roger J-B Wets. *Variational Analysis*. Springer, 2009.

[207] Roni Rosner. *Modular Synthesis of Reactive Systems*. PhD thesis, The Weizmann Institute of Science, 1991.

[208] Bartek Roszak and Mireille E Broucke. Necessary and sufficient conditions for reachability on a simplex. *Automatica*, 42(11):1913–1918, 2006.

[209] Halsey L Royen and Patrick M Fitzpatrick. *Real Analysis*. Prentice-Hall, 2010.

[210] Walter Rudin. *Principles of Mathematical Analysis*. McGraw-Hill, third edition, 1976.

[211] Matthias Rungger and Paulo Tabuada. Computing robust controlled invariant sets of linear systems. *IEEE Transactions on Automatic Control*, 62(7):3665–3670, 2017.

[212] S Safra. On the complexity of omega-automata. In *Annual Symposium on Foundations of Computer Science*, pages 319–327, 1988.

[213] Matthew Senesky, Gabriel Eirea, and T John Koo. Hybrid modelling and control of power electronics. In *International Conference on Hybrid Systems: Computation and Control*, page 450–465, 2003.

[214] Bruno Siciliano, Lorenzo Sciavicco, Luigi Villani, and Giuseppe Oriolo. *Robotics: Modelling, Planning and Control.* Springer, 2010.

[215] Hava T Siegelmann and Eduardo D Sontag. On the computational power of neural nets. *Journal of Computer and System Sciences*, 50(1):132–150, 1995.

[216] Eduardo D Sontag. A Lyapunov-like characterization of asymptotic controllability. *SIAM Journal on Control and Optimization*, 21(3):462–471, 1983.

[217] Eduardo D Sontag. A 'universal' construction of Artstein's theorem on nonlinear stabilization. *Systems & Control Letters*, 13(2):117–123, 1989.

[218] Eduardo D Sontag. *Mathematical Control Theory: Deterministic Finite Dimensional Systems.* Springer Science & Business Media, 2013.

[219] Olaf Stursberg and Bruce H Krogh. Efficient representation and computation of reachable sets for hybrid systems. In *International Workshop on Hybrid Systems: Computation and Control*, pages 482–497. Springer, 2003.

[220] Fei Sun, Necmiye Ozay, Eric M Wolff, Jun Liu, and Richard M Murray. Efficient control synthesis for augmented finite transition systems with an application to switching protocols. In *American Control Conference*, pages 3273–3280. IEEE, 2014.

[221] Jacek Szarski. *Differential Inequalities.* Instytut Matematyczny Polskiej Akademi Nauk (Warszawa), 1965.

[222] Paulo Tabuada. *Verification and Control of Hybrid Systems: A Symbolic Approach.* Springer Science & Business Media, 2009.

[223] Paulo Tabuada and George J Pappas. Linear time logic control of discrete-time linear systems. *IEEE Transactions on Automatic Control*, 51(12):1862–1877, 2006.

[224] Yuichi Tazaki and Jun-ichi Imura. Discrete abstractions of nonlinear systems based on error propagation analysis. *IEEE Transactions on Automatic Control*, 57(3):550–564, 2011.

[225] J G Thistle and W M Wonham. Control problems in a temporal logic framework. *International Journal of Control*, 44(4):943–976, 1986.

[226] John G Thistle and W Murray Wonham. Control of infinite behavior of finite automata. *SIAM Journal on Control and Optimization*, 32(4):1075–1097, 1994.

[227] John G Thistle and W Murray Wonham. Supervision of infinite behavior of discrete-event systems. *SIAM Journal on Control and Optimization*, 32(4):1098–1113, 1994.

[228] Wolfgang Thomas. Automata on infinite objects. In *Formal Models and Semantics*, pages 133–191. Elsevier, 1990.

[229] Bernd Tibken and O Hachicho. Estimation of the domain of attraction for polynomial systems using multidimensional grids. In *IEEE Conference on Decision and Control*, volume 4, pages 3870–3874, 2000.

[230] Claire Tomlin, John Lygeros, and Shankar Sastry. Synthesizing controllers for nonlinear hybrid systems. In *International Workshop on Hybrid Systems: Computation and Control*, pages 360–373. Springer, 1998.

[231] Claire J Tomlin, John Lygeros, and S Shankar Sastry. A game theoretic approach to controller design for hybrid systems. *Proceedings of the IEEE*, 88(7):949–970, 2000.

[232] Claire J Tomlin, Ian Mitchell, Alexandre M Bayen, and Meeko Oishi. Computational techniques for the verification of hybrid systems. *Proceedings of the IEEE*, 91(7):986–1001, 2003.

[233] Ufuk Topcu, Andrew K Packard, Peter Seiler, and Gary J Balas. Robust region of attraction estimation. *IEEE Transactions on Automatic Control*, 55(1):137–142, 2010.

[234] Marc Toussaint, Kelsey R Allen, Kevin A Smith, and Joshua B Tenenbaum. Differentiable physics and stable modes for tool-use and manipulation planning – extended abtract. In *International Joint Conference on Artificial Intelligence*, pages 6231–6235, 2019.

[235] AJ van der Schaft and JM Schumacher. *An Introduction to Hybrid Dynamical Systems*. Springer Verlag, 2000.

[236] Cristian Ioan Vasile and Calin Belta. Reactive sampling-based temporal logic path planning. In *IEEE International Conference on Robotics and Automation*, pages 4310–4315, 2014.

[237] Wolfgang Walter. *Differential and Integral Inequalities*. Springer, 1970.

[238] Peter Wieland and Frank Allgöwer. Constructive safety using control barrier functions. *IFAC Proceedings Volumes*, 40(12):462–467, 2007.

[239] Tichakorn Wongpiromsarn, Ufuk Topcu, and Richard M Murray. Receding horizon temporal logic planning for dynamical systems. In *IEEE Conference on Decision and Control held jointly with Chinese Control Conference*, pages 5997–6004. IEEE, 2009.

[240] Tichakorn Wongpiromsarn, Ufuk Topcu, and Richard M Murray. Receding horizon control for temporal logic specifications. In *ACM International Conference on Hybrid Systems: Computation and Control*, pages 101–110, 2010.

[241] Tichakorn Wongpiromsarn, Ufuk Topcu, and Richard M Murray. Receding horizon temporal logic planning. *IEEE Transactions on Automatic Control*, 57(11):2817–2830, 2012.

[242] Tichakorn Wongpiromsarn, Ufuk Topcu, Necmiye Ozay, Huan Xu, and Richard M Murray. TuLiP: A software toolbox for receding horizon temporal logic planning. In *International Conference on Hybrid Systems: Computation and Control*, pages 313–314, 2011.

[243] W Murray Wonham and Peter J Ramadge. On the supremal controllable sublanguage of a given language. *SIAM Journal on Control and Optimization*, 25(3):637–659, 1987.

[244] MingQing Xiao. Quantitative characteristic of rotating stall and surge for moore–greitzer PDE model of an axial flow compressor. *SIAM Journal on Applied Dynamical Systems*, 7(1):39–62, 2008.

[245] Xiangru Xu, Paulo Tabuada, Jessy W Grizzle, and Aaron D Ames. Robustness of control barrier functions for safety critical control. *IFAC-PapersOnLine*, 48(27):54–61, 2015.

[246] Liren Yang, Oscar Mickelin, and Necmiye Ozay. On sufficient conditions for mixed monotonicity. *IEEE Transactions on Automatic Control*, 64(12):5080–5085, 2019.

[247] Majid Zamani, Giordano Pola, Manuel Mazo, and Paulo Tabuada. Symbolic models for nonlinear control systems without stability assumptions. *IEEE Transactions on Automatic Control*, 57(7):1804–1809, 2012.

[248] Ye Zhao, Benito R Fernandez, and Luis Sentis. Robust optimal planning and control of non-periodic bipedal locomotion with a centroidal momentum model. *The International Journal of Robotics Research*, 36(11):1211–1242, 2017.

[249] Ye Zhao, Yinan Li, Luis Sentis, Ufuk Topcu, and Jun Liu. Reactive task and motion planning for robust whole-body dynamic locomotion in constrained environments. *The International Journal of Robotics Research*, May 2022, in press.

Index